International
REVIEW OF
Neurobiology
Volume 43

Neuromuscular Junctions IN Drosophila

International
REVIEW OF
Neurobiology
Volume 43

Neuromuscular Junctions in Drosophila

EDITED BY

VIVIAN BUDNIK
Department of Biology and Molecular and Cellular Biology Program
University of Massachusetts
Amherst, Massachusetts, USA

L. SIAN GRAMATES
Department of Biology and Molecular and Cellular Biology Program
University of Massachusetts
Amherst, Massachusetts, USA

ACADEMIC PRESS

San Diego London Boston New York Sydney Tokyo Toronto

Cover photo: Neuromuscular junction of muscles 12 and 13 of a third instar Drosophila larva, immunolabeled with α-HRP. Confocal micrograph by Dr. Michael Gorczyca

This book is printed on acid-free paper. ∞

Academic Press
a division of Harcourt Brace & Company
525 B Street, Suite 1900, San Diego, California 92101-4495, USA
http://www.apnet.com

Academic Press
24-28 Oval Road, London NW1 7DX, UK
http://www.hbuk.co.uk/ap/

International Standard Book Number: 0-12-366843-3 (case)
International Standard Book Number: 0-12-139370-4 (pb)

PRINTED IN THE UNITED STATES OF AMERICA
99 00 01 02 03 04 EB 9 8 7 6 5 4 3 2 1

To Kalpana White,
my mentor, my role model

CONTENTS

Contributors xi
Preface xiii

Early Development of the *Drosophila* Neuromuscular Junction: A Model for Studying Neuronal Networks in Development

Akira Chiba

I. Introduction 1
II. Neuronal Network Model System 3
III. Principles of Neuronal Network Development 10
IV. Future Challenges 17
V. Conclusions 19
References 19

Development of Larval Body Wall Muscles

Michael Bate, Matthias Landgraf, and Mar Ruiz Gómez Bate

I. Introduction 25
II. Origins of Muscle-Forming Cells from Embryonic Mesoderm 27
III. Muscle Specification 30
IV. The General Pathway of Myogenesis 37
V. Conclusions 40
References 40

Development of Electrical Properties and Synaptic Transmission at the Embryonic Neuromuscular Junction

Kendal S. Broadie

I. Introduction 45
II. Physiological Maturation of Electrical Properties in Embryonic Myotubes 46
III. Glutamate Receptors and Maturation of the Postsynaptic Glutamate-Gated Current in Embryonic Myotubes 50

IV. Neural Induction of Postsynaptic Specialization ... 53
V. Retrograde Induction of Presynaptic Specialization ... 55
VI. Physiological Maturation of Synaptic Transmission ... 57
VII. Development of Synaptic Modulation Properties ... 60
VIII. Methods for Electrophysiological Assays of the Embryonic Neuromuscular Junction ... 61
IX. Conclusions and Perspectives ... 64
References ... 65

Ultrastructural Correlates of Neuromuscular Junction Development

MARY B. RHEUBEN, MOTOJIRO YOSHIHARA, AND YOSHIAKI KIDOKORO

I. Introduction ... 69
II. Features and Development of the Junctional Aggregate ... 71
III. Differentiation of Postsynaptic Specializations ... 78
IV. Characterization and Development of Presynaptic Specializations ... 85
References ... 89

Assembly and Maturation of the *Drosophila* Larval Neuromuscular Junction

L. SIAN GRAMATES AND VIVIAN BUDNIK

I. Introduction ... 93
II. A Bit of Anatomy ... 94
III. A Whirlwind Tour of Development ... 97
IV. Mechanisms of Synapse Assembly ... 99
V. Structural Plasticity at the Neuromuscular Junction ... 104
VI. Concluding Remarks and Future Directions ... 110
References ... 111

Second Messenger Systems Underlying Plasticity at the Neuromuscular Junction

FRANCES HANNAN AND YI ZHONG

I. Introduction ... 119
II. Short-Term Plasticity ... 120
III. Neuromodulation ... 124
IV. Structural Plasticity ... 127
V. Long-Term Functional Plasticity ... 131
References ... 133

Mechanisms of Neurotransmitter Release

J. Troy Littleton, Leo Pallanck, and Barry Ganetzky

I. Introduction 139
II. The SNARE Hypothesis 140
III. Measuring Synaptic Function in *Drosophila* 142
IV. Mutational Analysis of the SNARE Complex 145
V. Synaptotagmin and Ca^{2+} Regulation of SNARE Function 148
VI. Additional Components of the Release Apparatus 151
VII. Conclusions 154
References 156

Vesicle Recycling at the *Drosophila* Neuromuscular Junction

Daniel T. Stimson and Mani Ramaswami

I. Introduction 163
II. Origins of Synaptic Vesicle Recycling Studies in *Drosophila* 164
III. Molecular and Phenotypic Analysis of *shi* Mutants 165
IV. Cell Biology of Synaptic Vesicle Recycling in *Drosophila* 167
V. Molecules Involved in Synaptic Vesicle Recycling 172
VI. Assays for Synaptic Vesicle Recycling in *Drosophila* 173
VII. Direct and Indirect Effects 181
VIII. Conclusions 182
References 183

Ionic Currents in Larval Muscles of *Drosophila*

Satpal Singh and Chun-Fang Wu

I. Introduction 191
II. Potassium Currents 193
III. Calcium Currents 209
IV. Conclusions 213
References 214

Development of the Adult Neuromuscular System

Joyce J. Fernandes and Haig Keshishian

I. Introduction 221
II. The Adult Musculature 223

III. Motor Neurons 228
IV. Formation of Adult Neuromuscular Junctions 230
V. Functional Development of the Adult Neuromuscular System 233
VI. Conclusions and Perspectives 233
References 235

Controlling the Motor Neuron

James R. Trimarchi, Ping Jin, and Rodney K. Murphey

I. Introduction 241
II. The Three Reflex Circuits 243
III. Future Analysis Using Intracellular Recordings from Central Neurons 256
IV. Using the GAL4/UAS System to Characterize Neural Circuits Underlying Behavior 257
V. Conclusions 259
References 260

Appendix 265
Index 267
Contents of Recent Volumes 285

CONTRIBUTORS

Numbers in parentheses indicate the pages on which the authors' contributions begin.

Mar Ruiz Gómez Bate (25), Department of Zoology, University of Cambridge, Cambridge CB2 3EJ, United Kingdom

Michael Bate (25), Department of Zoology, University of Cambridge, Cambridge CB2 3EJ, United Kingdom

Kendal S. Broadie (45), Department of Biology, University of Utah, Salt Lake City, Utah 84112

Vivian Budnik (93), Department of Biology and Molecular and Cellular Biology Program, University of Massachusetts, Amherst, Massachusetts 01003

Akira Chiba (1), Department of Cell and Structural Biology, University of Illinois at Urbana-Champaign, Urbana, Illinois 61801

Joyce J. Fernandes (221), Department of Biology, Yale University, New Haven, Connecticut 06520

Barry Ganetzky (139), Department of Genetics, University of Wisconsin, Madison, Wisconsin 53705

L. Sian Gramates (93), Department of Biology and Molecular and Cellular Biology Program, University of Massachusetts, Amherst, Massachusetts 01003

Frances Hannan (119), Cold Spring Harbor Laboratory, Cold Spring Harbor, New York 11724

Ping Jin (241), Department of Biology, Morrill Science Center, University of Massachusetts, Amherst, Massachusetts 01003

Haig Keshishian (221), Department of Biology, Yale University, New Haven, Connecticut 06520

Yoshiaki Kidokoro (69), Institute for Behavioral Sciences, Gunma University School of Medicine, 3-39-22 Showa-machi, Maebashi 371-8511, Japan

Matthias Landgraf (25), Department of Zoology, University of Cambridge, Cambridge CB2 3EJ, United Kingdom

J. Troy Littleton (139), Department of Genetics, University of Wisconsin, Madison, Wisconsin 53705

Rodney K. Murphey (241), Department of Biology, Morrill Science Center, University of Massachusetts, Amherst, Massachusetts 01003

Leo Pallanck (139), Department of Genetics, University of Washington, Seattle, Washington 98195

Mani Ramaswami (163), Department of Molecular and Cellular Biology and Arizona Research Laboratories Division of Neurobiology, University of Arizona, Tucson, Arizon 85721

Mary B. Rheuben (69), Department of Anatomy, College of Veterinary Medicine, Michigan State University, East Lansing, Michigan 48824

Satpal Singh (191), Department of Biochemical Pharmacology, SUNY at Buffalo, Buffalo, New York 14260

Daniel T. Stimson (163), Department of Molecular and Cellular Biology and Arizona Research Laboratories Division of Neurobiology, University of Arizona, Tucson, Arizona 85721

James R. Trimarchi (241), Department of Biology, Morrill Science Center, University of Massachusetts, Amherst, Massachusetts 01003

Chun-Fang Wu (191), Department of Biology, University of Iowa, Iowa City, Iowa 52242

Motojiro Yoshihara (69), Institute for Behavioral Sciences, Gunma University School of Medicine, 3-39-22 Showa-machi, Maebashi 371-8511, Japan

Yi Zhong (119), Cold Spring Harbor Laboratory, Cold Spring Harbor, New York 11724

PREFACE

This book is an attempt to gather the main contributions that research using the fruit fly *Drosophila melanogaster* has made in the area of synapse development, synapse physiology, and excitability of muscles and nerve cells. It was many decades ago that invertebrate model systems were recognized for the relative simplicity of their nervous systems, the accessibility of their nerve cells, and the feasibility of using them to address questions at the level of single identified neurons. However, it was not until this relative simplicity was combined with the fly's exceptional utility as a genetic system that perturbations of development and physiology could be routinely used to identify the genes and molecules underlying these processes.

The reviews in this book represent a synthesis of major advances in our understanding of neuronal development and synaptic physiology, which have been obtained using the approach just described. A major theme that emerges throughout the book is the great degree of evolutionary conservation that exists in the processes and molecules required to form a functional nervous system. Therefore, this approach has not only successfully expanded our understanding of these processes in the fly, but in many instances has complemented and guided the knowledge sought in other systems. This book is directed to the general neuroscience audience: researchers, instructors, graduate students, and advanced undergraduates who are interested in the mechanisms of synapse development and physiology. However, we envision that this book will also be a valuable resource for those that use the fruit fly as a model system in their laboratories.

The first four reviews are devoted to conveying our understanding of the early stages of neuromuscular junction formation. Akira Chiba discusses the major advantages of using the fruit fly neuromuscular junction as a model system and introduces the reader to the processes that culminate with synaptogenesis. He describes how motor neurons emerge from the ventral nerve cord and begin to extend the peripheral axons that will innervate the body wall muscles of the larva. Processes such as axon pathfinding, target recognition, synaptic attraction, and synaptic inhibition, as well as some of the genes and molecules involved in these processes, are presented. In their review, Michael Bate, Matthias Landgraf, and Mar Ruiz Gómez Bate discuss the processes of myogenesis and pattern formation that result in the highly organized body wall muscles. These muscles form the substrate over which growing axons develop and configure synapses. Reviews 3 and 4 recount major aspects of the very early stages of synaptogenesis. Kendal Broadie describes the major ion channel types present in the muscles before synaptogenesis and how they change during the first stages of motor neuron–muscle interactions. This review also recapitulates the main developmental landmarks that demarcate synapse formation, including neural induction of glutamate receptor aggregation, and the generation of the first synaptic currents. Additionally, it summarizes the main methodologies employed in the electrophysiological measurement of ionic and synaptic currents in late embryonic and early larval stages. Mary Rheuben, Motojiro Yoshihara, and Yoshiaki Kidokoro also examine early stages of synapse formation, but from a morphological and ultrastructural perspective. This review is a synthesis of the main morphological events that conclude in the formation of a mature synapse.

Reviews 5 and 6 proceed to stages beyond synaptogenesis and discuss the processes of synapse maturation and synapse modification. L. Sian Gramates and Vivian Budnik describe how neuromuscular synapses in fruit fly larvae are continuously changing in a surprisingly dynamic process and discuss some of the molecules and genes that participate in these events. Frances Hannan and Yi Zhong focus on plasticity of the neuromuscular junction and modulation of its activity by second messenger systems, as well as how genes identified in learning and memory mutants are also involved in the regulation of neuromuscular junction physiology and structure.

The next three reviews relate the main physiological attributes of the mature neuromuscular junction. Troy Littleton, Leo Pallanck, and Barry Ganetzky describe the molecules and mechanisms involved in the process of Ca^{2+}-dependent vesicle exocytosis. Daniel T. Stimson and Mani Ramaswami discuss how synaptic vesicles are recycled within the nerve terminal. The main ion channels and ionic currents at the body wall, as well as the genes underlying these currents, are described by Satpal Singh and Chun-Fang Wu.

The last two reviews of this volume deal with the formation and physiology of the adult neuromuscular junction. A remarkable process in holometabolous insects, such as fruit flies, is metamorphosis. In this process, a crawling eyeless maggot is transformed into an animal able to fly, walk, groom, and reproduce. This transition from larva to adult has dramatic consequences for the nervous system. Most larval muscles die and are colonized by new myoblasts that will form the adult musculature. Larval motor neurons retract their arborizations and establish new neuronal circuits. Joyce Fernandes and Haig Keshishian describe the processes involved in this reorganization of neuromuscular junctions. Finally, James Trimarchi,

Ping Yin, and Rodney Murphey examine some of the circuits underlying simple behaviors that are executed by the adult neuromuscular system. This review is an attempt to place the neuromuscular junction and the circuits that control it into a behavioral perspective—it is, after all, the climax of the developmental process—to build an organism that can operate, procreate, and behave in its environment.

Although much of what we know about neuromuscular junction development is represented in these reviews, this field is a vigorously pursued and rapidly developing area of research. We foresee that this will be just the first edition of this book.

Vivian Budnik
L. Sian Gramates

EARLY DEVELOPMENT OF THE *DROSOPHILA* NEUROMUSCULAR JUNCTION: A MODEL FOR STUDYING NEURONAL NETWORKS IN DEVELOPMENT

Akira Chiba

Department of Cell and Structural Biology,
University of Illinois at Urbana-Champaign,
Urbana, Illinois 61801

I. Introduction
  A. Brief History
II. Neuronal Network Model System
  A. High Resolution Cell Biology
  B. Accessible Genetics
III. Principles of Neuronal Network Development
  A. Motor Neuron Development
  B. Axon Pathfinding
  C. Synaptic Target Recognition
IV. Future Challenges
  A. Interfacing Specific Recognition to Growth Cone Motility Control
  B. Interfacing Specific Cell Recognition to Synaptogenesis Initiation
V. Conclusions
  References

I. Introduction

A major challenge in neuroscience today is to explain the genetic programs that direct development of the brain. The human brain is composed of sophisticated networks of neurons that self-organize specific connections during development. Furthermore, each neuron is molecularly unique and is capable of selectively establishing synaptic connections. How does a neuronal growth cone communicate with other cells while choosing specific pathways or settling on specific targets? What molecules are there to mediate such cell recognition and differentiation? In order to learn the basic organizational principles, developmental neurobiologists need a convenient model system in which both molecular and cellular components are easily dissected as well as reconstructed. A good analogy to such an ideal model system can be found in Bonsai gardening, which has a long history in Japan. A Bonsai garden, an ancient version of today's Biosphere

0074-7742/99 $30.00

II attempt, is a self-contained miniature universe. Its minimalistic style typically allows only one or two dwarf trees, a few blossoming grasses, and a small rock. They supposedly depict something far more complex and large, the real universe. The popularity of Bonsai, despite the profundity it represents, lies in its simplicity and accessibility.

This review illustrates the fly neuromuscular system as one such model system for biologists who study the genetic programming for constructing a functional neuronal network and, in particular, the organizational principles of axon guidance and synaptic target recognition. It also explains why this particular neuronal network occupies a unique position in the current neuroscience scene and summarizes the emerging principles that may be generalizable to the development of human and other animal brains.

A. Brief History

Long before the recent explosion of research on the fly nervous system, the forefathers of neurobiology, such as the 18th century naturalist Pierre Lyonnet, knew that the neuromuscular system of insects offered an accessible model nervous system.[1] In modern time, a series of electrophysiological studies in the mid-1970s first demonstrated that the larval neuromuscular synapses of the fruit fly *Drosophila melanogaster* serve as an excellent genetic model for studying synaptic functions.[2,3] About the same time, the power of single neuron analysis gained wide recognition, especially through the application of intracellular dye injection methods to insect nervous systems.[4–6]

Although implicit in these and other earlier studies, the idea that one can learn about the developmental regulation of synapses through the powerful genetic system of *Drosophila* had not been explicitly put forward until the late 1980s. In 1989, Haig Keshishian and colleagues at Yale University published the first of their series of papers.[7–9] They combined immunocytochemistry-based anatomy with electrophysiology and began to describe the details of the developing neuromuscular synapses in fly embryos and larvae. Soon after, systematic application of large-scale genetic screens started in the laboratory of Corey Goodman at the University of California.[10–12] Since the early 1990s, Keshishian, Goodman, and a number of others worldwide have amply demonstrated that the early development of the fly neuromuscular junction is the farm from which new genes are isolated and characterized in the specific contexts of axon guidance and synapse development. Advances in related fields have been fueling the success of these studies. Many insights are now available regarding the development of fly muscles, the synaptic partners of the motor neurons, a subject that is addressed by M. Bate *et al.* in this volume.[13,14] Other reviews discuss how electrophysiological studies of neuromuscular synapses have provided many illuminating insights about the molecular refinement mechanisms of the maturing synapse.[15–18]

Drosophila in general keeps evolving as an ever powerful *in vivo* genetic system, putting itself at the cutting edge of the functional analysis of developmental genes. Cloning homologous genes in mammals and other species has become so routine that directly applying what one finds in the fly nervous system to many other animals, including humans, is practical. As a result, the community of neuroscientists, especially those interested in the genetic programs of neural networking, are directing their attention to this simple neuronal network.

II. Neuronal Network Model System

Two critical advantages warrant the fly neuromuscular system a unique position as a leading *in vivo* model for developmental neuroscience research. The first is its simplicity as a multicellular system. The practical power of high resolution cell biology possible in this fly neuronal network cannot be overstressed, although the point is often overshadowed by another unparalleled advantage of the fly system, its genetic accessibility.

A. High Resolution Cell Biology

The overall organization of a fly embryo follows that of other segmented animals. The second through seventh abdominal segments develop through essentially the same molecular and genetic programming.[14,19–22] This means that one can observe the same sequence of events repeated many times within a single animal. These features greatly facilitate the quantitative analysis of the development of a single neuron.

Figure 1 illustrates the cellular organization of an abdominal hemisegment of the *Drosophila* embryo. In each embryonic/larval hemisegment, there are 30 skeletal muscles with various orientations and positions (see also Appendix).[23] Each muscle is a single cell resulting from the fusion of about a dozen myoblasts (see M. Bate *et al.,* this volume). Figure 1 also shows 31 motor neurons known to innervate specific muscles in late stage embryos.[23] The most crucial aspect of this system is that neurons and muscles are each uniquely identifiable and that their total number is vastly smaller

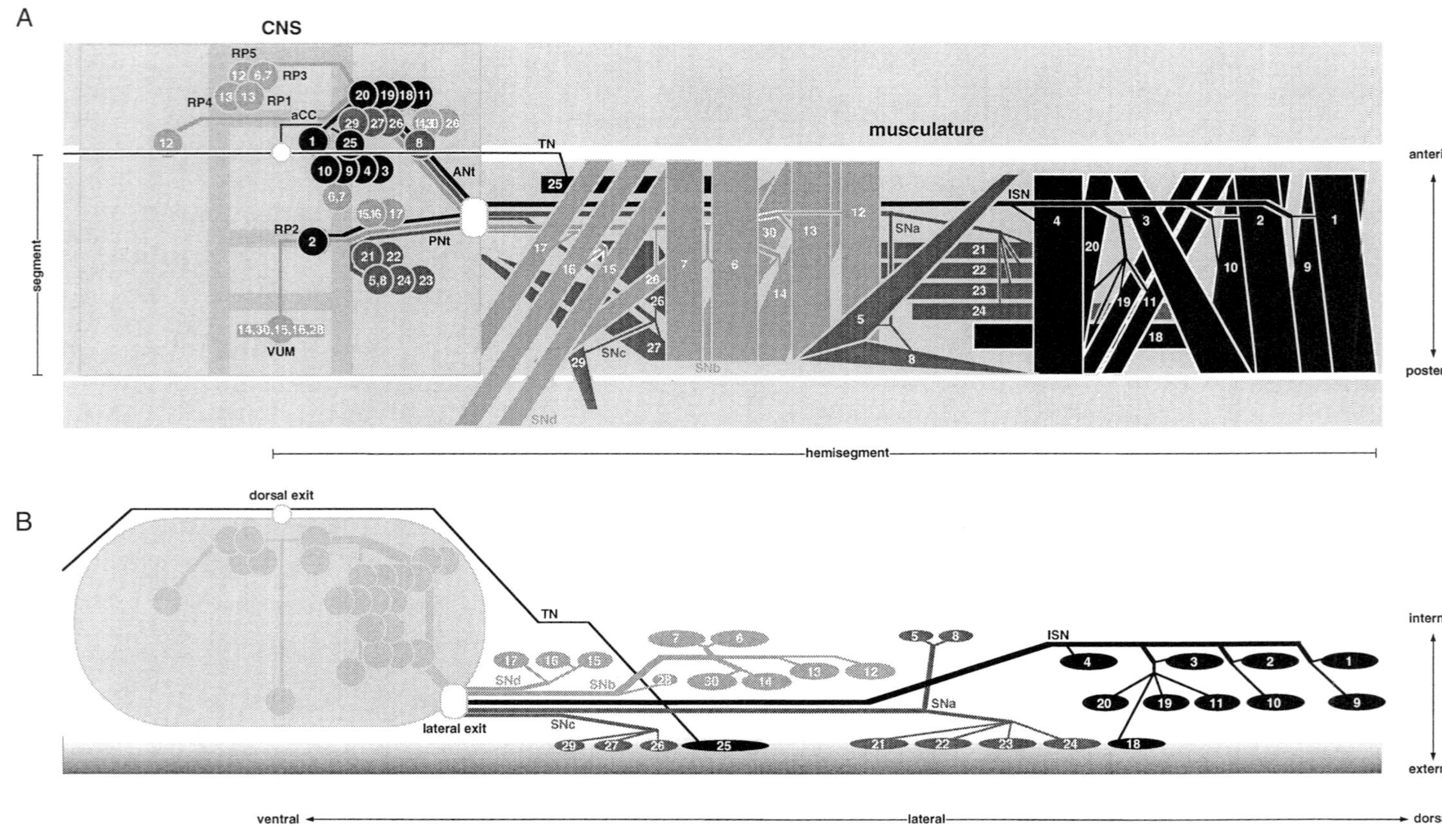
A
CNS
musculature
RP5
RP3
RP4
RP1
aCC
RP2
VUM
TN
ANt
PNt
ISN
SNa
SNb
SNc
SNd
anterior
posterior
segment
hemisegment
B
dorsal exit
lateral exit
internal
external
ventral
lateral
dorsal

than that in the vertebrate brain and musculature. Therefore, one can discuss, for example, how a particular motor neuron called the anterior corner cell (aCC) motor neuron interacts with a particular muscle cell known as muscle 1. This situation is entirely different from a number of other *in vitro* primary cultured neuron and *in vivo* nervous systems in which one can only discuss cellular interactions between a group of neurons and a group of muscles in more or less generic terms. In studies where one must deal with how the network specificity of neurons emerges, our ability to distinguish unique molecular personalities of the individual cellular components is essential.

A number of methods are available to monitor the developing neuromuscular system with high cellular resolution. They include a battery of antibodies that label the motor neurons and muscles either as a whole or as specific subsets.[12,20,25–28] The relative ease of uniquely identifying individual cells allows one to inject fluorescent intracellular dye, such as Lucifer yellow, that reveals the full morphology of individual axons most beautifully at the light microscopic level.[9,29–31] The identifiability of the cells is also extended to the electron microscopic level, allowing ultrastructural characterization of specific neurons' axon development and synapse formation.[32] The lipophilic dye DiI is not only useful in showing a single neuron fluorescently, but also in live visualization of its developing axon due to its relatively low phototoxicity.[24,29–31,33–35] Genetic engineering of flies now permits targeted expression of the clonable fluorescent molecule green fluorescent protein (GFP) in specific motor neurons and/or muscles.[36,37] These methods have facilitated the *in vivo* genetic analysis at the level of single neurons.

FIG. 1. The cellular organization of the embryonic neuromuscular system. The schematic shows a right half of the representative abdominal body wall musculature and their innervating motor neurons in a *Drosophila* embryo. Both the dorsal view (A) and the cross-sectional view (B) are shown. The 30 muscle cells are each uniquely identified based on their positions, orientations, and shapes, as well as molecular expression patterns. They are given identification numbers. The muscles are grouped into six groups according to the motor neuron groups that take the six different axon pathways to reach those muscles: TN, ISN, SNa, SNb, SNc, and SNd. Each of the 31 shown here innervates a specific muscle or group of muscles. Their target muscles are indicated with the numbers inside the cell bodies. Up to a dozen additional motor neurons are likely to be present but their identities are less well known. Note that some motor neurons arise from the contralateral side of the CNS and/or from a segment anterior to that of the muscles. Still others (e.g., VUM) are located on the midline and are not bilaterally paired as is the case with all the others. Two major axon pathways inside the CNS, ANt and PNt, converge at the lateral exit. A second exit, the dorsal exit, is used by only a small number of motor neurons. Information is extracted from Landgraf *et al.*[41] and the author's observations (A. Schmidt, A. Chiba, and C. Doe, in preparation).

I. Motor Neurons

Neurogenesis of the insect nervous system occurs along the ventral midline. This differs from vertebrates, who generate their central nervous system (CNS) at the dorsal midline. Otherwise, virtually the same general principles appear to apply to the generation of neural precursor cells from the ectodermal layer of the insect embryo.

The abdominal hemisegment of the CNS arises from 30 unique neuroblasts.[38] Of these, 10 neuroblasts give birth to the 31 motor neurons shown in Figure 1.[39] These motor neurons innervate the 30 muscles,[23] with each targeting a specific muscle(s) by the end of embryogenesis. Less well characterized are up to a dozen additional motor neurons that form synapses during the early larval stage.

II. Axon Pathways

Motor neurons extend their growth cone-tipped axons through specific axonal pathways (Fig. 1A). Pathways within the CNS reach one of two nerve exits, and from there the pathways diverge again to reach specific target regions of the body wall. These pathways are visualized easily with neuronal surface antibodies and more specifically with retrograde labeling from the motor neuron axons.[7,8,20,40,41]

Within each CNS hemisegment, all pathways of the motor neuron axons converge at either the lateral exit, located at the lateral side of the CNS, in the midpoint between the segmental borders, or the dorsal exit at the dorsal midline. The two major pathways that converge at the lateral exit are ANt, the anterior nerve tract (ANT or ISNt) and the posterior nerve tract (PNT or SNt) (Fig. 1A). They contain the bulk of the motor neuron axons that arise from both sides of the midline and some from the anterior segment. Axons that join the ANt from the contralateral CNS hemisphere cross the midline through either the anterior commissure or the posterior commissure, the major axon fascicles that connect the two sides of the brain (Fig. 1A). Pathways to the dorsal exit are less robust in appearance due to a smaller number of axons in them.

Outside the CNS, axon pathways are as specifically laid out as those in the CNS (Fig. 1A). Five major pathways extend out of the lateral exit.[8] A sixth one continues from the dorsal exit. Of the five lateral pathways, the intersegmental nerve (ISN) extends far out to the large and most dorsal (distal) muscle group. The segmental nerve a (SNa) reaches the lateral muscle group and diverges into two minor pathways. The SNb connects to the large ventrolateral muscle group. SNc specializes in the ventral muscles in the superficial layer. The SNd supplies innervation to the most ventral (proximal) muscles. Finally, from the dorsal exit extends a bilaterally symmetric transverse nerve (TN) that runs through the segmental border.

Within these axon pathways run individual motor neuron axons. Each axon takes a highly predictable path to reach its synaptic partner (Fig. 1). The situation may be likened to the commuter highways on which individual passenger cars merge, switch, and leave lanes, with each faithfully following an idiosyncratic route to get to their destiny.[19,20,42,43] The growth cone from an aCC motor neuron, for example, starts off within the CNS by taking the ANt route before reaching the lateral exit. From there, the axon extends through the ISN pathway all the way to the most distal muscle region. In contrast, the axon from the RP1 motor neuron first crosses the CNS midline and joins the ANt of the contralateral side. Upon reaching the lateral exit, the RP1 axon chooses to grow through the SNb pathway, eventually reaching the most internal layer of muscle.

III. Synaptic Target Selections

The matching between each motor neuron and its synaptic partner muscle(s) is also exact.[9,19,20,24,29,31] For example, the aCC motor neuron reproducibly forms a synapse on muscle 1, RP1 always settles on muscle 13, and the list goes on for all the other known motor neurons (see Fig. 1).

The system provides examples of both synaptic bifurcation and multiple innervation. The ventral unpaired median (VUM) motor neuron, for example, synapses with muscles 14, 15, 16, 28, and 30 (Fig. 1). However, the synapses for muscle 12 are supplied by the axons from the RP1 and RP4 motor neurons (Fig. 1).

Unlike in larger animals, synaptic targeting takes place in a short period of time, no longer than 3 hr, and within a short distance, less than 100 μm of the birth point of the motor neuron. Therefore, the fly neuromuscular system accommodates fast and repeated single cell analysis on synaptic target specificity, a subject technically difficult to study in most other systems.

B. Accessible Genetics

The fly system offers unparalleled access to its genome, and new genes can be isolated efficiently. Moreover, experimental techniques are available to genetically engineer the nervous system so that the development of an individual neuron is monitored *in vivo,* while its cellular and molecular microenvironment is specifically manipulated.

I. Efficient Gene Isolation

A straightforward strategy for isolating the genes that are involved in neuromuscular development is a systematic mutagenesis screen. This strategy assumes that when any of those genes are disrupted, the developmental program for the neuromuscular system deviates from its normal course. Detection of such developmental deviations depends critically on the resolution of one's analysis as the specific phenotypes resulting from losing one gene could be relatively subtle. This approach has so far isolated the *beaten path* gene among others (Table 1).[44]

A major drawback in applying such a classic mutagenesis strategy to a process such as neural development, which occurs rather late in embryogenesis, is that it fails to identify those genes that have essential functions during the earlier developmental stages. However, alternative screening strategies, looking for patterns of gene expression with enhancer traps, have proven successful. An assumption in these alternative screens is that those genes involved in axon guidance and/or synaptic target recognition are expressed in the motor neurons and/or muscles at the time of embryonic neuromuscular development. However, it is not safe to assume that gene expression in a particular cell type implies an essential and/or unique role of the gene in that particular cellular context. Even so, this approach has led to the isolation of a number of genes that have subsequently been shown to be directly involved in normal neuromuscular development, including *connectin* (Fig. 2). Antibodies generated against either neural or embryonic tissues have also been useful in screening for molecules (antigens) that show specific expression patterns within the neuromuscular system. The *fasciclins* and *Toll* are among those genes that were identified from such screens.[9,45–47] New fly genes can also be isolated purely on the basis of their molecular homologies to the genes that are implicated in neural development in other animals. This homology search strategy has isolated, for example, the tyrosine phosphatases DPTPs and DLAR.[48–50]

II. *In Vivo Functional Analysis*

Once candidate genes are isolated, their specific contributions to axon development and/or synaptic targeting need to be demonstrated *in vivo*. The most traditional analytical design for the functions of a specific gene is based on its deletion from the genome. Such a loss-of-function mutant analysis tests the necessity of the gene and is applied routinely to the genetic analyses of neural development. However, the strategy fails with those genes that have additional roles in either earlier stages of development or in other tissues because, for example, they can cause early lethality or affect synaptic development indirectly. Also, if the particular gene is widely expressed throughout the nervous system, its specific role in the specific context of axon guidance or synaptic targeting cannot be revealed easily. Several genetic approaches are used

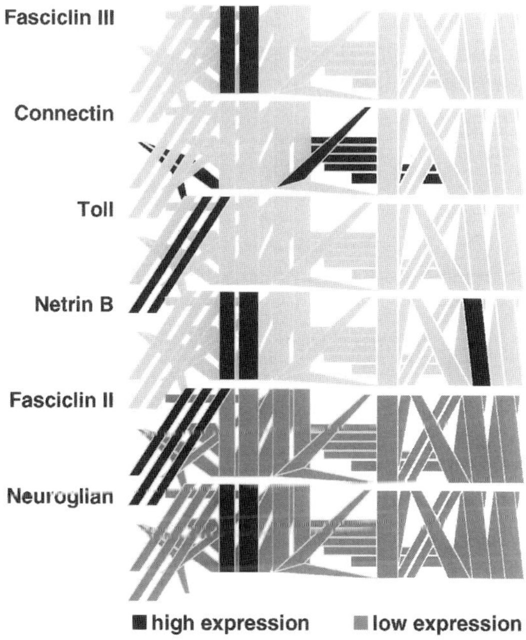

FIG. 2. Expression patterns of proposed synaptic target recognition molecules. The expression patterns of several putative synaptic target recognition molecules are shown. They exemplify the membrane-associated proteins of various structural families that are present in various subsets of muscles when motor neuron growth cones contact them.

to limit loss-of-function mutations to a particular time and/or a subset of cells. One is to isolate mutant alleles that exhibit temperature sensitivity.[51] Another is to induce genetic mosaicism in the nervous system.[52] A promising additional method is to knock out the gene's function through expression of a modified gene that would produce dominant negative effects on the endogenous gene product. An advantage of this approach is that such transgene expression can be targeted easily to a specific cell(s) or at specific times of development using the GAL4/UAS conditional gene activation system.[53] Using the same targeted gene expression system, another kind of functional analysis can be designed as well. For example, motor neuron X may normally contact muscles 1 and 2 but only synapses with muscle 1; gene Y may be normally expressed in muscle 1 but not 2. What will happen if the gene were to be ectopically expressed by muscle 2? Such an experiment tests the sufficiency of the gene and complements the loss-of-function experiments.

Thus, in this model system, one can attempt to both dissect the biological process of axon and synapse development down to single genes, as well as reconstruct the molecular events back into a whole neuronal network.

III. Principles of Neuronal Network Development

Many studies have taken advantage of these cell biological and genetic strategies. They are leading to new insights of the organizational principles that build a neuronal network.

A. Motor Neuron Development

Axonogenesis among the motor neurons occurs in several waves.[8] First, some of the axons that take the ISN pathway extend out from the CNS. They are destined for the most distal (dorsal) target region. Later, some of the SNa axons form a fascicle that is separated from the ISN. The two, however, extend through the same muscle layers until SNa reaches the lateral muscle group. The next pathway to form is SNb. The axons in this pathway first grow out by closely following the ISN axons, but the two do not fasciculate tightly to each other. The SNb axons then turn internally and diverge into four minor pathways within the ventrolateral region. The SNc and SNd pathways develop about the same time. The SNc axons separate from those in the SNa pathway and turn externally. The SNd axons, however, form as an offshoot to the SNb pathway. Within each axon pathway, specific axons develop first. In the ISN, for example, the aCC and RP2 axons serve the role of the pioneers, while in the SNb it is the RP1 axon.[9,29] Other motor neuron axons keep arriving at their target muscles at different times until the early larval stage. However, whether the latecomers are dependent on the pioneers for pathway and target guidance will have to be explicitly tested in the future.

We are beginning to determine the factors that best correlate with specific axon pathfinding and target selection (Table I). Neither the cell body positions nor the neuroblast origins predict the target selection.[29] Whereas the motor neuron axons that innervate a given muscle group all share the same peripheral pathways, their CNS pathways vary considerably (Fig. 1). In contrast, a correlation exists between the dendritic pattern and synaptic target selection.[41] This makes good sense in terms of organizing loci of information networks into distinct nuclei, or neuropils, within the CNS. Whether we can experimentally uncouple the axonal organizational

TABLE I
EXAMPLES OF PROPOSED AXON PATHFINDING AND SYNAPTIC TARGET RECOGNITION MOLECULES[a]

Molecule	Homolog	Domain	Expression	Ref.
1. Cell surface molecules				
Connectin		LRR	Neurons and muscles (subsets)	11, 60
DN-cadherin	Cadherin	Cadherin, TM	Neurons (all)	99
Derailed	RTK	TM, TK	Neurons	100
DLAR	LAR, PTP	lg, FN, TM, PTP	Neurons (all)	48
DPTP69D/DPTP99A	PTP, LAR	lg, FN, TM, PTP	Neurons (all)	49
Fasciclin II	NCAM	lg, FN, TM	Neurons and muscles (all)	57, 58
Fasciclin III	(lgCAM)	lg, TM	Neurons and muscles (subsets)	59, 73
Frazzled	DCC	lg, FN, TM	Neurons (all)	101
Late bloomer	Tetraspanin	TM	Neurons (all)	97
Neuroglian	L1	lg, FN, TM	Neurons and muscles (all)	102
Toll	(LRR)	LRR, TM, IL-1R	Muscles (subset)	74
2. Secreted molecules				
Beat			Neurons	44
D-semaphorin II	Semaphorin	Semaphorin, lg	Muscles (subset)	83
Netrin B	Netrin	Laminin, EGF	Muscles (subset)	103
3. Nuclear molecules				
Abrupt (clueless)		Zinc finger	muscles (all)	104

[a] This table is a partial list of a variety of molecules proposed to control axon guidance and/or synaptic target recognition in the fly neuromuscular system. The small numbers of molecules listed as secreted and nuclear molecules mostly reflect the fact that less is known about their roles in their specific contexts. Abbreviations: FN, fibronectin III; lg, immunoglobulin; IL-1R, interleukin 1 receptor; LRR, leucine-rich repeat; PTP; phosphotyrosine phosphatase; TK, tyrosine kinase; TM, transmembrane.

principles from the dendritic ones is one of the major questions for the future.

B. AXON PATHFINDING

1. Choice Point Hierarchy

Each axon grows through a unique series of choice points. A choice point is where multiple axons converge and/or diverge. It is analogous to a traffic intersection that has a stoplight. Just as the stoplight provides a signal to the cars in the intersection, so are the choice points thought to provide specific molecular cues to which individual axons respond specifically. In the case of the RP1 axon, there are at least five recognizable choice points before it approaches the future synaptic target, muscle 13 (Fig. 1).[9,31] The first choice point is at the CNS midline where its growth cone decides to cross the midline through the axon fascicle. Subsequently, the axon passes through a series of choice points within the contralateral side of the CNS, extending briefly through the longitudinal connective and then through the ANt. When it reaches the lateral exit, the RP1 axon takes the SNb pathway and, after navigating through the layers of muscles, reaches muscle 13. Other motor neuron axons go through similar numbers of choice points of their own. There is a unique hierarchy of choice points through which a given motor neuron axon grows (Fig. 1).

Mutagenesis studies have found several classes of choice point mutations. First is a set of mutations that cause midline decision errors, such as *roundabout* and *commissureless*.[12] Then there are mutation errors specific to the ISN/SNb diverging, such as *beaten path*.[10] Finally, there are those mutations that disrupt subbranching within the SNb pathway, such as *clueless* (*abrupt*).[10] Thus, at each choice point, there seems to be a set of genes that play crucial roles in proper cell recognition.

II. Mutual Independence of Choice Points

Does the choice point actively modulate the growth cone's molecular profile so that its subsequent responses are altered based on the choice point experience? Or is the main function of the choice point simply to direct the growth cone to a new microenvironment? How much of the axon's intrinsic guidance mechanism is reprogrammable through its interactions with extrinsic cues that surround it?

The first hint that mutually independent mechanisms are largely at play at different choice points comes from a study using midline mutants. In *commissureless* mutants, no motor neuron axons that would normally cross the CNS midline do so.[12] This phenotype is thought to be due to the loss of the Commissureless protein from the CNS midline glia.[54] The axons of RP1, RP3, RP5, and other motor neurons that normally cross the midline fail to do so in this mutant. They nevertheless follow the subsequent choice point sequence correctly in the ipsilateral side of the body, presenting the exact mirror image pathway of the normal one that would have been taken in the contralateral side.[55] This supports the idea that an early choice point decision at the midline does not reprogram the growth cone's responsiveness toward the cues experienced at later choice points.

This idea is supported in the peripheral axon pathways as well. Motor neuron axons normally approach their target muscles through specific pathways. However, when they are made to deviate from the usual pathways, they often take detours to reach their target muscles.[30,56] These situations tend to occur when the alternate axon pathways are relatively close to the normal pathways, presumably reflecting the radius in which the axon growth cones can still detect local cues. A number of independent cellular or molecular manipulations to the musculature can cause such detours. The growth cones are likely retaining their normal responsiveness to the cues from the target muscles, despite the incorrect decisions made at some earlier choice points. Once again, these observations suggest that molecular signaling systems that guide axons through a specific sequence of choice points operate largely independently of each other.

III. Fasciculation Control

Axons prefer to fasciculate with each other along the pathways. Fasciculation often occurs among those axons that share common cell adhesion molecules such as fasciclin II (FasII), fasciclin III (FasIII), connectin, and DN-cadherin.[11,57–62] However, the axons also switch fasciculation partners repeatedly. *In vivo,* when and where defasciculation occurs is highly predictable. For example, the SNa branches out into two minor pathways after it passes muscle 12 in the lateral region of the body wall (Fig. 1). Similarly, the SNb and SNd have their own defasciculation points within the ventral muscle regions (Fig. 1).

The underlying regulatory mechanisms of axon fasciculation control are beginning to be demonstrated. For motor neuron axons, one common adhesion molecule is FasII, a structural homolog of the vertebrate neural adhesion molecule NCAM. Levels of FasII have been shown to influence the fasciculation and defasciculation of motor neuron axons.[57,58]

In addition to the up- and downregulation of axon fasciculation molecules, one regulatory factor that influences axon adhesion in a locale-specific manner is a secreted molecule called beaten path. This motor neuron-derived protein accumulates at high levels at very specific choice points where the five major pathways diverge from each other. Beaten path's main role during neural development is to reduce cell–cell adhesion mediated by FasII as well as by connectin, two of the well-known axon fasciculation molecules in the fly nervous system.[63] This provides a good example of where the inhibition of cell adhesion is an actively regulated event.

C. Synaptic Target Recognition

Axon guidance and synaptic target recognition probably share many of the same molecular mechanisms. However, there are crucial differences.

Synaptic target recognition must involve the mechanisms that not only stop growth cone extension but also initiate extensive remodeling of the cytoskeleton and reallocation of the membranes.[43,64,65] These processes are necessary to convert a motile growth cone structure into a much more stable and highly differentiated presynaptic terminal. Studies using the fly neuromuscular system are rapidly advancing our knowledge of the molecular processes of presynaptic terminal differentiation, a subject that will be addressed in more detail by S. Gramates and V. Budnik in this volume.[66,67] Similarly, we are also beginning to unravel its very first step, a special cell recognition that occurs between presynaptic and postsynaptic partners.[68]

I. Targeting Accuracy

There is an amazing accuracy in synaptic targeting by motor neurons. It now appears that most of the precise connectivity that is established during embryonic neuromuscular development uses molecular mechanisms that are independent of neuronal electrical activity.[15,18,69] However, there is plenty of evidence that this electrical activity is a major modifier of the course of synapse maturation during larval development.[17,33,70–72] The situation mirrors the vertebrate brain development in which there is a two step synaptic connectivity development, with the first step being activity-independent axon targeting and the second step being the subsequent activity-dependent synaptic refinement.[42] In *Drosophila,* a number of studies using single cell dye injection show that the RP3 motor neuron nearly always selects muscles 6 and 7. The reported targeting rate in wild-type embryos is virtually 100%.[56,59,73,74] One specific region within the muscle membrane surfaces, the contact site between the two muscle membranes, is overwhelmingly the preferred target membrane domain. Other motor neurons demonstrate equally reproducible targeting accuracy.

When the target is experimentally removed, however, the axon often reaches the correct region but delays synaptogenesis.[30,56] However, when the targets are duplicated, the axons can innervate both simultaneously.[56] When muscle cells surrounding the target are ablated, the axons do not lose accuracy of targeting the correct targets.[56] These results support several conclusions. First, each axon is capable of positively recognizing the target. Second, even if it is capable of doing so, the axon does not form synapses ectopically unless the cues are correct. Third, positive recognition can occur between the axon and its target muscle with or without surrounding muscles.

II. Synaptic Attraction

Attraction is a major synaptic target recognition mechanism. Studies so far have identified a few molecules that mediate positive target cell recognition in this system. FasIII provides the first example. A short transmembrane protein with three immunoglobulin-like extracellular domains, FasIII is normally expressed by muscles 6 and 7, the two targets of the RP3 axon, as well as the axon of RP3 itself.[9,26,75,76] Thus, both axon and muscles, the synaptic partners, share the same cell surface molecules at the time of their contact. When FasIII is genetically eliminated from the muscles, the RP3 axon loses some of its targeting accuracy. However, when FasIII is placed on incorrect muscles that the RP3 axon normally ignores as it grows past, those muscles become acceptable alternative targets to RP3.[59] The same motor neuron hardly ever innervates these muscles in other experimental situations. The responsiveness to muscle-provided FasIII depends on the axon's own FasIII expression. The RP3 motor neuron that lost its own FasIII expression is no longer attracted to the FasIII-bearing alternative target muscles. This homophilic matchmaking idea has been further demonstrated in reciprocal experiments. Ectopic expression of FasIII in aCC and RP2, motor neurons that normally lack FasIII, induced these motor neurons to target FasIII-positive muscles. Therefore, FasIII functions as a positive synaptic target recognition molecule through its homophilic interaction.

A structurally unrelated molecule, connectin, is a cell surface associated adhesion molecule with leucine-rich repeats.[11,77] Connectin is normally expressed by the lateral muscle group before and during synaptogenesis on these muscles. Similar to FasIII, ectopic expression of this molecule on other muscles causes those axons that normally ignore them while targeting connectin-positive muscles to start synaptogenesis on the wrong muscles.[60,61] Connectin responsiveness is linked to those axons' own connectin expression. Thus, connectin, too, acts as a homophilic recognition molecule.

The fact that the two structurally unrelated cell adhesion molecules can mediate specific synaptic target recognition through homophilic cell adhesion raises the possibility that the microenvironment that permits synaptogenesis initiation requires a close and strong membrane apposition between the axon and the target muscle. Cell adhesion molecules with specifically regulated spatiotemporal expression patterns are primary candidates to serve such a purpose. Consistent with this, when the dosage of FasII, a NCAM-like cell adhesion molecule normally expressed by all motor neurons and muscles, is artificially raised in certain motor neurons and/or muscles, incorrect synaptic targeting between these FasII-enhanced cells is induced.[78]

Netrin B is another candidate-positive target recognition molecule for the RP3 growth cone.[79] This *Drosophila* gene is structurally homologous to the vertebrate netrins, which have been shown to attract certain growth cones while repelling others.[80,81] In *Drosophila,* netrin B is expressed in muscles 6 and 7 in the ventrolateral muscle group and in one additional muscle in the dorsal (most distal) muscle group.[79] Netrin B's putative DCC type receptor, Frazzled, is heavily enriched in the motor neuron axons.[82] Genetically altering netrin B expression in the musculature clearly affects the pathfinding and/or target selection of many motor neuron growth cones, likely including that of RP3.[79] Future experiments are still needed to specifically test whether netrin B is a factor that attracts the RP3 growth cone to its target muscles. If demonstrated, this will make a case for positive synaptic target recognition based on heterophilic interaction.

III. Synaptic Inhibition

Ample evidence supports the possibility that inhibitory cues also play important roles in accurate target recognition.[80] It is interesting to observe that inhibition effectively adds to the means to regulate complex organizational events in the developing neuronal network. One example of inhibitory signaling in the fly neuromuscular system involves D-semaphorin II, a molecule structurally related to collapsin and semaphorins in vertebrates.[16,83] D-semaphorin II is expressed at high levels in a specific ventral muscle in the thoracic region. This molecule, when ectopically placed in other muscles, inhibits the synaptogenesis of the motor neuron axons that normally innervate those muscles.[83] This experiment gives support to the hypothesis that under normal conditions, one reason that the axons do not mistakenly synapse with D-semaphorin II positive muscles is that they are actively repelled by the presence of D-semaphorin II. Such a mechanism can be called negative recognition.

Another example is Toll, a transmembrane molecule with an extracellular domain containing prominent leucine-rich repeats, similar to connectin.[84] Toll is normally expressed by the ventral-most muscles before they are innervated. During this time many axons navigate past the Toll-positive muscles without innervating them.[11,45,74] However, by the time the axons destined to innervate these ventral muscles arrive at the target regions, the Toll expression levels decrease dramatically. Is the early presence of Toll in the ventral muscles actively inhibiting the incorrect targeting? When muscles genetically lacked their Toll expression, they accepted innervation by the axons that should have normally ignored them.[74] In contrast, when Toll was placed on muscles that normally do not bear Toll at the time of motor neuron–muscle contacts, the axons failed to initiate synaptogenesis at those muscles.[74] Thus, this muscle membrane-associated molecule inhibits synaptogenesis through a very local interaction.

IV. Dynamism of the Target Cell Biology

Studies have begun to reveal that the muscles undergo dynamic cellular changes during the initiation of synaptogenesis. These changes involve endosome formation that is mediated by the muscle cell surface molecule Commissureless, the same molecule that is proposed to play a major role in CNS midline signaling.[54] In the muscles, Commissureless-dependent endocytosis is correlated with synaptogenesis initiation. When there is no Commissureless, synaptogenesis fails to initiate at high frequencies.[55] The axons reach the target correctly but then stall there. In a mutant that truncates the bulk of the Commissureless protein's cytoplasmic domain, the truncated Commissureless remains on the muscle surface. Synaptogenesis does not occur in this case.[55] Whether the endocytosis is the actual cause of, or merely correlated with, the initiation of synaptogenesis has yet to be determined. However, this suggests that there is reciprocally dynamic membrane remodeling between presynaptic motor neurons and postsynaptic muscles during the period of their interactions.

IV. Future Challenges

One general area in which much future progress is expected is the characterization of the molecular interface mechanisms that connect a context-dependent cell recognition process that takes place between a specific growth cone and a specific muscle to the cell machinery that is shared by many neurons.[65] How does an axonal growth cone convert recognition of a specific cue at a choice point to the cytoskeletal rearrangement that changes its direction of growth? Similarly, how does the activation of a specific cell recognition molecule link itself to the cytoplasmic cell biology of the growth cone that orchestrates its metamorphosis into a presynaptic terminal?

A. Interfacing Specific Recognition to Growth Cone Motility Control

Exactly how turning, stopping, and bifurcating of a growth cone occur in such a reproducible manner remains a major unresolved issue.[43,65,85–94] On one hand, as discussed in this review, we know that various recognition cue molecules are strategically positioned in the microenvironment. Each growth cone is thought to have a unique set of cue receptors that are localized on its filopodia, the dynamic microprocesses. Upon encountering a specific site of molecular cues, the receptors on a particular filopodium will be activated. Such a context-dependent activation of specific receptor molecules is thought to be the starting point of specific growth cone turning, or stopping, at that specific site. On the other hand, we know that there are networks of molecules within every growth cone that are responsible for the dynamic regulation of their motility. The molecules that compose the cytoskeleton (e.g., actins, microtubules) and those that regulate them (e.g., α-actinin, vinculin, rho subfamily small G proteins, myosin, integrins, calmodulin-dependent protein kinases) are believed to be rather widely found in most neurons. How the cytoskeletal control mechanisms interface to receptors activated in a specific cell context is one of the unresolved questions.

B. Interfacing Specific Cell Recognition to Synaptogenesis Initiation

Connectin, netrin B, FasII, and FasIII have been mentioned as a few of the putative, positive synaptic target recognition molecules in this system. The morphological changes accompanying the conversion of a growth cone to a presynaptic terminal are well documented.[64,68,77] There is even evidence that synaptic target recognition and presynaptic terminal differentiation can be genetically separable.[95] At the molecular level, however, it is poorly understood how synaptic target recognition events connect to the molecular networks responsible for assembling all that is needed to carry out synaptogenesis at an axon terminal. There seems to be a set of molecules that are commonly found in the majority of neurons during synaptogenesis, such as cytoskeletal control molecules (e.g., rho subfamily small G proteins, α-actinin, vinculin, nonmuscle myosin), vesicle transport molecules (e.g., kinesin, dynein, adaptin, dynamin), and cytoplasmic organizer molecules (e.g., PDZ family molecules, discussed by S. Gramates and V. Budnik, this volume).[66,67,96] Those general purpose cell dynamic molecules may be ready and, once triggered, may initiate construction of the synaptic transmission machinery autonomously. Molecules such as Late bloomer,[97] which are relatively small and associated with the membranes of motor neuron axons, may be involved in the initial steps of this mass molecular event.

There is an optimism among those of us who work on the fly nervous system. We now have a variety of genetic tools that allow us to manipulate many of the receptors as well as intracellular signaling molecules believed to be involved in dynamic neural differentiation. Defining the specific contributions of each and how they coordinate with others in axon guidance and synaptogenesis initiation will be one of the major challenges that await vigorous investigation.

V. Conclusions

This review described the unique advantages of the *Drosophila* neuromuscular system as a genetic model for studying neuronal network formation. It also briefly discussed a number of new findings and their implications that lead to new appreciation of the molecular nature of building of the brain. We live in an exciting age in which we are witnessing a research field that is beginning to crack the very codes of neuronal network formation. The major breakthroughs no doubt belong to the 21st century. The *Drosophila* neuromuscular system, with its Bonsai garden-like simplicity and accessibility, will allow neuroscientists to not only keep dissecting, but also start reconstructing the organizational principles that govern the development of the brain.

Acknowledgments

I am indebted to Haig Keshishian at Yale University for first introducing me to the fly neuromuscular system. I thank the members of my current laboratory at the University of Illinois for comments on this review manuscript.

References

1. Lyonnet, P. (1760). Traite anotomique de la chenille qui rounge le bois de saule.
2. Jan, L. Y., and Jan, Y. N. (1976). Properties of the larval neuromuscular junction in *Drosophila melanogaster. J Physiol (Lond.)* **262,** 189–214.
3. Jan, L. Y., and Jan, Y. N. (1976). L-glutamate as an excitatory transmitter at the *Drosophila* larval neuromuscular junction. *J Physiol (Lond.)* **262,** 215–236.
4. Murphey, R. K., Jacklet, A., and Schuster, L. (1980). A topographic map of sensory cell terminal argorizations in the cricket CNS: Correlation with birthday and position in a sensory array. *J. Comp. Neurol.* **191,** 53–64.
5. Bentley, D., and Keshishian, H. (1982). Pioneer neurons and pathways in insect appendages. *TINS* **5,** 364–367.
6. Levine, R. B. (1986). Reorganization of the insect nervous system during metamorphosis. *TINS* **9,** 315–319.
7. Johansen, J., Halpern, M. E., Johansen, K. M., and Keshishian, H. (1989). Stereotypic morphology of glutamatergic synapses on identified muscle cells of *Drosophila* larvae. *J Neurosci* **9,** 710–725.
8. Johansen, J., Halpern, M. E., and Keshishian, H. (1989). Axonal guidance and the development of muscle fiber-specific innervation in *Drosophila* embryos. *J. Neurosci.* **9,** 4318–4332.

9. Halpern, M. E., Chiba, A., Johansen, J., and Keshishian, H. (1991). Growth cone behavior underlying the development of stereotypic synaptic connections in *Drosophila* embryos. *J. Neurosci.* **11,** 3227–3238.
10. VanVactor, D., Sink, H., Fambrough, D., Tsoo, R., and Goodman, C. S. (1993). Genes that control neuromuscular specificity in *Drosophila. Cell* 73, 1137–1153.
11. Nose, A., Mahajan, V. B., and Goodman, C. S. (1992). Connectin: A homophilic cell adhesion molecule expressed on a subset of muscles and the motor neurons that innervate them in *Drosophila. Cell* **70,** 553–567.
12. Seeger, M., Tear, G., Ferres-Marco, D., and Goodman, C. S. (1993). Mutations affecting growth cone guidance in Drosophila: Genes necessary for guidance toward or away from the midline. *Neuron* **10,** 409–426.
13. Bate, M. (1993). The mesoderm and its derivatives. *In* "The Development of *Drosophila melanogaster,*" pp. 1013–1090. Cold Spring Harbor: CSHL Press,
14. Keshishian, H., Broadie, K., Chiba, A., and Bate, M. (1996). The *Drosophila* neuromuscular junction: A model for studying synaptic development and function. *Annu. Rev. Neurosci.* **19,** 545–575.
15. Broadie, K., and Bate, M. (1993). Activity-dependent development of the neuromuscular synapse during *Drosophila* embryogenesis. *Neuron* **11,** 607–619.
16. Davis, G. W., Schuster, C. M., and Goodman, C. S. (1996). Genetic dissection of structural and functional components of synaptic plasticity. III. CREB is necessary for presynaptic functional plasticity. *Neuron* **17,** 669–679.
17. Zhong, Y., Budnik, V., Wu, C. F. (1992). Synaptic plasticity in *Drosophila* memory and hyperexcitable mutants: Role of cAMP cascade. *J. Neurosci.* **12,** 644–651.
18. Kidokoro, Y., and Nishikawa, K. (1994). Miniature endplate currents at the newly formed neuromuscular junction in *Drosophila* embryos and larvae. *Neurosci. Res.* **19,** 143–154.
19. Keshishian, H., and Chiba, A. (1993). Neuromuscular development in *Drosophila:* Insights from single neurons and single genes. *TINS* **16,** 278–283.
20. Keshishian, H., Chiba, A., Chang, T. N., Halfon, M. S., Harkins, E. W., Jarecki, J., Wang, L., Anderson, M., Cash, S., and Halpern, M. E., *et al.* Cellular mechanisms governing synaptic development in *Drosophila melanogaster. J. Neurobiol.* **24,** 757–787.
21. Fletcher, T. L., De Camilli, P., and Banker, G. (1994). Synaptogenesis in hippocampal cultures: Evidence indicating that axons and dendrites become competent to form synapses at different stages of neuronal development. *J. Neurosci.* **14,** 6695–6706.
22. Thomas, J. B., Bastiani, M. J., Bate, M., and Goodman, C. S. (1984). From grasshopper to *Drosophila:* A common plan for neuronal development. *Nature* **310,** 203–207.
23. Crossley, C. A. (1978). The morphology and development of the *Drosophila* muscular system. *In* "The Genetics and Biology of *Drosophila,*" pp. 499–560. Academic Press, New York.
24. Sudhof, T. C. (1995). The synaptic vesicle cycle: A cascade of protein-protein interactions. *Nature* **375,** 645–653.
25. Jan, L. Y., and Jan, Y. N. (1982). Antibodies to horseradish peroxidase as specific neuronal markers in *Drosophila* and in grasshopper embryos. *Proc. Natl. Acad. Sci. USA* **79,** 2700–2704.
26. Patel, N. H., Snow, P. M., and Goodman, C. S. (1987). Characterization and cloning of fasciclin III: A glycoprotein expressed on a subset of neurons and axon pathways in *Drosophila. Cell* **48,** 975–988.
27. Fujita, S. C., Zipursky, S. L., Benzer, S., Ferrus, A., and Shotwell, S. L. (1982). Monoclonal antibodies against the *Drosophila* nervous system. *Proc. Natl. Acad. Sci. USA* **79,** 7929–7933.
28. Kania, A., Han, P. L., Kim, Y. T., and Bellen, H. (1993). Neuromusculin, a *Drosophila* gene expressed in peripheral neuronal precursors and muscles, encodes a cell adhesion molecule. *Neuron* **11,** 673–687.

29. Sink, H., and Whitington, P. M. (1991). Location and connectivity of abdominal motor neurons in the embryo and larva of *Drosophila melanogaster. J. Neurobiol.* **22,** 298–311.
30. Sink, H., and Whitington, P. M. (1991). Early ablation of target muscles modulates the arborisation pattern of an identified embryonic *Drosophila* motor axon. *Development* **113,** 701–707.
31. Sink, H., and Whitington, P. M. (1991). Pathfinding in the central nervous system and periphery by identified embryonic *Drosophila* motor axons. *Development* **112,** 307–316.
32. Barinaga, M. (1997). NGF signals ride a trolley to nucleus. *Science* **277,** 1037.
33. Cash, S., Chiba, A., and Keshishian, H. (1992). Alternate neuromuscular target selection following the loss of single muscle fibers in *Drosophila. J Neurosci* **12,** 2051–2064.
34. Bossing, T., Technau, G. M., and Doe, C. Q. (1996). *Huckbein* is required for glial development and axon pathfinding in the neuroblast 1-1 and neuroblast 2-2 lineages in the *Drosophila* central nervous system. *Mech. Dev.* **55,** 53–64.
35. Bossing, T., and Technau, G. M. (1994). The fate of the CNS midline progenitors in *Drosophila* as revealed by a new method for single cell labeling. *Development* **120,** 1895–1906.
36. Yeh, E., Gustafson, K., and Boulianne, G. L. (1995). Green fluorescent protein as a vital marker and reporter of gene expression in *Drosophila. Proc. Natl. Acad. Sci. USA* **92**, 7036–7040.
37. Brand, A. (1995). GFP in *Drosophila. Trends Genet.* **11,** 324–325.
38. Robinson, M. S. (1997). Coats and vesicle budding. *Trends Cell Biol.* **7,** 99–102.
39. Bossing, T., Udolph, G., Doe, C. Q., and Technau, G. M. (1996). The embryonic central nervous system lineages of *Drosophila melanogater.* I. The neuroblast lineages derived from the ventral half of the neuroectoderm. *Dev. Biol.* **179,** 41–64.
40. Broadie, K. S., and Bate, M. (1993). Development of the embryonic neuromuscular synapse of *Drosophila melanogaster. J. Neurosci* **13,** 144–166.
41. Landgraf, M., Bossing, T., Technau, G. M., and Bate, M. (1997). The origin, location, and projections of the embryonic abdominal motor neurons of *Drosophila. J. Neurosci.* **17,** 9642–9655.
42. Goodman, C. S., and Shatz, C. J. (1993). Developmental mechanisms that generate precise patterns of neuronal connectivity. *Cell* **72**(Suppl.), 77–98.
43. Goodman, C. S. (1996). Mechanisms and molecules that control growth cone guidance. *Annu. Rev. Neurosci.* **19,** 341–377.
44. Fambrough, D., and Goodman, C. S. (1996). The *Drosophila beaten path* gene encodes a novel secreted protein that regulates defasciculation at motor axon choice points. *Cell* **87,** 1049–1058.
45. Halfon, M. S., Hashimoto, C., and Keshishian, H. (1995). The *Drosophila Toll* gene functions zygotically and is necessary for proper motor neuron and muscle development. *Dev Biol* **169,** 151–167.
46. Zinn, K., McAllister, L., and Goodman, C. S. (1988). Sequence analysis and neuronal expression of fasciclin I in grasshopper and *Drosophila. Cell* **53,** 577–587.
47. Grenningloh, G., Bieber, A. J., Rehm, E. J., Snow, P. M., Traquina, Z. R., Hortsch, M., Patel, N. H., and Goodman, C. S. (1990). Molecular genetics of neuronal recognition in *Drosophila:* Evolution and function of immunoglobulin superfamily cell adhesion molecules. *CSH Symp. Quant. Biol.* **60,** 327–340.
48. Krueger, N. X., Van Vactor, D., Wan, H. I., Gelbart, W. M., Goodman, C. S., and Saito, H. (1996). The transmembrane tyrosine phosphatase DLAR controls motor axon guidance in *Drosophila. Cell* **84,** 611–622.
49. Desai, C. J., Gindhart, J. G., Goldstein, L. S. B., and Zinn, K. (1997). Receptor tyrosine phosphatases are required for motor axon guidance in the *Drosophila* embryo. *Cell* **84,** 599–609.

50. Desai, C. J., Krueger, N. X., Saito, H., and Zinn, K. (1997). Competition and cooperation among receptor tyrosine phosphatases control motor neuron growth cone guidance in *Drosophila. Development* **124,** 1941–1952.
51. Hall, S. G., and Bieber, A. L. (1997). Mutations in the *Drosophila* neuroglian cell adhesion molecule affect motor neuron pathfinding and peripheral nervous system patterning. *J. Neurobiol.* **32,** 325–340.
52. Xu, T, and Harrison, S. D. (1994). Mosaic analysis using FLP recombinase. *In* "Methods in Cell Biology," pp. 655–681. Academic Press, San Diego.
53. Brand, A., Manoukian, A. S., and Perrimon, N. (1994). Ectopic expression in *Drosophila. In* "Methods in Cell Biology," pp. 683–696. Academic Press, San Diego.
54. Tear, G., Harris, R., Sutaria, S., Kilomanski, K., Goodman, C. S., and Seeger, M. A. (1996). Commissureless controls growth cone guidance across the CNS midline in *Drosophila* and encodes a novel membrane protein. *Neuron* **16,** 501–514.
55. Wolf, B., Seeger, M. A., and Chiba, A. (1998). Commissureless endocytosis is correlated with initiation of neuromuscular synaptogenesis. *Development* **125,** 3853–3863.
56. Chiba, A., Hing, H., Cash, S., and Keshishian, H. (1993). Growth cone choices of *Drosophila* motor neurons in response to muscle fiber mismatch. *J. Neurosci.* **13,** 714–732.
57. Lin, D. M., Fetter, R. D., Kopczynski, C., Grenningloh, G., and Goodman, C. S. (1994). Genetic analysis of fasciclin II in *Drosophila:* Defasciculation, refasciculation, and altered fasciculation. *Neuron* **13,** 1055–1069.
58. Lin, D. M., and Goodman, C. S. (1994). Ectopic and increased expression of fasciclin II alters motor neuron growth cone guidance. *Neuron* **13,** 507–523.
59. Chiba, A., Snow, P., Keshishian, H., and Hotta, Y. (1995). Fasciclin III as a synaptic target recognition molecule in *Drosophila. Nature* **374,** 166–168.
60. Nose, A., Umeda, T., and Takeichi, M. (1997). Neuromuscular target recognition by a homophilic interaction of connectin cell adhesion moleucles in *Drosophila. Development* **124,** 1433–1441.
61. Nose, A., Takeichi, M., and Boodman, C. S. (1994). Ectopic expression of connectin reveals a repulsive function during growth cone guidance and synpase formation. *Neuron* **13,** 525–539.
62. Iwai, Y., Usui, T., Hirono, S., Steward, R., Takeichi, M., and Uemura, T. (1997). Axon patterning requries DN-cadherin, a novel neuronal adhesion receptor, in the *Drosophila* embryonic CNS. *Neuron* **19,** 77–89.
63. Hoffmann, J. A., and Reichhart, J. (1997). *Drosophila* immunity. *Trends Cell Biol* **7,** 309–316.
64. Haydon, P. G., and Drapeau, P. (1995). From contact to connection: Early events during synaptogenesis. *TINS* **18,** 196–201.
65. Chiba, A., and Keshishian, H. (1996). Neuronal pathfinding and recognition: Roles of cell adhesion molecules. *Dev. Biol.* **180,** 424–432.
66. Budnik, V., Koh, Y. H., Guan, B., Hartman, B., Hough, C., Woods, D., and Gorczyca, M. (1996). Regulation of synaptic structure and function by the *Drosophila* tumor suppressor gene *dlg. Neuron* **17,** 627–640.
67. Budnik, V. (1996). Synaptic maturation and structural plasticity at *Drosophila* neuromuscular junctions. *Curr. Opin. Neurobiol.* **6,** 858–867.
68. Chiba, A., and Rose, D. (1998). "Painting" the target: How local molecular cues define synaptic relationships. BioEssays 20, *in press.*
69. Broadie, K., and Bate, M. (1993). Innervation directs receptor synthesis and localization in *Drosophila* embryo synaptogenesis. *Nature* **361,** 350–353.
70. Jarecki, J., and Keshishian, H. (1995). Role of neural activity during synaptogenesis in *Drosophila. J. Neurosci.* **15,** 8177–8190.

71. Budnik, V., Zhong, Y., Wu, C. F. (1990). Morphological plasticity of motor axons in *Drosophila* mutants with altered excitability. *J. Neurosci.* **10,** 3754–3768.
72. Chang, T. N., and Keshishian, H. (1996). Laser ablation of *Drosophila* embryonic motor neurons causes ectopic innervation of target muscle fibers. *J. Neurosci.* **16,** 5751–5726.
73. Cremona, O., and De Camilli, P. (1997). Synaptic vesicle endocytosis. *Curr. Opin. Cell Biol.* **7,** 323–330.
74. Rose, D., Zhu, X., Kose, H., Hoang, B., Cho, J., and Chiba, A. (1997). Toll, a muscle cell surface molecule, locally inhibits synaptic initiation of the RP3 motor neuron growth cone in *Drosophila. Development* **124,** 1561–1571.
75. Snow, P. M., Bieber, A. J., and Goodman, C. S. (1989). Fasciclin III: A novel homophilic adhesion molecule in *Drosophila. Cell* **59,** 313–323.
76. Snow, P. M., Zinn, K., Harrelson, A. L., McAllister, L., Schilling, J., Bastiani, M. J., Makk, G., and Goodman, C. S. (1988). Characterization and cloning of fasciclin I and fasciclin II glycoproteins in the grasshopper. *Proc. Natl. Acad. Sci. USA* **85,** 5291–5295.
77. Yoshihara, M., Rheuben, M. B., and Kidokora, Y. (1997). Transition from growth cone to functional motor nerve terminal in *Drosophila* embryos. *J. Neurosci.* **17,** 8408–8426.
78. Davis, G. W., Schuster, C. M., and Goodman, C. S. (1997). Genetic analysis of the mechanisms controlling target selection: Target-derived fasciclin II regulates the pattern of synapse formation. *Neuron* **19,** 561–573
79. Mitchell, K. J., Doyle, J. L., Serafini, T., Kennedy, T. E., Tessier-Lavigne, M., Goodman, C. S., and Dickson, B. J. (1996). Genetic analysis of netrin genes in *Drosophila:* Netrins guide CNS commissural axons and peripheral motor axons. *Neuron* **17,** 203–215.
80. Tessier-Lavigne, M., and Goodman, C. S. (1996). The molecular biology of axon guidance. *Science* **274,** 1123–1133.
81. Ming, G., Song, H., Berninger, B., Holt, C. E., Tessier-Lavigne, M., and Poo, M. (1997). cAMP-dependent growth cone guidance by netrin-1. *Neuron* **19,** 1225–1235.
82. Kolodziej, P. A., Timpe, L. C., Mitchell, K. J., Fried, S. R., Goodman, C. S., Jan, L. Y., and Jan Y. N. (1996). *frazzled* encodes a *Drosophila* member of the DCC immunoglobulin subfamily and is required for CNS and motor axon guidance. *Cell* **87,** 197–204.
83. Matthes, D. J., Sink, H., Kolodkin, A. L., and Goodman, C. S. (1995). Semaphorin II can function as a selective inhibitor of specific synaptic arborizations. *Cell* **81,** 631–639.
84. Hashimoto, C., Hudson, K. L., and Anderson, K. V. (1988). The *Toll* gene of *Drosophila,* required for dorsal-ventral embryonic polarity, appears to encode a transmembrane protein. *Cell* **52,** 269–279.
85. Bentley, D., and O'Connor, T. P. (1994). Cytoskeletal events in growth cone steering. *Curr. Opin. Neurobiol.* **4,** 43–48.
86. Burden-Gulley, S. M., Payne, H. R., and Lemmon, V. (1995). Growth cones are actively influenced by substrate-bound adhesion molecules. *J. Neurosci.* **15,** 4370–4381.
87. de la Torre, J. R., Hopker, V. H., Ming, G., Poo, M., Tessier-Lavigne, M., Hemmati-Brivanlou, A., and Holt, C. E. (1997). Turning of retinal growth cones in a netrin-l gradient mediated by the netrin receptor DCC. *Neuron* **19,** 1211–1224.
88. Chang, H. Y., Takei, K., Sydor, A. M., Born, T., Rusnak, F., and Jay, D. G. (1995). Asymmetric retraction of growth cone filopodia following focal inactivation of calcineurin. *Nature* **376,** 686–690.
89. Dodd, J., and Schuchardt, A. (1995). Axon guidance: A compelling case for repelling growth cones. *Cell* **81,** 471–474.
90. Grabham, P. W., and Goldberg, D. J. (1997). Nerve growth factor stimulates the accumulation of B1 integrin at the tips of filopodia in the growth cones of sympathetic neurons. *J. Neurosci.* **17,** 5455–5465.

91. Klinz, S. G., Schachner, M., and Maness, P. F. (1995). L1 and N-CAM antibodies trigger protein phosphatase activity in growth cone-enriched membranes. *J. Neurochem.* **65,** 84–95.
92. Kuhn, T. B., Schmidt, M. F., and Kater, S. B. Laminin and fibronectin guideposts signal sustained but opposite effects to passing growth cones. *Neuron* **14,** 275–285.
93. Lin C.-H., and Forscher, P. (1995). Growth cone advance is inversely proportional to retrograde F-actin flow. *Neuron* **14,** 763–771.
94. Zheng, J. Q., Felder, M., Connor, J. A., and Poo, M. M. (1994). Turning of nerve growth cones induced by neurotransmitters. *Nature* **368,** 140–144.
95. Prokop, A., Landgraf, M., Rushton, E., Broadie, K., and Bate, M. (1996). Presynaptic development at the *Drosophila* neuromuscular junction: Assembly and localization of presynaptic active zones. *Neuron* **17,** 617–626.
96. Sone, M., Hoshino, M., Suzuki, E., Kuroda, S., Kaibuchi, K., Nakagoshi, H., Saigo, K., Nabeshima, Y., and Hama, C. (1997). Still life, a protein in synaptic terminals of *Drosophila* homologous to GDP-GTP exchangers. *Science* **275,** 543–547.
97. Kopczynski, C. C., Davis, G. W., and Goodman, C. S. (1996). A neural tetraspanin, encoded by late bloomer, that facilitates synapse formation. *Science* **271,** 1867–1870.
98. Raghavan, S., and White, R. A. H. (1997). Connectin mediates adhesion in *Drosophila. Neuron* **18,** 873–880.
99. Senger, D. L., and Campenot, R. B. (1997). Rapid retrograde tyrosine phosphorylation of trkA and other proteins in rat sympathetic neurons in compartmented cultures. *J. Cell Biol.* **138,** 411–421.
100. Callahan, C. A., Muralidhar, M. G., Lundgren, S. E., Scully, A. L., and Thomas, J. B. (1995). Control of neuronal pathway selection by a *Drosophila* receptor protein-tyrosine kinase family member. *Nature* **376,** 171–174.
101. Marks, M. S., Ohno, H., Kirchhausen, T., and Bonifacino, J. S. (1997). Protein sorting by tyrosin-based signals: Adapting to the ys and wherefores. *Trends Cell Biol.* **7,** 124–128.
102. Schuster, C. M., Davis, G. W., Fetter, R. D., and Goodman, C. S. (1996). Genetic dissection of structural and functional components of synaptic plasticity. II. Fascilin II controls presynaptic structural plasticity. *Neuron* **17,** 655–667.
103. Kramer, H., and Phistry, M. (1996). Mutations in the *Drosophila* hook gene inhibit endocytosis of the boss transmembrane ligand into multivesicular bodies. *J. Cell Biol.* **133,** 1205–1215.
104. Cagan, R. L., Kramer, H., Hart, A. C., and Zipursky, S. L. (1992). The bride of sevenless and sevenless interaction: Internalization of a transmembrane ligand. *Cell* **69,** 393–399.

DEVELOPMENT OF LARVAL BODY WALL MUSCLES

Michael Bate, Matthias Landgraf, and Mar Ruiz Gómez Bate

Department of Zoology, University of Cambridge, Cambridge CB2 3EJ, United Kingdom

I. Introduction
II. Origins of Muscle-Forming Cells from Embryonic Mesoderm
III. Muscle Specification
  A. Segregation of Muscle Founder Cells and Precursors of Adult Muscles
  B. Founder Cells, Fusion-Competent Cells, and the Formation of Muscles
  C. Progenitor Cell Lineages and the Significance of Founder Cell Gene Expression
IV. The General Pathway of Myogenesis
  A. *Nautilus* and *Dmef2* as Regulators of Muscle Differentiation
  B. Cell Fusion, Migration of Precursors; Formation of Attachments
V. Conclusions
  References

I. Introduction

Larval muscles in *Drosophila* form a complex pattern of contractile fibers that attach to the inner surface of the developing epidermis. Each segment has its own, highly specific set of muscles, but from the point of view of the neuromuscular junction, most attention has been focused on the muscles of the abdominal segments. An equivalent set of 30 muscles is repeated in each abdominal hemisegment from A2 to A7 (Fig. 1; Appendix), with variations on this set in A1 and in segments posterior to A7.[1–3]

Much of our understanding of muscle development comes from the study of myogenesis in vertebrates, especially mammals.[4] For this reason, it is important to realize at the outset that while the body wall muscles of *Drosophila* resemble those of higher vertebrates in some ways (they are striated, syncytial fibers formed by the fusion of myoblasts), there is nonetheless a significant difference that has consequences for the development, the innervation, and the function of the musculature. This difference has to do with the way in which muscles are formed from myotubes that are themselves the products of myoblast fusion. In *Drosophila*, each larval muscle is formed from a single myotube that differentiates to form a very specific contractile element in the overall muscle pattern: each muscle (myotube) has a distinctive size, shape, sites of attachment to the epidermis, and stereotyped patterns of connection with innervating motor neurons. In

0074-7742/99 $30.00

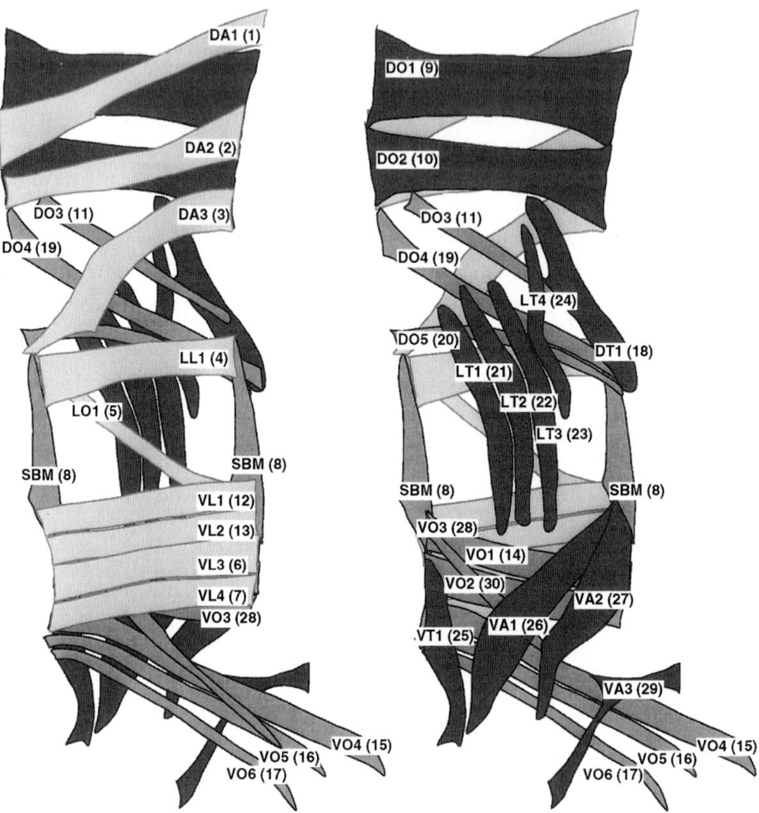

FIG. 1. Diagram showing internal (left) and external (right) views of the larval muscle pattern in abdominal segments A2–A7. External muscles are shown in dark gray, with more internal muscles in lighter shades of gray. Each muscle is identified according to the scheme given in Refs. 2 and 3 and by the number given by Crossley.[1] DA, dorsal acute; DO, dorsal oblique; DT, dorsal transverse; LO, lateral oblique; LT, lateral transverse; LL, lateral longitudinal; VA, ventral acute; VO, ventral oblique; VT, ventral transverse; VL, ventral longitudinal; SBM, segment border muscle.

higher vertebrates, however, each muscle is formed from a bundle of myotubes that aggregate together to form an identifiable unit in the neuromuscular system. This bundle of fibers has a recognizable size, shape, sites of attachment, and is innervated by a pool of motor neurons, each of which is connected to a subset of the myotubes that make up the muscle. Thus the mechanical unit on which the motor neurons operate in *Drosophila* is a single myotube, as against the bundle of myotubes that constitutes a muscle in a vertebrate. It is important to bear this distinction in mind when considering how the muscles are formed and in particular how they acquire their specific properties. In *Drosophila,* the diversification of myotubes is extreme: 30 different myotubes (muscles) in a hemisegment (Fig. 1). In vertebrates there is a smaller number of different myotube classes: primary, secondary, fast, and slow are the obvious categories, and these are bundled together to make different muscles with distinctive properties.[5] Thus the specification of elements in the muscle pattern in *Drosophila* revolves around the way in which differences are made between myotubes, and this may be different from the way in which muscles are patterned in a vertebrate.

Nonetheless, it is also true that myogenesis appears, like many other developmental mechanisms, to be a conserved process. This conservation is particularly clear in the details of the general pathway of myogenic differentiation to which all muscles in the developing pattern conform and on which specific characteristics are then overlaid. This distinction between specific and general muscle properties underlies the structure of this review because it is important for the way in which we think about the construction of the neuromuscular junction: all muscles and motor neurons make neuromuscular synapses, but each muscle is innervated by specific motor neurons.[6] Thus the general pathway of myogenic differentiation has to be implemented for a neuromuscular junction to form, but each muscle must develop unique characteristics to allow it to be recognized as an appropriate target for innervation by its partner motor neuron.

This survey begins with the origin of the muscle-forming cells of the embryo, moves on to consider how specific muscles are formed from these cells, and then deals with the general pathway of myogenic differentiation, which will culminate in the synthesis of the contractile apparatus and the formation of the mature neuromuscular junction.

II. Origins of Muscle-Forming Cells from Embryonic Mesoderm

All the muscles in the fly (larval and adult) are ultimately derived from a set of cells that invaginates early in embryogenesis to form an internal cell layer, the mesoderm.[3,7] These cells are set aside as a midventral sector of the embryonic blastoderm by their response to high intranuclear levels of the transcription factor Dorsal, which in turn is distributed along the future dorsoventral axis of the embryo by the action of maternally acting coordinate genes.[8,9] Early mesodermal cells are characterized by the expression of two transcription factors: Twist, a bHLH protein,[10] and Snail, a zinc-finger protein.[11] Whereas *snail* largely (but not exclusively) acts to repress the expression in the mesoderm of genes characteristic of more lateral sectors of the blastoderm (notably nervous system-specific genes),[12,13] *twist* activates many (but not all) mesoderm-specific genes.[14] These include *tinman* (*tin*), a gene required for the formation of dorsal cell types in the mesoderm,[15–17] and *myocyte enhancer factor* (*Dmef-2*), the homolog of vertebrate *Mef2* genes, which is an essential gene for the general pathway of myogenic differentiation in the fly.[18–20]

Mesodermal cells that have invaginated at gastrulation spread over the inner face of the embryonic ectoderm to form a monolayer of *twist*-expressing cells.[7] Cells in this layer divide four times, during gastrulation and in the course of the next 5 hr of embryogenesis.[3] As a result of the increase in cell numbers and because of specific cell rearrangements, the original monolayer breaks down to produce a more complex, multilayered internal structure from which different mesodermal derivatives, including the muscles, will be formed.[21] Despite its increasing complexity, there is a clear division of the mesoderm in every segment into two domains (Fig. 2). This division is based on modulated patterns of *twist* expression and the expression of patterning (pair rule) genes that underlie this modulation. As

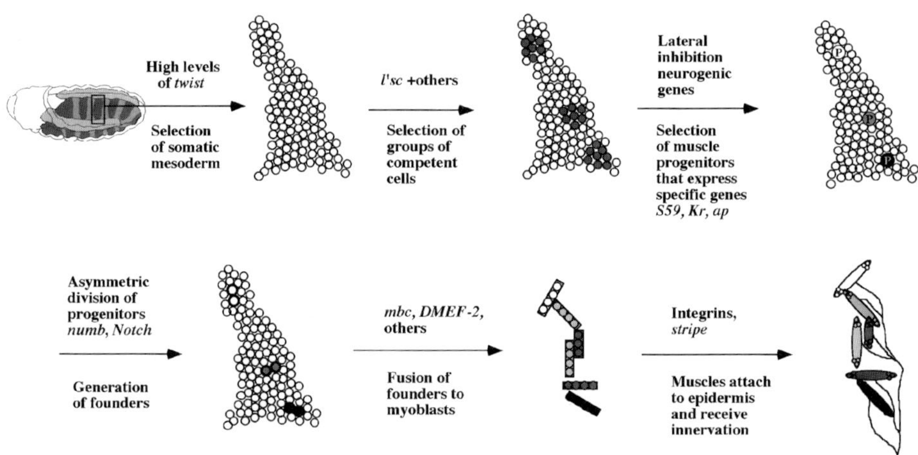

FIG. 2. Sequence of events leading from muscle forming mesoderm to differentiated muscle. At top left an embryo at stage 10 showing modulation of *twist* in the mesoderm to give domains of high (dark gray) and low (light gray) expression. Domains of high and low *twist* expression approximate the *slp* and *eve* domains, respectively (see text for details). From cells of the high *twist* domain, muscle progenitors are segregated by lateral inhibition from groups of competent cells that express *l'*(*sc*). Progenitors divide to give rise to sibling founder cells that have specific patterns of gene expression. Differences between siblings depend on the asymmetric distribution of Numb. Founders are surrounded by fusion-competent myoblasts, with which they fuse to form syncytial precursors of individual muscles. The precursors differentiate as mature, innervated muscles that are anchored to the larval epidermis.

far as we know the somatic (body wall) muscles are derived from only one of these domains, where *twist* expression remains high during the period following gastrulation and mesodermal migration (Fig. 2).

During the blastoderm stage that precedes gastrulation, the embryo is divided up along the anteroposterior axis by a series of dorsoventral stripes of pair rule gene expression. The partially overlapping expression patterns of these genes are an essential part of the machinery that leads to the subdivision of the fly into a reiterated pattern of segmental units.[22] Pair rule gene expression extends to the cells of the future mesoderm and is carried over into the internalized mesoderm at gastrulation. For our purposes, two of these genes are particularly important: *even-skipped* (*eve*) and *sloppy paired* (*slp*). The requirement for these genes divides each future mesodermal segment into two domains: an anterior domain where *eve* is required and a posterior domain where *slp* is required.[23,24] It is from the domain of *slp* expression that the somatic muscles will be formed.[24] This mesodermal subdivision in the anteroposterior axis of each segment is reflected in the changing patterns of *twist* expression that appear after gastrulation. Cells in the *eve* domain gradually lose *twist* expression, whereas those of the *slp* domain maintain relatively high levels of *twist,* leading to a characteristic modulation of the pattern of *twist* expression, with a sequence of high and low levels (Fig. 2) repeated in every segment along the body axis.[21,23,24]

Experimentally altering the levels of *twist* in the early mesoderm, either by targeting high levels of expression to both domains or, using a temperature-sensitive combination of *twist* alleles, by lowering expression in both domains shows that *twist* is a crucial determinant for the entry of cells to myogenesis.[25] Overexpression of *twist* blocks the formation of derivatives such as the gut musculature that comes from the *eve* domain of normally low *twist* expression and propels some of these cells into aberrant somatic myogenesis, normally characteristic of the *slp* domain of high *twist* expression. Conversely, lowering *twist* levels throughout the mesoderm severely disrupts the formation of body wall muscles from the *slp* domain while leaving the differentiation of structures from the *eve* domain of normally low *twist* expression intact. Experiments of this kind lead to the conclusion that high levels of *twist* are essential for cells to enter somatic myogenesis and that the modulation of *twist* expression under the influence of the pair rule genes defines a sector of the mesoderm in every segment from which such muscle-forming cells will be formed.[25]

Cells in this muscle-forming population are acted on by at least two signals emanating from the ectoderm that are likely to be important in regulating the spatial organization of myogenesis: Decapentaplegic (Dpp), a transforming growth factor β family member,[26] and Wingless (Wg), the homolog of Wnt signaling molecules in vertebrates.[27] *dpp* is expressed in a dorsal sector of the blastoderm and postblastoderm embryo and is required both for the maintenance of *tin* expression in the underlying dorsal mesoderm and for the repression in the same cells of more ventrally expressed mesodermal genes such as *pox meso.*[28,29] In this way, *dpp* expression in the ectoderm subdivides the mesoderm into dorsal and ventral sectors by an inductive signal. While there is a demonstrable requirement for the Dpp signal in the formation of dorsal derivatives of the *eve* domain, it is likely that it will prove to be an important factor in determining the formation of more dorsal muscles from the *slp* domain as well. *wg* is expressed in a dorso/ventral stripe of cells just anterior to the stripe of ectodermal *engrailed* expression that marks the most posterior cells of the future segment. The posterior margin of *slp* expression (abutting the anterior margin of the *eve* domain) lies at the junction of *wg* and *engrailed* expressing cells: thus *wg* expressing cells are adjacent to muscle-forming cells of the *slp* domain.[24] *wg* expression is required for at least two characteristics of these cells: the maintenance of an enhanced level of *twist* expression by cells immediately adjacent to the *wg* stripe and for the formation of the majority of somatic muscles.[30,31] The exact role of *wg* in myogenesis remains to be elucidated, as does the question of whether its influence on myogenesis is actually mediated through its effects on *twist* expression; nonetheless, for the majority of muscles, myogenesis fails in the absence of the *wg* signal.

III. Muscle Specification

A. Segregation of Muscle Founder Cells and Precursors of Adult Muscles

The earliest sign of muscle differentiation is the appearance of groups of cells in muscle-forming mesoderm that express the proneural gene, *lethal of scute* (*l'sc*) (Fig. 2).[32] As in the nervous system, where clusters of cells give rise to single neuroblasts,[33] each cluster of proneural gene-expressing cells in the mesoderm represents a group of apparently equivalent cells from which a single muscle-forming cell (a muscle progenitor) will segregate. This segregation in turn depends on a process of lateral inhibition, which is mediated by neurogenic genes.[33] The Notch signaling pathway is activated in cells surrounding the segregating progenitor and these cells are then blocked from adopting the progenitor fate. Unlike the muscle progenitor itself, these cells lose *l'sc* expression.[32] Current thinking is that they probably go on to form a second essential class of muscle-forming cells, the fusion competent myoblasts (see further below). A corollary of these events is that in embryos mutant for any of the neurogenic genes, lateral inhibition fails, high levels of *l'sc* expression are maintained throughout the muscle-forming mesoderm,[32] and excess muscle progenitors are formed. In such embryos myogenesis is severely deranged.[34,35]

Muscle progenitors each divide once to give rise to two muscle-forming cells.[32] In all cases at least one of these sibling cells is a so-called muscle founder cell (Figs. 2–4). Muscle founder cells have an essential role in the patterning and specification of the larval muscles, which is considered further later. However, to understand fully the lineages generated by muscle progenitors it is necessary to appreciate that there is a second class of cell in the embryo that will later (at the end of larval life) initiate the formation of adult muscles.[36] These cells are known as adult muscle precursors (APs). Cells of the larval somatic muscle lineage lose *twist* expression as muscle differentiation begins in the embryo during stage 12, whereas APs maintain *twist* expression throughout embryogenesis and during larval life, when they divide to form pools of adult myoblasts from which adult muscles will be formed.[36] It is only as they assemble to form the adult muscles that these cells in turn lose *twist* expression and begin to differentiate.[37,38] Thus maintained *twist* expression in the late embryo labels a special set of adult muscle-forming cells that, like the founders of larval muscles, are formed as a consequence of the specification and division of muscle progenitors.

B. Founder Cells, Fusion-Competent Cells, and the Formation of Muscles

As emphasized at the beginning of this review, all muscles share common characteristics of contractility and excitability that result from the activation of a general pathway of myogenic differentiation. However, each muscle has unique properties such as size, shape, attachment sites, and innervation that reflect a specific program of differentiation that is executed solely by the cells that contribute to that particular fiber. Thus while it is necessary to understand how myogenesis generally is initiated in the somatic mesoderm, it is equally essential to understand how groups of myoblasts within the somatic mesoderm gain access to the properties necessary to form a specific muscle in the pattern. The founder cell model to explain this process was originally based on observations both of early muscle development in the fly embryo and of early patterns of gene expression in muscle-forming mesoderm,[39,40] but it is now strongly supported by evidence of muscle differentiation in embryos where myoblast fusion fails.[41]

One way of explaining the specification of muscles is to assume that myoblasts are selected in groups and assigned to fuse and form a particular kind of fiber. While this is a straightforward explanation, there is so far no compelling evidence in its favor. In fact, early patterns of gene expression suggest to the contrary that single cells are specified to express genes characteristic of individual muscles or muscle groups and that neighboring myoblasts are then recruited to these patterns of expression as they fuse with the originally expressing cells. The type example for such expression is the homeobox containing gene *S59,* which is finally expressed in a very few muscles in each segment but whose expression begins in a small number of specific cells in muscle-forming mesoderm.[40] Many such genes have now been documented.[3,42] Like *S59,* their expression begins in a few cells at predictable locations in muscle-forming mesoderm and is maintained in a small number of muscles or muscle subsets. There is increasing evidence (see below) that it is the selective expression of such genes that endows each muscle in the pattern with its distinctive properties.

The founder cell model suggests that the muscle pattern is seeded, and muscle-specific patterns of gene expression are initiated, as single cells are specified as founder myoblasts in muscle-forming mesoderm (Fig. 2). Because the development of each muscle begins with the segregation of such a founder, there should be 30 muscle founders in each hemisegment of the abdomen, seeding the formation of 30 different muscles. Work using a P[LacZ] insertion that marks muscle progenitors and founders suggests that this is indeed the case.[43] Founders fuse with neighboring, nonfounder myoblasts that are designated as fusion-competent cells and these cells are thereby recruited to the founder's own characteristic pattern of gene expression. Together, the fused group of cells forms the syncytial precursor of a specific muscle (Fig. 2).

This hypothesis, among other things, puts a great deal of weight on the founders as special cells that have privileged access to the information necessary to form specific muscles. In contrast, fusion-competent cells are seen as "naive" myoblasts that cannot form muscles independently of founder cells. A critical test of this hypothesis is provided by embryos where myoblast fusion fails. What is the behavior of founders and fusion-competent cells under these conditions? The answer, as it turns out, is very clear: in the absence of fusion, the founders form at appropriate locations, express their normal complement of genes, and then go on to differentiate as tiny, mononucleate muscles.[41] These miniature fibers are properly innervated and contractile[44]; in other words, they appear to differentiate perfectly normally. Fusion-competent cells, however, which are unable to fuse with the founders, express muscle myosin, but remain rounded and otherwise undifferentiated; they develop none of the specific characteristics of the muscles they would normally contribute to and eventually many of them degenerate.[41] It is important to note that the only defect in such embryos is a block to myoblast fusion[45]—there is no other defect in the myogenic pathway, which is apparently completed normally by the founders.[41] Thus, where myoblast fusion fails, founder myoblasts are revealed as a special class of cells that uniquely have access to the information necessary, (a) to complete myogenesis and (b) to execute the specific program of differentiation characteristic of the muscles whose formation they seed.

As founders are uniquely capable of completing myogenesis, they gate the process of muscle development—it cannot proceed without them (this assertion can be tested by blocking the formation of founders with activated Notch—under these circumstances no muscles are formed). Because this is so the specific genes (such as *S59*) that individual founders express can condition the general pathway of myogenesis so that the individual properties of particular muscles can be superimposed upon it. Given the key function of founders in the construction of the muscle pattern, it is necessary to return to the muscle progenitors and the lineages they generate and understand how each founder acquires its distinctive pattern of gene expression and how these expression patterns actually affect muscle development.

C. Progenitor Cell Lineages and the Significance of Founder Cell Gene Expression

The progenitor cell division that gives rise to muscle founder cells is characteristically asymmetric, in that each such division gives rise to two different classes of cells (Fig. 3). Each sibling cell either maintains or loses the expression of genes that were originally expressed by the parent progenitor.[46] Thus muscle founder cells can be classified as either + or − according to these expression patterns. The unequal outcome of the progenitor division depends on an asymmetric distribution of the cytoplasmic protein Numb between the two daughter cells.[46–49] The cell that receives Numb maintains progenitor gene expression, whereas the one lacking Numb loses expression (Fig. 3). This consequence of the presence or absence of Numb depends (at least in part) on the action of Numb in blocking activation of the Notch signaling pathway,[46] presumably by its previously demonstrated capacity to bind to the intracellular domain of Notch.[50] Thus activation of Notch in one founder downregulates the expression of genes previously expressed in the progenitor, whereas Numb blocks this in the other founder and progenitor gene expression is maintained in this cell (Fig. 3). All APs arise as siblings of muscle founders and the

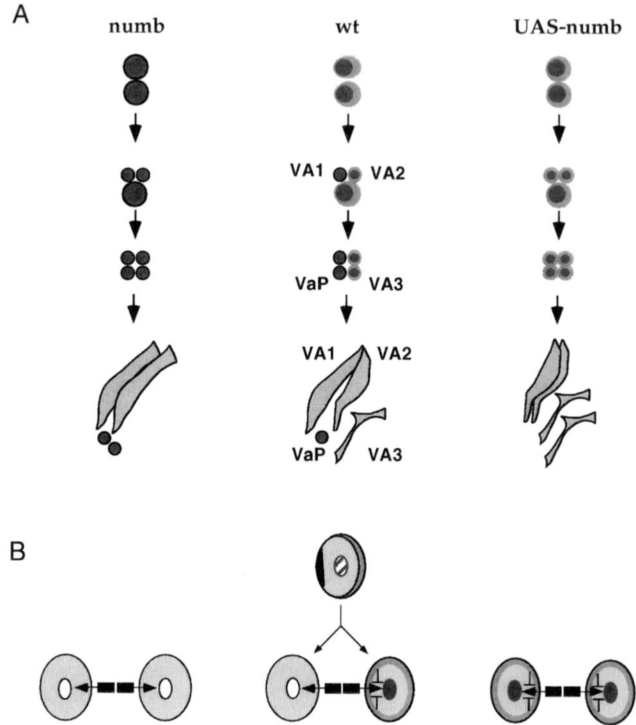

FIG. 3. Diagrams illustrating effects of loss and gain of *numb* expression on the segregation of lineages in the mesoderm. See text and Refs. 46 and 58 for details. (A) Development of VA1-3 and VaP in wild-type (wt), *numb*[1] (left) and UAS-*numb* (right) embryos. During normal development two ventral *S59*-expressing muscle progenitors divide sequentially to give rise to four cells, the founders for muscles VA1-3 and the ventral adult precursor (VAP). Numb is asymmetrically distributed in both progenitors and the generation of the four distinct cells requires the segregation of Numb to one cell in each pair of siblings. Thus, after division of the dorsalmost progenitor, Numb is segregated to the VA2 founder and excluded from VA1. Similarly the division of the second progenitor generates a VA3 founder containing Numb and VaP that lacks Numb. In *numb* mutant embryos each sibling adopts the fate of the cell that does not receive Numb in the wild type, so producing duplicate VA1s and VaPs. When Numb is provided ectopically in the mesoderm, the opposite fates are adopted by the two sibling cells of each pair, giving rise to two VA2 and two VA3. The presence of Numb is indicated by the lighter shading in progenitors and founder cells. (B) Implementation of alternative fates in the sibling cells produced by muscle progenitors. Numb and Inscuteable (Insc)[46,58] are asymmetrically distributed in muscle progenitors (Numb, gray and Insc black sector). Progenitors differ in the combination of marker genes that they express: e.g., DA1 progenitor coexpresses *eve* and *Kr*, VA2 progenitor *S59* and *Kr*, indicated by crosshatched nucleus. Segregation of Numb to one of the founders after division ensures that the two sibling cells will adopt different fates. Implementation of the Notch pathway (arrows) in the sibling cell that does not receive Numb results in the repression of genes such as *S59*, *Kr*, and *eve* (empty nuclei) and allows the maintenance of *twist* expression in adult precursor

formation of APs requires Notch activation and is blocked by the presence of Numb. Thus APs are the siblings of founder myoblasts of the + class.[46]

Because subsets of muscles and the founder cells that give rise to them express particular transcription factors such as *eve,*[51] *S59*[40] *Krüppel* (*Kr*),[52,53] and *apterous* (*ap*),[54] it has been suggested that it is the expression of these genes that is responsible for the development of specific muscle characteristics.[3,42] Studies confirm that this is so in the case of *Kr* and *S59*[53] and indicate that it may also be true for the *Drosophila* homolog of vertebrate MyoD genes, *nautilus* (*nau*),[55] and for the LIM homeodomain protein encoded by *ap.*[54] *Kr* is expressed in several muscle progenitors and in founders and muscles derived from these. However, its function in muscle development has been studied in detail in just two ventral muscles, VA1 and VA2, which are derived from a single progenitor that coexpresses *Kr* and *S59*[32,53] (Fig. 4). During normal development, both *Kr* and *S59* expression are lost from the VA1 founder, whereas both are maintained in VA2. As expected, these expression patterns can be switched by manipulating the distribution of Numb: in *numb* mutants, both cells lose expression of *Kr* and *S59,* whereas if Numb is present in both cells, both maintain expression as Notch-mediated signaling is blocked.[46] The specific role of *Kr* can be tested by looking at mutations that remove *Kr* from the mesoderm.[53] Under these conditions, both cells not only lack *Kr,* but lose expression of *S59.* In addition, both muscles whose development is initiated by the two sibling founders now differentiate identically to form duplicates of VA1, the muscle that normally does not express either gene (Fig. 4). This experiment suggests that a function of *Kr* is to maintain the expression of *S59* in the founder in which it is expressed and that, together, the expression of both genes regulates differences between muscles VA1 and VA2. This can be confirmed by ectopically expressing *Kr* throughout the mesoderm, using the GAL4/UAS targeted expression system.[56] Under these conditions, both founders maintain *Kr* and *S59* expression, and the two muscles they form both differentiate as VA2 (Fig. 4).[53] Presumably the specific characteristics of muscles such as VA1 and VA2 depend on the activation of many genes that are downstream of selectively expressed transcription factors such as *Kr* and S59. One such gene appears to be *knockout* (*ko*), which is a downstream target of *Kr* and is required for the development of a normal pattern

---

cells. In the sibling cell receiving Numb, the Notch signaling pathway is blocked and the expression of progenitor genes such as *S59, Kr,* and *eve* is maintained (black nuclei). In the absence of Numb, activation of Notch results in the repression of *S59, Kr,* and *eve* in both sibling cells, whereas ectopic activation of Numb in both sister cells interferes with the activation of Notch and produces two cells that adopt the alternative fate.

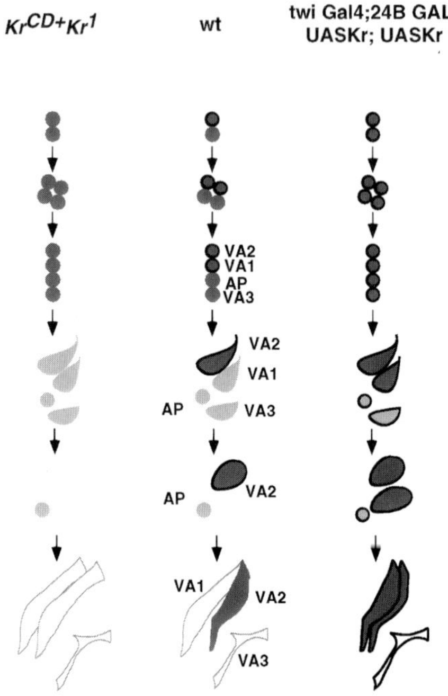

FIG. 4. Diagram showing the consequences of lack and excess of function of *Kr* for the development of muscles VA1–VA3 and the adult ventral muscle precursor (VAP). See text and Ref. 53 for further details. Light and dark shading indicates levels of *S59, Kr* expression is represented by a black outline. Only cells that can be identified by *S59* or myosin (muscles, bottom line) expression are represented. During normal development (center: wt), two *S59*-positive progenitors give rise to the three *S59*-positive founders that will seed the formation of muscles VA1-VA3 and to the AP. Only the more dorsal progenitor and its two founders will express *Kr*. *Kr* is lost in VA1 founder and *S59* decays in precursors VA1 and VA3 and in AP, whereas both *S59* and *Kr* are uniquely maintained in the VA2 precursor. In the absence of *Kr* (left) the segregation of *S59*-positive progenitors and founders is not affected. However, *S59* expression declines in all the precursors by stage 13, indicating that the maintenance but not the initiation of *S59* expression in VA2 is dependent on *Kr*. In these conditions VA2 is transformed toward its non-*S59*-expressing sibling VA1. When *Kr* is ectopically provided in the whole mesoderm (right), the segregation of *S59*-expressing cells is unaffected, confirming that *Kr* is unable to initiate *S59* expression. Furthermore, there is no effect on *S59* expression in those cells where it is not normally expressed (VA3 and VAP). However, *S59* expression is maintained in VA1 precursor and muscle, which is transformed toward the *S59*-expressing VA2 fate. wt, wild type.

of muscle innervation by motorneurons.[57] While loss of *Kr* disrupts the development of many specific muscle attributes, loss of *ko* probably affects only one of these, namely innervation.

Because *Kr* is expressed in many muscles, but its effects on VA1 and VA2 are rather specific, it seems likely that individual muscle phenotypes are the result of the coincident activation of several different factors in the same founder cell. This expectation is borne out by the expression of other transcription factors: they are almost without exception expressed in partially overlapping subsets, rather than in specific muscles. Indeed, VA1/2 founders and muscles must be subject to controls other than *Kr* because, while loss and gain of *Kr* can switch the phenotypes of VA1/VA2, it cannot alter the overall characteristics of this pair, namely that they are external, ventral acute muscles (Fig. 1). *Kr* simply regulates the differences between these sibling cells.

Numb clearly functions as a master switch for distinctions between siblings, and in accordance with this, loss and gain of Numb in the mesoderm are sufficient to produce, with high penetrance, global switches of the muscle pattern as founders (and APs) are switched between alternative fates[16,58] (Fig. 3). However, some genes expressed by progenitors, such as *ap*, are not expressed differentially by founders, but are maintained in both and it may be that genes of this kind are responsible for the differentiation of common, local attributes shared by groups of muscles. An example of this would be the four LT muscles (Fig. 1), all of which express *ap*. In *ap* mutants the number of such muscles may be reduced, whereas if *ap* is overexpressed using heat shock the number of LT fibers can be increased.[54]

The most plausible model for the moment is that specific muscle characteristics are regulated by the selective expression of transcription factors in particular muscles, with a number of such factors controlling the distinctive features of any one muscle in a hierarchical fashion. The clearest insight that we have at present is into the way in which distinctions between sibling muscles and between larval founders and APs are controlled by the asymmetric distribution of Numb and its effect on the expression of factors such as *Kr* and *S59*. One implication of experiments with Numb is that altering the specific pattern of transcription factor expression in a muscle precursor does not block myogenesis. Under these conditions a muscle is formed, but it is a different muscle. Thus factors such as *Kr* and *S59* do not control myogenesis, they condition it to produce specific outcomes.

IV. The General Pathway of Myogenesis

A. *Nautilus* AND *Dmef-2* AS REGULATORS OF MUSCLE DIFFERENTIATION

In vertebrates, the MyoD family of transcription factors has been identified as key factors in regulating myogenic differentiation.[59] In flies, the MyoD homolog *nau* was isolated by homology and was found not to be expressed throughout the myogenic lineage, but rather specifically in a small number of cells that would give rise to a subset of the final muscles.[60,61] The exact role of *nau* in myogenesis has still not been clarified, but it seems clear from its limited pattern of expression that it cannot have the pivotal role attributed to vertebrate MyoD. Indeed it may be more helpful to think of *nau* as another factor controlling the specific characteristics of founders and muscles in which it is expressed. Overexpression experiments tend to support this view.[55]

Ectopic activation of *twist,* however, is sufficient to cause mesodermal and probably ectodermal cells to enter the pathway of somatic myogenesis,[25] suggesting that genes downstream of *twist* are essential activators of myogenic differentiation. One of these genes is clearly the MADS box containing gene *Dmef-2. Dmef-2* is a target of *twist* activation and has a complex and dynamic pattern of expression in the mesoderm from gastrulation onward.[20] Loss of *Dmef-2* blocks the formation of the body wall muscles completely, without, interestingly, affecting the selection and specification of founder cells. However, the founder cells in *Dmef-2* mutants fail to complete differentiation as they would in embryos mutant for genes such as *myoblast city* (*mbc*) that are simply blocked in myoblast fusion.[41,44,45] In fact, *Dmef-2* seems to be required for the proper differentiation of many different, contractile cell types in the mesoderm.[18,19] In the case of body wall muscles, this program of differentiation includes the fusion of myoblasts, the synthesis of the contractile apparatus, the normal differentiation of the neuromuscular junction, and the formation of anchorage sites on the larval epidermis.

B. Cell Fusion, Migration of Precursors; Formation of Attachments

Fusion between muscle founders and neighboring myoblasts occurs progressively after the segregation of founders during late stage 11 and stage 12.[39,62] The final size of any muscle is a specific characteristic that is presumably (although this has never been tested) set by an intrinsic property of the founding myoblast. Fusion is completed rapidly for smaller muscles such as VA3 that consist of two or three fused cells, but continues at least until stage 13 for larger muscles, which may incorporate 10 to 20 myoblasts. Several genes have now been characterized that are essential to the fusion process: *mbc,*[41,45] *blown fuse* (*blow*),[62] and *rolling stone* (*rost*).[63,64] Interestingly, the expression of one of these, *rost,* reportedly in muscle founders,[64] gives some support to the notion that myoblast fusion may be an inherently asymmetrical process, requiring the presence of both founders and fusion-competent cells if it is to occur. However, a detailed analysis of the fusion process at the ultrastructural level reveals no sign of any such asymmetry, suggesting instead that equivalent cellular processes are matched and synchronized between partner myoblasts as they align and coalesce.[62] Failure of myoblast fusion is one of the most common phenotypes seen in mutant screens for lethal embryos that affect myogenesis[65] and it is likely that the identification of further genes will rapidly advance our understanding of this essential but as yet poorly understood feature of the myogenic pathway.

Founder myoblasts and the fusing precursors of individual muscles undergo two migrations in the embryo. The first of these are the small-scale migrations of individual precursors to their appropriate locations on the developing larval epidermis. These local events can be exemplified by the orderly alignment of the four myoblasts that give rise to the VA1,2, and 3 muscles and the ventral adult muscle progenitor on the ventral posterior margin of the segment during stage 12.[32,40] The second migration allows the developing muscles to span their future territories. During this phase, muscle precursors elongate, extending processes toward their future attachment sites. These movements are executed by motile extensions of the cell that resemble growth cones as they move over the inner surface of the larval epidermis.[39] Each precursor extends over a defined territory, and their growth cones may cross segment borders to form attachments in neighboring segments. Both local migration and extension are characteristics that are manifested by founder myoblasts in the absence of fusion, whereas in the same embryos, neighboring, nonfounder myoblasts remain rounded and make no obviously directed movements.[41] Once again suggesting that there are aspects to the muscle differentiation program that are intrinsic to founders, both in their specific characteristics and in their general implementation.

The muscle pattern is completed by the formation of a mechanical coupling between muscles and the body wall at sites of attachment and between adjacent muscles, via muscle/muscle junctions. Integrins are expressed and required at these junctions for proper anchorage to be established.[66] The location of attachment sites in the developing epidermis is signaled by the expression of several proteins in the attachment cells, including β_1 tubulin,[67] the basic helix-loop-helix protein encoded by *delilah,*[68] *groovin,*[69] and *stripe* (*sr*).[70–72] *sr* encodes a zinc finger transcription factor and is required for the expression of other genes characteristic of attachment cells.[71, 72] While proteins such as β_1 tubulin and the integrins are required for proper differentiation of the anchorage site, evidence from the ectopic expression of *sr* in the epidermis shows that it is sufficient to activate a signal in cells in which it is expressed. This signal acts at a distance to guide growing myotubes toward their proper sites of attachment.[72,73] In addition, once they reach their attachment sites, the muscles have an inductive function in maintaining characteristic patterns of gene expression in epidermal attachment cells.[73]

V. Conclusions

The formation of the muscle pattern depends on an interaction between a general pathway of myogenic differentiation and local controls that endow individual muscles with their special characteristics. Each muscle synthesizes a contractile apparatus and anchorages to the epidermis and forms a neuromuscular junction with the terminals of innervating motor neurons. These functions are common to all muscles and, at least in part, are downstream of control factors such as *Dmef-2.* However, each muscle in the pattern has its own highly distinctive properties, including size, shape, orientation, sites of attachment, and innervation by particular motor neurons. These individual properties depend on the local expression of transcription factors such as *S59* and *Kr.* This combination of general and specific regulatory mechanisms in the myogenic pathway leads to the formation of a specialized set of contractile elements on which motor neurons can act to produce the coordinated movements of larval behavior.

Acknowledgment

The authors' work is supported by grants from the Wellcome Trust.

References

1. Crossley, A. C. (1978). The morphology and development of the *Drosophila* muscular system. *In* "The Genetics and Biology of *Drosophila*" (M. Ashburner and T. Wright, eds.), Vol. 2b, pp. 499–560. Academic Press, New York.
2. Campos-Ortega, J. A., and Hartenstein, V. (1997). "The Embryonic Development of *Drosophila melanogaster*," 2nd Ed. Springer Verlag, Berlin.
3. Bate, M. (1993). The mesoderm and its derivatives. *In* "The Development of *Drosophila melanogaster*" (M. Bate and A. Martinez-Arias, eds.), Vol. 2, pp. 1013–1090. Cold Spring Harbor Laboratory Press, Cold Spring Harbor, NY.
4. Yun, K., and Wold, B. (1996). Skeletal muscle determination and differentiation: Story of a core regulatory network and its context. *Curr. Opin. Cell Biol.* **8,** 877–889.
5. Donoghue, M. J., and Sanes, J. R. (1994). All muscles are not created equal. *TIGS* **10,** 396–401.

6. Keshishian, H., Broadie, K., Chiba, A., and Bate, M. (1996). The *Drosophila* neuromuscular junction: A model system for studying synaptic development and function. *Annu. Rev. Neurosci.* **19,** 545–575.
7. Leptin, M., and Grunewald, B. (1990). Cell shape changes during gastrulation in *Drosophila. Development* **110,** 73–84.
8. Ray, R. P., Arora, K., Nüsslein Volhard, C., and Gelbart, W. M. (1991). The control of cell fate along the dorsal ventral axis of the *Drosophila* embryo. *Development* **113,** 35–54.
9. St Johnston, R. D., and Nüsslein Volhard, C. (1992). The origin of pattern and polarity in the *Drosophila* embryo. *Cell* **68,** 201–219.
10. Thisse, B., Stoetzel, C., Gorostiza-Thisse, C., and Perrin-Schmitt, F. (1988). Sequence of the *twist* gene and nuclear localization of its protein in endomesodermal cells of early *Drosophila* embryos. *EMBO J.* **7,** 2175–2183.
11. Boulay, J.-L., Dennefeld, C., and Alberga, A. (1987). The *Drosophila* developmental gene *snail* encodes a protein with nucleic acid binding fingers. *Nature* **330,** 395–398.
12. Leptin, M. (1991). *twist* and *snail* as positive and negative regulators during *Drosophila* mesoderm development. *Genes Dev.* **5,** 1568–1576.
13. Kosman, D., Ip, Y. T., Levine, M., and Arora, K. (1991). Establishment of the mesoderm-neuroectoderm boundary in the *Drosophila* embryo. *Science* **254,** 118–122.
14. Casal, J., and Leptin, M. (1996). Identification of novel genes in *Drosophila* reveals the complex regulation of early gene activity in the mesoderm. *Proc. Natl. Acad. Sci. USA* **93,** 10327–10332.
15. Bodmer, R. (1993). The gene *tinman* is required for specification of the heart and visceral muscles in *Drosophila. Development* **118,** 719–729.
16. Azpiazu, N., and Frasch, M. (1993). *tinman* and *bagpipe:* Two homeo box genes that determine cell fate in the dorsal mesoderm of *Drosophila. Genes Dev.* **7,** 1325–1340.
17. Yin, Z., Xu, X.-L., and Frasch, M. (1997). Regulation of the Twist target gene tinman by modular cis-regulatory elements during early mesoderm development. *Development* **124,** 4971–4982.
18. Bour, B. A., O'Brien, M. A., Lockwood, W. L., Goldstein, E. S., Bodmer, R., Taghert, P. H., Abmayr, S. M., and Nguyen, H. T. (1995). *Drosophila* MEF2, a transcription factor that is essential for myogenesis. *Genes Dev.* **9,** 730–741.
19. Lilly, B., Zhao, B., Ranganayakulu, G., Paterson, B. M., Schulz, R. A. and Olson, E. N. (1995). Requirement of MADS domain transcription factor D-MEF2 for muscle formation in *Drosophila. Science* **267,** 688–693.
20. Taylor, M. V., Beatty, K. E., Hunter, H. K., and Baylies, M. K. (1995). Drosophila MEF-2 is regulated by *twist* and is expressed in both the primordia and differentiated cells of the embryonic somatic, visceral and heart musculature. *Mech. Dev.* **50,** 29–41.
21. Dunin Borkowski, O. M., Brown, N. H., and Bate, M. (1995). Anterior-posterior subdivision and the diversification of the mesoderm in *Drosophila. Development* **121,** 4183–4193.
22. Akam, M. (1987). The molecular basis for metameric pattern in the *Drosophila* embryo. *Development* **101,** 1–22.
23. Azpiazu, N., Lawrence, P. A., Vincent, J.-P., and Frasch, M. (1996). Segmentation and specification of the *Drosophila* mesoderm. *Genes Dev.* 3183–3194.
24. Riechmann, V., Irion, U., Wilson, R., Grosskortenhaus, R., and Leptin, M. (1997). Control of cell fates and segmentation in the *Drosophila* mesoderm. *Development* **124,** 2915–2922.
25. Baylies, M. K., and Bate, M. (1996). *twist:* A myogenic Switch in *Drosophila. Science* **272,** 1481–1484.
26. Kingsley, D. M. (1994). The TGF-β superfamily: New members, new receptors, and new genetic tests of function in different organisms. *Genes Dev.* **8,** 133–146.

27. McMahon, A. P. (1992). The Wnt family of developmental regulators. *Trends Genet.* **8,** 236–242.
28. Staehling-Hampton, K., Hoffmann, F. M., Baylies, M. K., Rushton, E., and Bate, M. (1994). *dpp* induces mesodermal gene expression in *Drosophila. Nature* **372,** 22–29.
29. Frasch, M. (1995). Induction of visceral and cardiac mesoderm by ectodermal Dpp in the early *Drosophila* embryo. *Nature* **374,** 464–467.
30. Bate, M., and Rushton, E. (1993). Myogenesis and muscle patterning in *Drosophila. C. R. Acad. Sci. Paris* **316,** 1055–1061.
31. Baylies, M. K., Martinez-Arias, A., and Bate, M. (1995). *wingless* is required for the formation of a subset of muscle founder cells during *Drosophila* embryogenesis. *Development* 3829–3837.
32. Carmena, A., Bate, M., and Jiménez, F. (1995). *lethal of scute,* a proneural gene, participates in the specification of muscle progenitors during *Drosophila* embryogenesis. *Genes Dev.* **9,** 2373–2383.
33. Campos-Ortega, J. A, (1993). Early neurogenesis in *Drosophila melanogaster:* "The development of *Drosophila melanogaster*" (M. Bate and A. Martinez-Arias, eds.), Vol. 2, pp. 1091–1129. Cold Spring Harbor Laboratory Press, Cold Spring Harbor, NY.
34. Corbin, V., Michelson, A. M., Abmayr, S. M., Neel, V., Alcamo, E., Maniatis, T., and Young, M. W. (1991). A role for the *Drosophila* neurogenic genes in mesoderm differentiation. *Cell* **67,** 311–323.
35. Bate, M., Rushton, E., and Frasch, M. (1993). A dual requirement for neurogenic genes in *Drosophila* myogenesis. *Development Suppl.* 149–161.
36. Bate, M., Rushton, E., and Currie, D. A. (1991). Cells with persistent *twist* expression are the embryonic precursors of adult muscles in *Drosophila. Development* **113,** 79–89.
37. Currie, D. A., and Bate, M. (1991). The development of adult abdominal muscles in *Drosophila:* Myoblasts express *twist* and are associated with nerves. *Development* 91–102.
38. Fernandes, J., Bate, M., and Vijayraghavan, K. (1991). Development of the indirect flight muscles of *Drosophila. Development* **113,** 67–77.
39. Bate, M. (1990). The embryonic development of larval muscles in *Drosophila. Development* **110,** 791–804.
40. Dohrmann, C., Azpiazu, N., and Frasch, M. (1990). A new *Drosophila* homeobox gene is expressed in mesodermal precursor cells of distinct muscles during embryogenesis. *Genes Dev.* **4,** 2098–2111.
41. Rushton, E., Drysdale, R., Abmayr, S. M., Michelson, A. M., and Bate, M. (1995). Mutations in a novel gene, *myoblast city,* provide evidence in support of the founder cell hypothesis for *Drosophila* muscle development. *Development* **121,** 1979–1988.
42. Abmayr, S. M., Erickson, M. S., and Bour, B. A. (1995). Embryonic development of the larval body wall musculature of *Drosophila melanogaster. Trends Genet.* **11,** 153–159.
43. Nose, A., Isshiki, T., and Takeichi, M. (1998). Regional specification of muscle progenitors in *Drosophila:* The role of the *msh* homeobox gene. *Development* **125,** 215–223.
44. Prokop, A., Landgraf, M., Rushton, E., Broadie, K., and Bate, M. (1996). Presynaptic development at the *Drosophila* neuromuscular junction: The assembly and localization of presynaptic active zones. *Neuron* **17,** 617–626.
45. Erickson, M. R. S., Galletta, B. J., and Abmayr, S. M. (1997). *Drosophila myoblast city* encodes a conserved protein that is essential for myoblast fusion, dorsal closure, and cytoskeletal organization. *J Cell Biol.* **138,** 589–603.
46. Ruiz-Gómez, M., and Bate, M. (1997). Segregation of myogenic lineages in *Drosophila* requires Numb. *Development* **124,** 4857–4866.
47. Uemura, T., Shepherd, S., Ackerman, L., Jan, L. Y., and Jan, Y. N. (1989). *numb,* a gene required in determination of cell fate during sensory organ formation in *Drosophila* embryos. *Cell* **58,** 349–360.

48. Spana, E., Kopczynski, C., Goodman, C. S., and Doe C. Q. (1995). Asymmetric localization of Numb autonomously determines sibling neuron identity in *Drosophila* CNS. *Development* **121,** 3489–3494.

49. Rhyu, M. S., Jan, L. Y., and Jan, Y. N. (1994). Asymmetric distribution of Numb protein during division of the sensory organ precursor cell confers distinct fates to daughter cells. *Cell* **76,** 477–491.

50. Guo, M., Jan, L. Y., and Jan, Y. N. (1996). Control of daughter cell fates during asymmetric division: Interaction of Numb and Notch. *Neuron* **17,** 27–41.

51. Frasch, M., Hoey, T., Rushlow, C., Doyle, H., and Levine, M. (1987). Characterization and localization of the *even-skipped* protein of *Drosophila. EMBO J.* **6,** 749–759.

52. Gaul, U., Seifert, E., Schuh, R., and Jackle, H. (1987). Analysis of *Krüppel* protein distribution during early *Drosophila* development reveals posttranscriptional regulation. *Cell* **50,** 639–647.

53. Ruiz-Gómez, M., Romani, S., Hartmann, C., Jäckle, H., and Bate, M. (1997). Specific muscle identities are regulated by *Krüppel* during *Drosophila* embryogenesis. *Development* **124,** 3407–3414.

54. Bourgouin, C., Lundgren, S. E., and Thomas, J. B. (1992). *apterous* is a *Drosophila* LIM domain gene required for the development of a subset of embryonic muscles. *Neuron* **9,** 549–561.

55. Keller, C. A., Erickson, M. S., and Abmayr, S. M. (1997). Misexpression of *nautilus* induces myogenesis in cardioblasts and alters the pattern of somatic muscles fibers. *Dev. Biol.* **181,** 197–212.

56. Brand, A. H., and Perrimon, N. (1993). Targeted gene expression as a means of altering cell fates and generating dominant phenotypes. *Development* **118,** 401–415.

57. Hartmann, C., Landgraf, M., Bate, M., and Jäckle, H. (1997). *Krüppel*-dependent *knockout* activity is required for proper innervation of a specific set of *Drosophila* larval muscles. *EMBO J.* **16,** 5299–5309.

58. Carmena, A., Murugasu-Oei, B., Menon, D., Jimenez, F., and Chia, D. (1998). *inscuteable* and *numb* mediate asymmetric muscle progenitor divisions during *Drosophila* myogenesis. *Genes Dev.,* in press.

59. Weintraub, H., Davis, R., Tapscott, S., Thayer, M., Krause, M., Benezra, R., Blackwell, T. K., Turner, D., Rupp, R., Hollenberg, S., Zhuang, Y., and Lassar, A. (1991). The *MyoD* gene family: Nodal point during the specification of the muscle cell lineage. *Science* **251,** 761–766.

60. Paterson, B. M., Walldorf, U., Eldridge, J., Dubendorfer, A., Frasch, M., and Gehring, W. J. (1991). The *Drosophila* homologue of vertebrate myogenic-determination genes encodes a transiently expressed nuclear protein marking primary myogenic cells. *Proc. Natl. Acad. Sci. USA* **88,** 3782–3786.

61. Michelson, A. S., Abmayr, S., Bate, M., Arias, A. M., and Maniatis, T. (1990). Expression of a *Myo-D* family member prefigures muscle pattern in *Drosophila* embryos. *Genes Dev.* **4,** 2086–2097.

62. Doberstein, S. K., Fetter, R. D., Mehta, A. Y., and Goodman, C. S. (1997). Genetic analysis of myoblast fusion: *blown fuse* is required for progression beyond the prefusion complex. *J. Cell Biol.* **136,** 1249–1261.

63. Paululat, A., Burchard, S., and Renkawitz-Pohl, R. (1995). Fusion from myoblasts to myotubes is dependent on the *rolling stone* gene (*rost*) of *Drosophila. Development* **121,** 2611–2620.

64. Paululat, A., Goubeaud, A., Damm, C., Knirr, S., Burchard, S., and Renkawitz-Pohl, R. (1997). The mesodermal expression of *rolling stone* (*rost*) is essential for myoblast fusion in *Drosophila* and encodes a potential transmembrane protein. *J. Cell Biol.* **138,** 337–348.

65. Drysdale, R., Rushton, E., and Bate, M. (1993). Genes required for embryonic muscle development in *Drosophila melanogaster:* A survey of the X chromosome. *Roux's Arch. Dev. Biol.* **202,** 276–295.
66. Brown, N. H. (1993). Integrins hold *Drosophila* together. *BioEssays* **15,** 383–390.
67. Buttgereit, D., Leiss, D., Michiels, F., and Renkawitz-Pohl, R. (1991) During *Drosophila* embryogenesis the β_1 tubulin gene is specifically expressed in the nervous system and the apodemes. *Mech. Dev.* **33,** 107–118.
68. Armand, P., Knapp, A. C., Hirsch, A. J., Wieschaus, E. F., and Cole, M. D. (1994). A novel basic helix-loop-helix protein is expressed in muscle attachment sites of the *Drosophila* epidermis. *Mol. Cell. Biol.* **14,** 4145–4154.
69. Volk, T., and VijayRaghavan, K. (1994). A central role for epidermal segment border cells in the induction of muscle patterning in the *Drosophila* embryo. *Development* **120,** 59–70.
70. Lee, J. C., VigayRaghavan, K., Celniker, S. E., and Tanouye, M. A. (1995). Identification of a *Drosophila* muscle development gene with structural homology to mammalian early growth response transcription factors. *Proc. Natl. Acad. Sci. USA* **92,** 10344–10348.
71. Frommer, G., Vorbrüggen, G., Pasca, G., Jäckle, H., and Volk, T. (1996). Epidermal egr-like zinc finger protein of *Drosophila* participates in myotube guidance. *EMBO J.* **15,** 1642–1649.
72. Vorbrüggen, G., and Jäckle, H. (1997). Epidermal muscle attachment site-specific target gene expression and interference with myotube guidance in response to ectopic *stripe* expression in the developing *Drosophila* epidermis. *Proc Natl. Acad. Sci. USA* **94,** 8606–8611.
73. Becker, S., Pasca, G., Strumpf, D., Min, L., and Volk, T. (1997). Reciprocal signaling between *Drosophila* epidermal muscle attachment cells and their corresponding muscles. *Development* **124,** 2615–2622.58.

DEVELOPMENT OF ELECTRICAL PROPERITIES AND SYNAPTIC TRANSMISSION AT THE EMBRYONIC NEUROMUSCULAR JUNCTION

Kendal S. Broadie

Department of Biology, University of Utah, Salt Lake City, Utah 84112

I. Introduction
II. Physiological Maturation of Electrical Properties in Embryonic Myotubes
III. Glutamate Receptors and Maturation of the Postsynaptic Glutamate-Gated Current in Embryonic Myotubes
IV. Neural Induction of Postsynaptic Specialization
V. Retrograde Induction of Presynaptic Specialization
VI. Physiological Maturation of Synaptic Transmission
VII. Development of Synaptic Modulation Properties
VIII. Methods for Electrophysiological Assays of the Embryonic Neuromuscular Junction
IX. Conclusions and Perspectives
References

I. Introduction

The synapse is a specialized intercellular junction devoted to communication between two electrically excitable cells. In most organisms, synapse formation, or synaptogenesis, begins during early to mid embryogenesis when the primary neural circuits are formed, but often continues postembryonically and, in specialized circumstances, can be maintained for the lifetime of the organism. For practical purposes, this review defines synaptogenesis as the formation of the intercellular communication link: downstream of neuronal pathfinding and target recognition, but upstream of the established mechanism of synaptic transmission and the maintained synaptic developmental potential known as synaptic plasticity. The process of synaptogenesis defines the site of the intercellular junction and provides for the precise alignment of the presynaptic signaling apparatus and the postsynaptic receptor field. Inductive communication between synaptic partners determines both the molecular nature and the transmission strength of the assembling synapse.

Transmission of information at the synapse depends on precisely orchestrated electrical signals coupled to the activity of an array of pre- and postsynaptic proteins. To accomplish high fidelity information transfer,

0074-7742/99 $30.00

tightly concentrated protein assemblies must be localized on both sides of the developing synaptic cleft in strict spatial and temporal synchrony. Postsynaptically, this requires the synthesis, assembly, and localization of a variety of ion channels, including voltage-gated, Ca^{2+}-gated, and ligand (neurotransmitter)-gated channels, and the cytoskeletal framework required to maintain their concentration in the postsynaptic membrane. Delayed specializations often include the formation of postsynaptic morphological characteristics, most prominently membrane folding, the purpose of which is to maximize the membrane area surrounding the postsynaptic transmitter receptor field. Presynaptically, the challenge is even greater, requiring not only a similar assembly of transmembrane ion channels, but also the assembly of the synaptic vesicle exocytosis and endocytosis complexes composed of transmembrane, membrane-associated, and cytosolic proteins. In addition, a variety of presynaptic cytosolic components must be marshaled at the developing membrane fusion sites; these include mitochondria, synaptic vesicles, and the organelle intermediates of the synaptic vesicle cycle and the presynaptic cytoskeletal network required to tether and regulate the movement of all these components. Delayed specializations often include the swelling of the nerve terminal surrounding presynaptic active zones to form varicosities, known as boutons, which serve as storehouses for the reserve synaptic vesicle population.

Because of its large size, relative simplicity, and accessibility for experimental manipulation, the leading model of synaptogenesis has always been the neuromuscular junction. The attraction of the *Drosophila* neuromuscular junction is that it combines these features in a system amenable also to systematic molecular genetic analyses. Thus, we are able to harness genetic mutations as tools to dissect the mechanisms of synaptogenesis as well as to screen directly for new genes/molecules required for synapse formation. This review focuses exclusively on the development of electrical properties and synaptic transmission at the embryonic *Drosophila* neuromuscular junction. The discussion includes the development of muscle electrical properties, pre- and postsynaptic maturation of neuromuscular junction transmission characteristics, and a brief discussion of experimental methods used to obtain electrophysiological recordings from the *Drosophila* embryonic preparation. Related aspects of morphological synaptogenesis can be found in earlier and later reviews of this volume.

II. Physiological Maturation of Electrical Properties in Embryonic Myotubes

Prior to neuromuscular target recognition, myoblast fusion is complete and the newly formed syncitial myotubes have differentiated their final attachments to the epidermis, generating the mature larval muscle pattern.[1] At this stage [12–13 hr after fertilization (AEL) at 25°C], the myotubes are extensively electrically and dye coupled together into large multifiber networks.[2–4] This coupling includes myotubes of different positional classes (longitudinal, transverse, oblique) within a hemisegment as well as extensive coupling across segmental borders. Analysis of muscle electrical properties at this stage is difficult as the extensive coupling prevents voltage clamp recordings of ionic conductances.[5] However, the myotubes abruptly uncouple during early stage 16 (13–13.5 hr AEL; Fig. 1).[6] At this point,

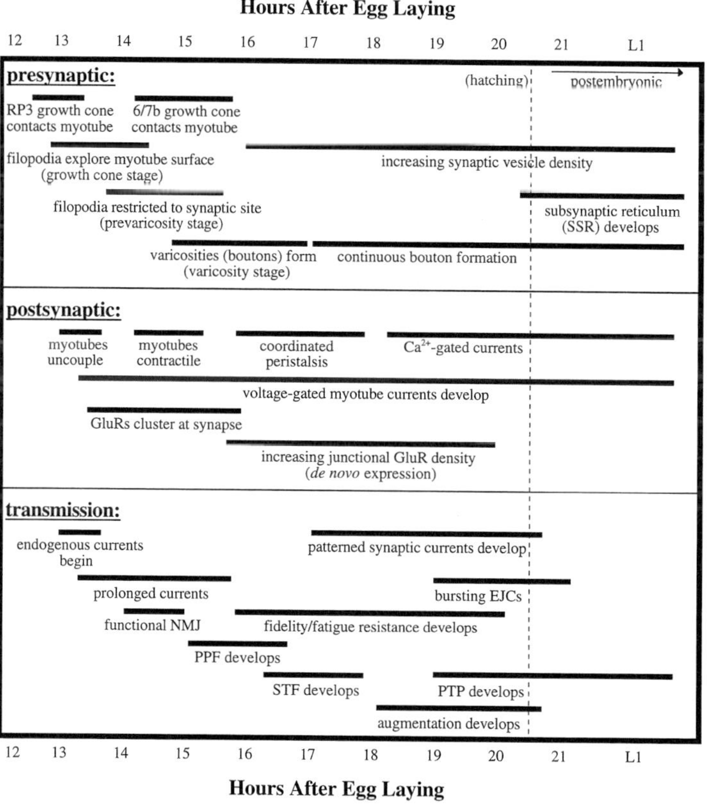

FIG. 1. A time line of neuromuscular junction (NMJ) synaptogenesis on ventral longitudinal muscle 6 from wild-type strain Oregon-R. Time is given in hours after fertilization at 25°C for the 8- to 9-hr period encompassing embryonic synaptogenesis. For clarity, events are segregated into morphological events in the presynaptic terminal (top), postsynaptic development (middle), and synaptic transmission characteristics (bottom). Due to a dorsal–ventral development gradient, synaptogenesis at dorsal neuromuscular junctions is delayed relative to the timing shown here.

dye coupling is restricted to <10% of dye-injected myotubes; a figure that remains constant for the remainder of embryonic and larval development.[2,4,7] Although relatively strong electrical coupling between myotubes remains detectable, a tight voltage clamp of the myotubes is now readily accomplished. At maturity, the intersegmental coupling coefficients between muscles 6 and 7 remain high (range of 0.33–0.43), although intrasegmental coupling between these muscles is lower (coupling coefficient of 0.16) and coupling with other classes of muscles is greatly reduced or absent.[7] The purpose of this electrical coupling may be to aid the coordination of muscle contraction during locomotion.

Five prominent voltage-activated currents are observed in the mature larval muscle: a voltage-gated calcium current (I_{ca}), two voltage-gated potassium currents (I_A, I_K), and two calcium-gated potassium currents (I_{CF}, I_{CS}).[8–10] These currents appear and mature in embryonic myotubes in a consistent order: (1) I_{ca}, (2) I_A /I_K, and (3) I_{CF}/I_{CS}.[5,11] The first to appear when the myotubes uncouple (13–13.5 hr AEL) is a small inward, voltage-gated I_{Ca}. This current clearly precedes the development of any other detectable voltage-gated current and is present immediately after myotube uncoupling, suggesting that it is also present prior to uncoupling (Fig. 1).[5] Second, the two outward, voltage-gated K^+ currents appear within 1 hr of myotube uncoupling at 13–14 hr AEL: (1) a fast, inactivating A current (I_A) and (2) a delayed, noninactivating K current (I_K).[11] Both currents appear in the muscle at nearly the same time and rapidly mature through later myogenesis.[5] Third, very late in embryogenesis (mid-late stage 17), the outward, calcium-dependent K^+ currents appear (Fig. 1). The rapid, inactivating I_{CF} is first detected at 17–18 hr AEL and the delayed noninactivating I_{CS} appears shortly thereafter at 18–19 hr AEL.[5] Thus, all five voltage-dependent currents are present in the myotubes prior to hatching at 20–21 hr AEL.

The development of current density is specific for each class of current. When the myotubes uncouple, the I_{Ca} whole cell current is already prominent and the amplitude increases very gradually throughout embryogenesis so that at hatching the current has only doubled its peak amplitude.[5] Voltage-gated I_A and I_K appear after uncoupling and rapidly increase in amplitude during later myogenesis. I_K displays a rapid increase in amplitude early in myogenesis (13.5–15 hr AEL) and then a continued, gradual increase for the rest of embryogenesis. I_A displays a dramatic increase in amplitude for most of embryogenesis (13.5–18 hr AEL) but then actually decreases in amplitude prior to hatching.[5] The late-developing calcium-gated K^+ currents both show a slow, gradual rise in amplitude during the last 4 hr of embryogenesis.[5] As a result of these developmental profiles, voltage-gated K^+ currents dominate the whole cell current profile throughout late myogenesis, with I_A having the largest amplitude and I_K the second largest. The inward calcium current is largely masked by these outward K^+ currents but has the third largest amplitude. Calcium-gated K^+ currents contribute only marginally to the whole cell current profile, with I_{CF} making a slightly larger contribution than I_{CS} at embryonic maturity.

The only additional K^+ current documented in embryonic myotubes is a stretch-activated, outward K^+ current.[11] Stretch-activated channels are apparent throughout myogenesis and appear to be present immediately after myotube uncoupling, suggesting that they are also present prior to uncoupling.[5] Concentrations of the stretch receptors have not been reported and seem to be distributed widely throughout the myotube membrane. These channels have not been specifically studied and so their properties and physiological relevance remain to be tested.

In general, the voltage-dependent properties of embryonic myotube currents are similar to mature currents in the larval muscle. The only exception is the early embryonic I_A current (14 hr AEL), which has a midpoint of steady-state voltage inactivation that is 40 mV more negative than in the mature embryo (21 hr AEL) or the larva.[5] As myogenesis proceeds, the I_A steady-state inactivation curve develops a biphasic character, suggesting that a low voltage inactivation I_A channel is present in early development and is replaced by the mature high voltage inactivated form as development proceeds. The I_A current at all development stages can be completely eliminated in *Shaker* K^+ channel mutants (Sh^{KS133}), suggesting that the Shaker I_A channel may be present in two forms, embryonic and mature, during myogenesis.[5] Interestingly, the prominent I_A current present in early myogenesis appears to be almost entirely inactivated at the physiological resting potential of −60 mV. A prepulse at very negative potentials (e.g., −100 mV) is required to completely activate this current.[5] The significance and mechanisms of this developmental shift in channel properties remain unclear and have not been sufficiently investigated.

The developmental progression of voltage-gated current maturation is reflected in dynamic alterations in the voltage response of the myotube membrane. Prior to myotube uncoupling (13–13.5 hr AEL), the myotubes show a completely passive response to current injection.[5] After uncoupling and during most of early electrogenesis (14–18 hr AEL) the myotubes show only a small positive rectification in response to the depolarizing current. By 18 hr AEL, a larger outward rectification is apparent, but voltage responses appear restricted below 0 mV. At hatching (21 hr AEL), a regenerative action potential is often detected in response to depolarizing current injection.[5] This response has a threshold of -10 to -20 mV and peaks at 20 to 30 mV, and the regenerating potential continues for the duration of the current injection. Similar muscle action potentials have been observed in mature larval muscle, although independent investigators debate the existence and physiological relevance of such responses.[10,12] Interestingly, dynamic muscle voltage responses appear only after the developmental shift of the I_A K^+ current inactivation properties and the appearance of calcium-gated K^+ currents, suggesting that these currents may also play an important role in the development of mature muscle voltage responses.

III. Glutamate Receptors and Maturation of the Postsynaptic Glutamate-Gated Current in Embryonic Myotubes

Synaptic transmission at the neuromuscular junction is mediated by L-glutamate-gated neurotransmitter receptors (GluRs). The receptors are nonselective cation channels most permeable to monovalent cations, especially Na^+, and are markedly less permeable to divalent cations such as Ca^{2+}.[13–15] Thus, these channels are unlikely to activate muscle contraction directly but rather do so through the activation of voltage-dependent Ca^{2+} channels. The GluR channels are synthesized and expressed very early in myogenesis (<10 hr AEL) and are detectable by immunohistological staining in single myoblasts prior to fusion.[16] It is uncertain how early functional channels are present in the membrane, but they are detectable by electrophysiology recordings immediately after myotube uncoupling (13–13.5 hr AEL), suggesting that they are also functionally present earlier.[4,17] Two muscle-specific GluR genes, DGluRIIA and DGluRIIB, have been identified and their products have been shown to be expressed in embryonic myotubes.[18,19] Likewise, two distinct functional classes of GluRs have been reported in the embryo, although the relationship between genes and functional receptors has yet to be investigated.[20] Interestingly, the two GluR classes appear to segregate exclusively to the junctional and extrajunctional myotube membrane (Fig. 2), unlike the distribution of acetylcholine receptor classes in the vertebrate neuromuscular junction.

GluR properties have been assayed with cell-attached patches from myotubes both in culture and *in vivo* and appear to show similar characteristics in both systems.[20,21] Extrajunctional GluRs have a ~7-pA amplitude at −60 mV, a linear conductance of ~120 pS, and are highly subject to desensitization.[21] Junctional GluRs are not detectable using cell-attached patches outside of the junctional membrane domain, even in the perijunctional membrane, and therefore have only been assayed as components of synaptic currents in whole cell recordings.[20] Junctional receptors have two-step amplitudes at 9.5 and 18.5 pA at −60 mV, a reversal potential (E_{rev}) ~10 mV more positive than the extrajunctional receptor and are more

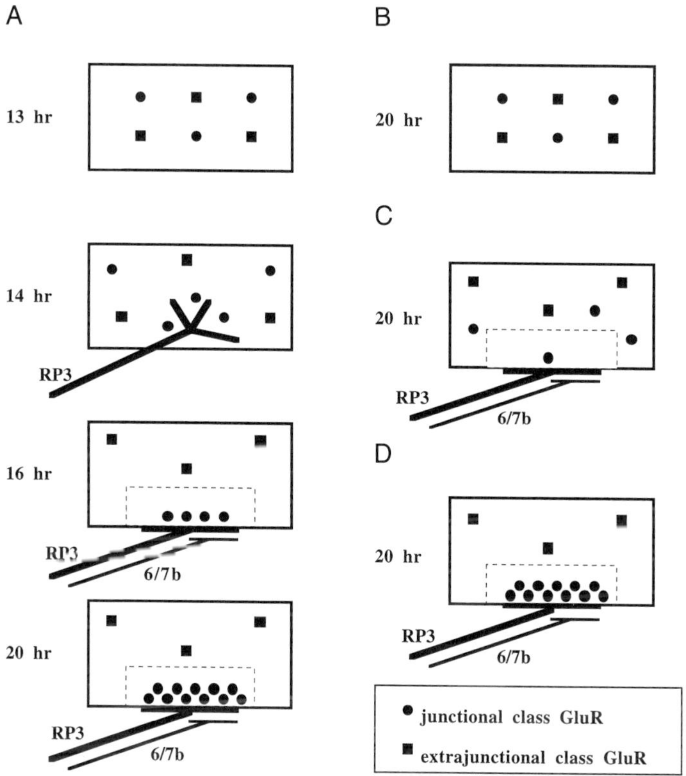

FIG. 2. Formation of the postsynaptic GluR field on muscle 6. (A) Time course of normal development. Distinct origins for junctional and extrajunctional GluRs are assumed, although it is possible that junctional GluRs represent a functionally modified version of the extrajunctional class in the postsynaptic membrane. Initially (13 hr AEL), GluRs appear widely distributed in the myotube. Following sequential innervation by motor neurons RP3 and 6/7b, GluRs cluster in the postsynaptic membrane and are segregated into junctional and extrajunctional classes. In late embryogenesis (20 hr AEL), GluR density increases dramatically in the postsynaptic membrane. (B and C) Preventing innervation (aneural myotube; B) or blocking action potential (AP) propagation (C) disrupts formation of the postsynaptic GluR field. However, blocking glutamatergic transmission by the transgenic expression of tetanus toxin (D) does not perturb GluR field formation.

resistant to desensitization. Thus, the two classes of GluRs differ in (1) mean unitary current amplitude, (2) reversal potential, and (3) desensitization properties. These two distinct types of GluRs are present throughout myogenesis at least from 14 hr AEL onward. It is presently unclear whether, as in vertebrates, these receptors represent distinct gene products (e.g., DGluRIIA and DGluRIIB) or result from a rapid modification of junctional type receptors following recruitment to the postsynaptic membrane. To date, glutamate channels are the only ligand-gated channels reported in embryonic myotubes. In particular, assays for inhibitory input, such as GABA receptor channel activity, have failed to find any detectable response, at least in muscle 6, suggesting that myotubes receive only excitatory glutamate innervation during embryogenesis.[4]

The distribution of GluRs in the myotube membrane during development has been followed with anti-GluR immunohistochemistry[22] and localized L-glutamate iontophoresis[4] onto discrete areas of the myotube surface (Fig. 2A). The GluR immunohistochemistry does not, as far as is known, distinguish between junctional and extrajunctional GluR types detected with electrophysiology. However, the iontophoresis technique appears to reveal only the larger amplitude junctional class of GluRs,[4] presumably due to the prominent desensitization properties of the lower amplitude extrajunctional receptors.[14,21] During early myogenesis (<13.5 hr AEL), functional glutamate receptors detected by iontophoresis are distributed broadly throughout the myotube membrane. Immunohistological studies suggest that GluRs may form clusters near myotube nuclei prior to innervation, although iontophoresis records have not detected such GluR hot spots.[4,22] Following innervation, (13–14 hr AEL), GluRs begin to concentrate in the developing synaptic cleft and are progressively lost from extrajunctional regions of the myotube membrane (Fig. 2). The redistribution of GluRs continues for the next several hours so that by 15–16 hr AEL functional receptors are rarely detected outside of the synaptic cleft region.[4] The cell-attached patch recording shows that the extrajunctional class of GluRs remains present at a low density in regions removed from the neuromuscular junction (Fig. 2A).[20] The increasing density of junctional GluRs can be assayed by glutamate iontophoresis at the synaptic cleft during progressive stages of development.[4] When the myotubes uncouple (13–13.5 hr AEL), usually only single channel openings are detected. Over the next several hours (13–16 hr AEL) this response quickly increases to an average of 25 massed receptor openings and accumulation continues through later stages of synaptogenesis until at hatching (20–21 hr AEL) several hundred GluRs can be activated in the synaptic zone. The amplitude of the glutamate-evoked postsynaptic current compares well with the maximum nerve-evoked excitatory junctional current (EJC) throughout synaptogenesis.[4]

Development of the glutamate-gated postsynaptic current appears to occur in two stages (Figs. 1 and 2A). First, at 13–16 hr AEL, preexisting GluRs migrate to the synaptic cleft through lateral diffusion and are selectively stabilized in this domain.[4] The result is that GluRs are rapidly concentrated in the postsynaptic membrane. A low density of extrajunctional GluRs remains in the myotube surface, but the relationship of these receptors to the junctional GluRs and their physiological role remain unclear. Second, at 16–21 hr AEL, *de novo* synthesis of new GluRs dramatically increases the density of GluRs in the postsynaptic membrane (Figs. 1 and 2A).[4] It is likely that these receptors are targeted nonselectively to the plasma membrane and are later trapped in the postsynaptic zone through a mechanism similar to that used to concentrate preexisting GluRs4 and other postsynaptic components.[23] However, if this hypothesis is correct, the synaptic localization of new junctional receptors must be very rapid, as this class of receptors cannot be detected in recordings from the extrajunctional membrane during late embryogenesis.[4,20]

IV. Neural Induction of Postsynaptic Specialization

The arrival of the presynaptic growth cone correlates closely with the onset of physiological maturation in the target myotube. At the time of initial contact, myotubes abruptly uncouple and commence the rapid program of membrane electrogenesis described earlier (Fig. 1).[4,5] However, despite this temporal coincidence, most aspects of myotube physiological maturation occur independently of neural induction. This was demonstrated with the use of a mutation in *prospero,* a gene involved in neuronal fate decisions. In *prospero* mutant embryos, peripheral innervation is delayed or eliminated, allowing the assay of innervation-dependent aspects of myotube differentiation.[24] In the absence of innervation, myotubes uncouple at the correct time. Furthermore, normal electrical properties appear and develop to maturity with the normal time course.[24] This normal development extends to all aspects of muscle electrical properties, including the developmental shift in I_A inactivation properties. By the end of embryogenesis, aneural muscle has normal morphology, contracts normally in response to a depolarizing current injection, and has the full range of mature physiological electrical properties.[24] Thus, myotube differentiation per se appears to occur independently of neural direction.

The site of the putative synaptic zone also appears to be correctly specified by the myotube in the absence of innervation.[25] Homophilic cell adhesion molecules are normally expressed on the neuronal growth cone and the muscle synaptic domain prior to contact, supporting the hypothesis that specific adhesive interaction may guide the selection of appropriate synaptic sites. In *prospero* mutants that lack innervation, these adhesive molecules are still expressed on the myotube at the correct developmental stage and localize to delineate the normal synaptic site. For example, the cell adhesion molecule fasciclin III is expressed only in the synaptic domain in the central cleft between muscles 6 and 7 in aneural muscle.[25] These findings suggest that the muscle may define the future synaptic domain of the neuromuscular junction and is capable of localizing transmembrane proteins to this specific membrane domain in the absence of neural direction.

In contrast, the maturation of the postsynaptic receptor field is critically dependent on neural induction, similar to the inductive interaction during development of the vertebrate neuromuscular junction. In aneural muscle, functional GluRs are expressed in the muscle normally but fail to aggregate to the synaptic domain (Figs. 2A, and 2B).[25] Moreover, the density of GluRs in the membrane does not increase substantially in the absence of innervation. These observations suggest that the presynaptic terminal is required to induce GluR aggregation in the postsynaptic membrane and trigger the later wave of accelerated GluR synthesis. This conclusion is supported by assays of *prospero* mutants where myotubes are often innervated at abnormally late developmental stages or at aberrant, ectopic locations. Under these conditions, GluRs are recruited to the synaptic site only on innervation, regardless of its location.[25] These observations suggest that the development of the postsynaptic GluR field is wholly dependent on presynaptic direction, which is both necessary and sufficient to induce postsynaptic maturation.

The molecular nature of the presynaptic inducing signal(s) has not been demonstrated. However, one clue to the possible induction mechanism stems from the observation that GluR aggregation is dependent on presynaptic electrical activity.[22,26] This suggestion was demonstrated by reducing or blocking presynaptic action potentials through (1) pharmacological means such as tetrodotoxin (TTX), which interferes with Na^+ channels, or (2) mutations in the *paralytic* gene, which encodes a voltage-gated Na^+ channel.[22,26] Results showed that a complete blockade of action potential propagation could eliminate formation of the postsynaptic GluR field (Fig. 2C). However, this blockade had to be complete to show this effect. For example, *paralytic* null mutants at physiological temperatures show only minimal neuromuscular junction transmission but display a normally patterned postsynaptic GluR field.[26] Only in *paralytic* mutants reared at high temperatures (>30°C), which show a complete loss of action potential propagation, is GluR field formation effectively prevented (Fig. 2C). This effect seemed to result directly from the action potential blockade, as mutant embryos that were downshifted to permissive temperatures late in embryogenesis initiated postsynaptic GluR field formation with the normal time course.[26] It should be stressed that there appears to be no close correlation between the degree of electrical activity and the size or density of the GluR field. For example, mutant embryos with activity levels <10% of normal (*paralytic* null mutants, permissive temperature) or >200% of normal (*eag; Sh* double mutants) do not show a pronounced developmental alteration in their postsynaptic GluR field during embryogenesis.[26] Consequently, the physiological relevance of electrical activity as a direct postsynaptic inducing signal is questionable.

One possible mechanism for the regulation of GluR aggregation and synthesis is the action of the ligand, glutamate. However, blockade of the neuromuscular junction transmission does not strongly affect postsynaptic GluR field formation (Fig. 2D). For example, mutations of synaptotagmin or syntaxin, which result in severe impairment or loss of synaptic transmission, do not perturb the postsynaptic response to glutamate iontophoresis.[27–29] Likewise, the transgenic presynaptic expression of tetanus neurotoxin, which completely blocks synaptic transmission, also does not perturb GluR field formation in the myotube (Fig. 2D).[30] From these studies, it can be concluded that normal glutamatergic synaptic transmission is not required for induction of the postsynaptic receptor field and that the requirement for a minimal level of presynaptic action potential propagation must be affecting an independent induction pathway. A caveat to this conclusion is that none of these experiments have blocked GluRs directly and that nonsynaptic or nonvesicular glutamate release, which is not readily detectable, cannot be excluded as a possible presynaptic signal. The nature of the presynaptic inducing signal remains unknown at this time.

V. Retrograde Induction of Presynaptic Specialization

Retrograde signals from the developing myotube also induce specialization in the presynaptic terminal. Retrograde signaling was demonstrated by using mutations in genes encoding transcription factors involved in successive stages of myogenesis.[31] These include *twist* mutants, which show no mesodermal differentiation, *myocyte-enhancer factor-2* (*mef-2*) mutants, which develop an undifferentiated array of myoblasts but no muscle, and *myoblast city* (*mbc*) mutants, which lack myoblast fusion but otherwise develop fully differentiated mononucleate muscles (Fig. 3). These mutants were used to assay presynaptic maturation in the complete absence of postsynaptic targets or under conditions where postsynaptic maturation was blocked at specific points in the myogenesis program.

Fully differentiated presynaptic boutons developed in all these mutants, demonstrating that target-derived signals are not involved in triggering morphological specializations.[28,31] These boutons contained structures iden-

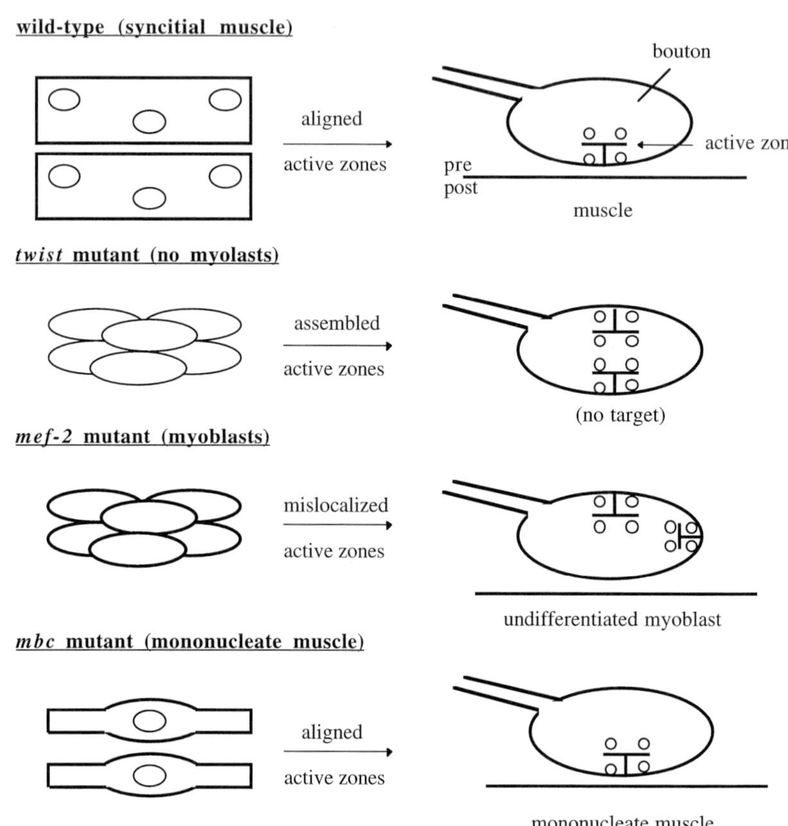

FIG. 3. A muscle-derived retrograde signal is required for presynaptic active zone localization. Wild-type embryos possess syncytial myotubes and precisely align presynaptic active zones with the postsynaptic target. In *twist* mutants, no mesoderm is formed but presynaptic terminals differentiate normally. In *mef-2* mutants, myoblasts fail to differentiate and these immature targets are unable to specify presynaptic active zone localization. In *mbc* mutants, myoblast fusion is disrupted but differentiated mononucleate myotubes are able to specify presynaptic active zone localization. Thus, a muscle-derived retrograde signal is required to direct the appropriate placement of presynaptic active zones; this signal depends on *mef-2* but not mbc function.

tical to presynaptic active zones, including electron-dense membrane containing characteristic membrane-associated T bars, which are believed to be the site of synaptic vesicle fusion (Fig. 3).[31] In the absence of postsynaptic targets, synaptic vesicles were docked at these active sites and clustered in the cytoplasm in small clouds surrounding these regions. Thus, target-derived signals are not required for the basic processes of presynaptic maturation, including presynaptic bouton formation, synthesis of active zone components, or assembly of active zones and associated synaptic vesicle pools.

In contrast, the correct localization and alignment of presynaptic active zones appears to be critically dependent on a muscle-derived retrograde signal. For example, in *twist* or *mef-2* mutants, active zones appear to be distributed randomly in the neuronal membrane and, in the case of *mef-2* mutants, fail to localize opposite myoblast membranes (Fig. 3).[31] In *mbc* mutants, however, active zones always exactly align with an underlying postsynaptic membrane of the mononucleate muscle. Moreover, *mbc* active zones are functional and display apparently normal synaptic vesicle release and GluR activation in the muscle.[31] These results show that a mef-2-dependent retrograde localization signal is required for the correct presynaptic placement of active zones (Fig. 3). Thus, pre- and postsynaptic cells have similar inductive requirements; in both, the basic elements of the signaling pathway assemble independently of intercellular interaction (transmitter receptors in the muscle, active zones in the neuron) but only localize appropriately to the synaptic site in response to inductive signals from the synaptic partner (Figs. 2 and 3).[25,31] The identity of the mef-2-dependent retrograde signal remains unknown at this time.

VI. Physiological Maturation of Synaptic Transmission

A time table of comprehensive physiological maturation has been established for the muscle 6/7 neuromuscular junction (Fig. 1).[4,32,33] Comparable analyses have been performed on the ventral oblique muscles 15–17 and the dorsal longitudinal muscles 1 and 9.[17,20,34] Similar conclusions have been derived from these studies, although the timing of events differs considerably due to a pronounced delay in dorsal development relative to ventral events. The sequence of events below is based on development of the ventral muscle 6/7 neuromuscular junction.[4] In larvae, these two muscles are coinnervated by two motorneurons, RP3 and 6/7b, both of which differentiate the larger type I glutamatergic terminals. Embryonic synaptogenesis of this junction commences at 12–13 hr AEL with initial neuromuscular contact and continues through hatching at 21 hr AEL (Fig. 1).[4] The initial pioneer contact is made by RP3 followed within 1–2 hr by 6/7b, which forms an overlapping terminal at the predefined junctional zone (Fig. 2).[4] Electrophysiological recordings have been taken from the larger muscle 6 in all analyses.

Synaptic transmission of the neuromuscular junction is detectable within a few minutes of motor neuron contact of the target muscle. When the myotubes uncouple (13–13.5 hr AEL), activation of isolated GluRs by spontaneous transmission from the growth cone is immediately detected, suggesting that the growth cone is synthesizing and releasing the transmitter at or prior to neuromuscular contact (Fig. 1).[4,17] This conclusion is supported by the observation that strong glutamate immunoreactivity is detected in the growth cone at, or prior to, neuromuscular contact.[4] The initial period (~1 hr) of synaptic transmission is highly labile/variable, reflecting the absence of a strong GluR field as well as a very low level of glutamate release from the immature presynaptic terminal. The amplitude of initial transmission events consists of quantal multiples of one to several GluR channel openings.[4] Initial transmission is almost certainly limited by the low density of postsynaptic GluRs. Indeed, throughout synaptogenesis, it appears that postsynaptic receptor capacity always limits the degree of transmission as the amplitude of spontaneous and evoked transmission correlates with current amplitude obtained from direct glutamate iontophoresis.[4] Because of this restriction, it cannot be determined whether initial transmitter release is quantal, nonquantal, or a combination of the two. One indication that nonquantal release may occur in the early neuromuscular junction is the appearance of prolonged (hundreds of milliseconds to seconds) transmission events interspersed with "normal" synaptic transmission events of <10 msec (Fig. 1).[4] These long time course currents are characteristic of the very early synapse (13–16 hr AEL) but become increasingly rare and are essentially absent by mid-synaptogenesis (17 hr AEL).

The neuromuscular junction first becomes fully functional at 14.5–15.5 hr AEL when direct electrical stimulation of the motor nerve results in muscle contraction.[4] This developmental period also corresponds to the first endogenous movements by the embryo, although early movement is sporadic and does not display the coordinated peristalsis of locomotory movements (Fig. 1). Endogenous synaptic transmission through the early period (up to 16–18 hr AEL) does not appear patterned. Rather a range of currents is observed from single GluR channel openings through to the largest macroscopic currents, the amplitude of which increases rapidly and steadily as development proceeds.[4] Coordinated peristaltic contractions are first observed at 16–18 hr AEL and become increasingly precise and robust as development proceeds. Concurrently, synaptic transmission events begin to show a rough pattern with larger currents clustered into bursts of activity bracketed by periods of relative quiescence (Fig. 1). Toward the end of embryogenesis (18–21 hr AEL), synaptic transmission is refined to periodic bursts of high-frequency (5–20 Hz) synaptic currents separated by periods nearly free of transmission.[4,35] Each burst of synaptic current underlies a single muscle contraction and is precisely orchestrated between neuromuscular junctions in the same and adjacent segments. Persistent electrical coupling between myotubes is revealed by the existence of small currents with slow rise and fall times that arise from synaptic events in coupled myotubes.[17,28] The physiological relevance of this coupling is uncertain but it may play some role in coordinating muscular activity.

An interesting feature of embryonic transmission is the reduced calcium dependence of synaptic vesicle release relative to the mature terminal.[27] At all synapses, the likelihood of synaptic vesicle release is strongly related to extracellular [Ca^{2+}], which appears to be the limiting factor of transmission and is linearly related to evoked current amplitude at low [Ca^{2+}] (i.e., 0.1 to 0.4 m*M*). In low [Ca^{2+}], transmission amplitude at the mature *Drosophila* neuromuscular junction, like other synaptic systems, shows a three to four power dependence on external [Ca^{2+}]. This observation has led to the hypothesis that four Ca^{2+} ions must act cooperatively to trigger the fusion of a single synaptic vesicle. In contrast, the *Drosophila* embryonic neuromuscular junction displays Ca^{2+} dependence half that of the mature terminal (power dependence of one to two) at the end of embryogenesis.[27] Therefore, the Ca^{2+} dependence of transmission must alter during postembryonic maturation. This alteration may reflect a change in the basic Ca^{2+}-sensing mechanisms coupled to synaptic vesicle fusion or some other developmental modification of the presynaptic signaling apparatus, but the physiological significance of this alteration is not clear.

The early terminal shows very weak, low fidelity transmission properties.[4] Prior to 15 hr AEL, the terminal rapidly fatigues in response to even very low stimulation frequencies (<5 Hz). This fatigue is manifest as a complete loss of transmission within a few seconds of maintained stimuli, which require up to many tens of seconds to fully recover. The strengthening of fatigue-resistance properties is a slow process that occurs over several hours (Fig. 1).[4] After 16 hr AEL, the terminal is able to show maintained transmission at low stimulation frequencies (<5 Hz), and fatigue induced by higher stimulation frequencies (>10 Hz) recovers within a few seconds. Moreover, during this period the terminal shows very poor transmission fidelity in that a given stimulus results in a wide variation of transmission amplitude events. For example, at 15 hr AEL, a stimulus train of 1 Hz (subfatigue frequency) results in a roughly equal frequency of transmission events from single GluR channel openings through the maximum macroscopic amplitude, as well as 15–20% transmission failures.[4] Transmission fidelity refines at the same time as fatigue resistant properties (Fig. 1). This increase in fidelity probably reflects both the increasing density of the postsynaptic GluR field and the reliability of synaptic vesicle release mechanisms in the presynaptic terminal.[4] By 18–20 hr AEL, the neuromuscular junction shows robust, high-fidelity transmission with few or no transmission failures at stimulation frequencies up to 10 Hz and evoked synaptic current amplitudes clustered near the maximum current amplitude.

The maturation of fidelity and fatigue-resistance properties correlates with the development of reserve synaptic vesicle pools in the presynaptic terminal. Prior to 15 hr AEL, ultrastructural examination of the newly forming terminal reveals a low density of synaptic vesicles, and immunohistological staining at the light microscope level for synaptic vesicle proteins, such as synaptotagmin, shows only a weak signal.[34] However, both synaptic vesicle density and synaptotagmin labeling increases dramatically beginning at 16 hr AEL. Ultrastructural images show a rapid increase in synaptic vesicle density and presynaptic cluster size at this developmental stage.[34] This synaptic vesicle accumulation correlates closely with the strengthening and refinement of neuromuscular junction transmission properties (Fig. 1). Therefore, it appears that maturation of fidelity and fatigue-resistance properties is critically dependent on the development of adequate presynaptic synaptic vesicle pools.

VII. Development of Synaptic Modulation Properties

Synaptic modulation properties are the last aspect of transmission characteristics to mature during synaptogenesis (Fig. 1). All forms of facilitory synaptic modulation are effectively absent prior to 15 hr AEL and mature only slowly during late synaptogenesis (K. S. Broadie, unpublished observations). There are two possible explanations for this delayed maturation. First, modulation properties are likely to be obscured by immature transmission properties such as fatigue and low-fidelity transmission, as the high-frequency stimulation used to induce most forms of facilitation is likely to rapidly deplete limiting synaptic vesicle pools.[34] Moreover, in general, most forms of facilitation at the embryonic neuromuscular junction can only be detected in low external [Ca^{2+}] (i.e., 0.2 m*M*) as synaptic depression or fatigue tends to dominate in higher [Ca^{2+}].[36] This observation is true to a lesser extent even at the mature larval neuromuscular junction. Second, the molecular mechanisms involved in synaptic modulation may be slow developing. For example, a protein called Leonardo, which encodes a 14-3-3 protein known to regulate long-term synaptic modulation properties at the neuromuscular junction, is first detected in the embryonic terminal at 16–18 hr AEL.[36] This expression pattern follows by several hours the establishment of basic excitation–secretion properties at the developing synapse. Thus, Leonardo and probably similar modulatory proteins display a delayed expression time course relative to proteins involved in more basic features of synaptic function, suggesting that the maturation of synaptic modulation properties may be a delayed developmental event.

Four forms of synaptic modulation are routinely assayed at the *Drosophila* neuromuscular junction. In order of duration these are (1) paired-pulse facilitation (PPF), (2) short-term facilitation (STF), (3) long-term augmentation, and (4) posttetanic potentiation (PTP). Simple rapid forms of modulation such as paired-pulse facilitation, occurring when a control stimulation is followed at short intervals by a second stimulation, are the first to appear during synaptogenesis and can be routinely detected as early as 15–16 hr AEL (Fig. 1). The interval for effective paired-pulse facilitation remains constant during synaptogenesis; significant PPF begins at a 100-msec interval and the degree of facilitation increases with a decreasing stimulation interval. By 18–20 hr AEL, PPF of 200+% can be detected at the most effective interval of 20 msec. The second form of facilitation, short-term facilitation, occurs during short, high-frequency stimulation trains. STF becomes apparent slightly later than PPF following the maturation of robust fatigue-resistant properties (16–18 hr AEL); in the earlier synapse, fatigue always dominates over any form of facilitation so that STF is rarely observed.[4] STF can be effectively measured during short (20 pulse) stimulation trains at 5–10 Hz during the later stages of synaptogenesis. By 20 hr AEL, STF of 200+% can be detected during short stimulus trains at 10 Hz. Augmentation is a form of facilitation that occurs over a longer period of time during prolonged, high frequency stimulation trains. Like other forms of facilitation, augmentation is first detected after a delay following the acquisition of fatigue-resistance properties and usually cannot be elicited prior to 18–20 hr AEL (Fig. 1). By 20 hr AEL, long trains of 5 Hz stimulation will result in 200+% augmentation of transmission amplitude, which develop over the first few seconds (STF) and persist for the duration of the stimulus train.[36] The longest-term modulation, posttetanic potentiation, is the last to develop and usually cannot be elicited prior to 18–20 hr AEL (Fig. 1). PTP is a form of potentiation that results as a consequence of a prolonged, high-frequency stimulus train (5–10 Hz for >30 sec). Following such a train, basal frequency (0.5 Hz) stimulation will result in amplitudes potentiated by 200+% for several minutes, often followed by a prolonged, low-level (25%) potentiation lasting for the duration of the recording. In the embryonic neuromuscular junction, maximum PTP of 200+% is only present during the last few hours of embryonic synaptogenesis.[36]

VIII. Methods for Electrophysiological Assays of the Embryonic Neuromuscular Junction

The *Drosophila* larval neuromuscular junction preparation approaches the ideal for electrophysiological synaptic recording and, as a consequence, has been used extensively for the last two decades.[37] Experiments are almost always performed on the wandering third instar larva approximately 4 days following hatching. This animal is relatively large (5 mm in length) and can be easily dissected with surgical scissors and pinned open using insect pins. Records are standardly made from a specific large, ventral longitudinal muscle (muscle 6) in the anterior abdomen (A2–A4) either with a simple intracellular recording electrode or with a two-electrode voltage clamp. The large size of the larval muscle makes two electrode impalement straightforward. Typically, the presynaptic terminal is stimulated with a glass suction electrode after severing the nerve from the ventral nerve cord. Using standard salines, the dissected preparation remains healthy for one to a few hours, although to the author's knowledge longer time course experiments have not been attempted. Recordings can be routinely made using only a dissection microscope, although some detailed manipulations benefit from use of a compound manipulation microscope similar to that used with embryonic preparations (see below). All of these techniques are fairly standard and can be mastered by the beginner in a relatively short period of time.

Recording from the embryonic preparation is considerably more challenging, primarily due to the small size (1 mm in length) of the embryo. Dissection of the embryo is usually performed under a good dissection microscope at a magnification of 60× or greater. First, the embryo must be freed from the extraembryonic membranes. Typically, the chorion is removed with a short (1–2 min) bath in commercial bleach and the vitelline membrane is then removed by mechanical dissection. Second, the embryo must be adhered to a coverslip surface. Younger embryos (< mid-stage 16; 15 hr AEL)[6] adhere readily to clean class slides coated with polylysine.[4] Older embryos (>15 hr AEL) have formed a tough, nonadhesive cuticle and must be attached by other methods. A variety of techniques have been employed, including pinning with sharpened steel pins to a plastic coverslip or gluing to a sylgard-coated glass coverslip using a water-polymerizing surgical glue such as acrylamide.[4,17] Both techniques involve attaching the head and tail to the substrate following immersion of the embryo in saline. Third, the embryo must be dissected along the length of the dorsal midline. Once again, a variety of techniques have been employed, including laser incision, dissection with sharpened tungsten needles, and dissection with glass microelectrodes, either the same as those pulled for recording (see below) or needles pulled from solid glass. Finally, the internal organs (other than the ventral nerve cord) are removed and the embryo is attached flat to the coverslip. The internal organs can be removed with a pair of fine forceps or, preferentially, by suction using a broken microelectrode attached to thin tubing. The embryo can be secured flat using extremely fine pins, dragline spider silk from the common Japanese spider *Nephila clavata*, or surgical glue.[4,17] At the end of this dissection, the embryo should be lying flat on a coverslip with the epidermis down and the prominent ventral nerve cord and internal muscles uppermost. The preparation can most conveniently be viewed with a 40× water-immersion lens using Nomarski optics on a compound, fixed-stage micromanipulation microscope.

Most embryonic recordings have been made from acutely dissected preparations. Embryos can be accurately staged in humid chambers at 25°C to the desired age either from the time of egg laying (in a healthy, heavily laying population) or, preferentially, from a defined morphological age marker such as onset/completion of gastrulation or dorsal epidermal closure. Alternatively, records can be made from dissected embryos cultured for prolonged periods of time.[38] Typically, for neuromuscular junction recording, embryos are dissected at or just prior to initial neuromuscular target recognition (11–13 hr AEL) and then cultured for the appropriate period. Cultured preparations develop neuromuscular junctions that are morphologically, ultrastructurally, and functionally similar to those developing in intact embryos.[38] However, the rate of development is significantly reduced relative to *in vivo* development and can be somewhat variable. Therefore, the precise stage of developmental events is less certain in cultured preparations, and it is suggested that acute dissection of timed embryos be used whenever possible.

Intracellular recording in the embryo is possible but quite difficult due to the fragile nature of the embryonic tissues. Electrophysiological records from the embryonic muscle are typically made in a whole cell patch-clamp configuration, which offers superior, low-noise recording conditions in these small cells.[4,17] The primary target muscles have been the ventral longitudinal 6 and the dorsal longitudinal muscle 1 and 9 in anterior abdominal segments (A2–A4). The properties of the neuromuscular junctions on these muscles appear similar, although the initial development of the dorsal neuromuscular junctions is delayed by approximately 30 min and appears to be further retarded in later stages of embryogenesis. In very young embryos (<13 hrs AEL) the muscles are extensively dye-coupled and exceedingly difficult to voltage-clamp. In older embryos (13.5–16 hr AEL), muscle coupling is dramatically diminished, excellent voltage clamp conditions are possible, and muscles can be rapidly patched without need of enzymatic treatment. In stage 17 embryos (16 hr to hatching), the muscle increasingly develops the sheath typical of the mature muscle and must be briefly enzymatically treated (e.g., 1 mg/ml collagenase; 1 min) to remove the sheath prior to recording. This enzymatic treatment has no detectable impacts on the neuromuscular junction synapse as judged by recording and ultrastructural analyses. On the clean muscle surface, patch-clamp recording techniques are highly successful. Any standard patch recording configuration is possible, although synaptic transmission and whole muscle properties are best monitored in a whole cell configuration. Both perforated patch and standard whole cell access have been successfully employed. The recording electrode can be pulled from a variety of soft and hard glasses and should have a final resistance of 2–10 MΩ; smaller electrodes (>5 MΩ) allow more stable recording but often lose access. Fire-polishing electrodes also greatly increase success. Steady-state input resistance of embryonic muscle 6 is 1–3 GΩ but declines rapidly during postembryonic development so that resistance in the third instar is in the range of 5–10 MΩ. The presynaptic motor nerve is stimulated with a glass suction electrode with a tip diameter just greater than the normal diameter of the nerve. For complete stimulus control it is advisable to sever the nerve trunk where it exits the ventral nerve cord; however, in practice the nerve attachment can often be left intact. Optimal recordings are made at temperatures of 14–16°C, although records are typically made in the range of 18–23°C. Recording at temperatures greater than 25°C for prolonged periods is extremely difficult due to the increased fluidity of the muscle membrane and the inability to maintain whole cell access. At 18°C, recordings in standard salines can be obtained for one to a few hours and longer recording sessions (up to at least 24 hr) are possible in culture conditions.

IX. Conclusions and Perspectives

Synaptogenesis at the *Drosophila* neuromuscular junction is similar in most regards to the thoroughly studied development of the vertebrate neuromuscular junction.[32,33,39] Although these neuromuscular junctions are functionally and morphologically similar, the two terminals are chemically and molecularly distinct; the vertebrate neuromuscular junction is cholinergic and the *Drosophila* neuromuscular junction is glutamatergic. Therefore, the similarity of the developmental programs at these two synapses is encouraging for two reasons. First, it suggests that different types of synapses develop using similar mechanisms. Second, it suggests that synaptogenesis is an evolutionarily conserved process with striking similarities across a wide range of animal species. Consequently, it seems probable that the study of synaptogenesis in the *Drosophila* embryo should be directly relevant to synaptogenesis processes at both central and peripheral synapses in higher animals.

The only justification for study of the *Drosophila* embryonic neuromuscular junction is to exploit the powerful molecular genetics of the system in order to assay the molecular and genetic mechanisms of synaptogenesis. Progress using the mature larval neuromuscular junction discussed elsewhere in this volume demonstrates the validity of this goal and the huge potential of the *Drosophila* system for identifying and characterizing the function of specific synaptic components. However, nearly all progress to date has arisen through the isolation of viable synaptic mutations or fortunate conditional mutants, both of which can be revealed by screening for adult behavioral defects. The weakness of this approach is that it may not allow isolation of many genes essential for synaptogenesis and synaptic function which, when mutated, will invariably result in early lethality, either in the late embryo or very early larva. Moreover, screens that rely on viability or rare conditional mutant phenotypes may not allow the identification of genes essential for earlier developmental processes. Therefore, exploitation of the *Drosophila* neuromuscular junction has been somewhat limited to date and has specifically failed to target the process of synaptogenesis

The present challenge is to employ the same types of systematic mutageneses used to address other *Drosophila* development processes to assay the mechanisms of neuromuscular junction synaptogenesis. The difficulty is that target phenotypes will exhibit functional impairment rather than an obvious and readily scoreable morphological character such as has been subject to prior *Drosophila* screening techniques. Thus, genetic analysis of synaptogenesis requires employing difficult and time-consuming behavioral and functional screening assays targeted at uncovering neuromuscular junction functional defects. Such screens are currently being conducted in a number of laboratories, but it has not yet been determined whether this approach will provide a feasible method of systematically investigating the mechanisms of synaptogenesis. The next few years should be exciting in the field as new mutants are characterized, genes cloned, and the nature of the encoded products revealed.

Acknowledgments

I am particularly grateful to Jeff Rohrbough and Emma Rushton for insightful comments on this manuscript. The author is supported by the National Institutes of Health, National Science Foundation, Muscular Dystrophy Association, Office of Naval Research, an Alfred P. Sloan Fellowship, and a Searle Scholar Award.

References

1. Bate, M. (1990). The embryonic development of larval muscles in *Drosophila. Development* **110,** 791–804.

2. Gho, M. (1994). Voltage-clamp analysis of gap junctions between embryonic muscles in *Drosophila*. *J. Physiol.* (*Lond.*) **481,** 371–383.
3. Johansen, J., Halpern, M., and Keshishian, H. (1989). Axonal guidance and development of muscle fiber-specific innervation in *Drosophila* embryos. *J. Neurosci.* **9,** 4318–4332.
4. Broadie, K. S., and Bate, M. (1993). Development of the embryonic neuromuscular synapse of *Drosophila melanogaster*. *J. Neurosci.* **13,** 144–166.
5. Broadie, K. S., and Bate, M. (1993). Development of larval muscle properties in the embryonic myotubes of *Drosophila melanogaster*. *J. Neurosci.* **13,** 167–180.
6. Campos-Ortega, J. A., and Hartenstein, V. (1985). The Embryonic Development of *Drosophila melanogaster*. Springer Verlag, Berlin.
7. Ueda, A., and Kidokoro, Y. (1996). Longitudinal body wall muscles are electrically coupled across the segmental boundary in the third instar larva of *Drosophila melanogaster*. *Invertebr. Neurosci.* **1,** 315–322.
8. Wu, C.-F., and Haugland, F. (1985). Altered potassium conductances in larval muscle fibers of Shaker mutants of *Drosophila*. *J. Neurosci.* **5,** 2626–2640.
9. Ganetzky, B., and Wu, C.-F. (1986). Neurogenetics of membrane excitability in *Drosophila*. *Annu. Rev. Genet.* **20,** 13–44.
10. Singh, S., and Wu, C.-F. (1990). Properties of potassium currents and their role in membrane excitability in *Drosophila* larval muscle fibers. *J. Exp. Biol.* **152,** 59–76.
11. Zagotta, W., Brainard, M., and Aldrich, R. (1988). Single-channel analysis of four distinct classes of potassium channels in *Drosophila* muscle. *J. Neurosci.* **8,** 4765–4779.
12. Yamaoka, K., and Ikeda, K. (1988). Electrogenic responses elicited by transmembrane depolarizing current in aerated body wall muscles of *Drosophila melanogaster* larvae. *J. Comp. Physiol.* **163,** 705–714.
13. Jan, L. Y., and Jan, Y. N. (1976). L-glutamate as an excitatory transmitter at the *Drosophila* larval neuromuscular junction. *J. Physiol.* (*Lond.*) **262,** 215–236.
14. Kidokoro, Y., and Chang, H. (1991). Kinetic properties of glutamate receptor channels in embryonic *Drosophila* myotubes in culture. *Biomed. Res.* **12,** 73–76.
15. Chang, H., Ciani, S., and Kidokoro, Y. (1994). Ion permeation properties of the glutamate receptor channel in cultured embryonic *Drosophila* myotubes. *J. Physiol.* (*Lond.*) **476,** 1–16.
16. Currie, D. A., Truman, J. W., and Burden, S. J. (1995). Drosophila glutamate receptor RNA expression in embryonic and larval muscle fibers. *Dev. Dyn.* **203,** 311–316.
17. Kidokoro, Y., and Nishikawa, K.-I. (1994). Miniature endplate currents at the newly formed neuromuscular junction in *Drosophila* embryos and larvae. *Neurosci. Res.* **19,** 143–154.
18. Schuster, C. M., Ultsch, A., Schloss, P., Cox, J. A., Schmidt, B., and Betz., H. (1991). Molecular cloning of an invertebrate glutamate receptor subunit expressed in *Drosophila* muscle. *Science* **254,** 112–114.
19. Petersen, S. A., Fetter, R. D., Noordermeer, J. N., Goodman, C. S., and DiAntonio, A. (1997). Genetic analysis of glutamate receptors in *Drosophila* reveals a retrograde signal regulating presynaptic transmitter release. *Neuron* **19,** 1237–1248.
20. Nishikawa, K.-I., and Kidokoro, Y. (1995). Junctional and extrajunctional glutamate receptor channels in *Drosophila* embryos and larvae. *J. Neurosci.* **15,** 7905–7915.
21. Chang, H., and Kidokoro, Y. (1996). Kinetic properties of glutamate receptor channels in cultured embryonic *Drosophila* myotubes. *Jpn. J. Physiol.* **46,** 249–264.
22. Saitoe, M., Tanaka, S., Takata, K., and Kidokoro, Y. (1997). Neural activity affects distribution of glutamate receptors during neuromuscular junction formation in *Drosophila* embryos. *Dev. Biol.* **184,** 48–60.
23. Zito, K., Fetter, R. D., Goodman, C. S., and Isacoff, E. Y. (1997). Synaptic clustering of Fasciclin II and Shaker: Essential targeting sequences and role of Dlg. *Neuron* **19,** 1007–1016.

24. Broadie, K., and Bate, M. (1993). Muscle development is independent of innervation during *Drosophila* embryogenesis. *Development* **119,** 397–418.
25. Broadie, K., and Bate, M. (1993). Innervation directs receptor synthesis and localization in *Drosophila* embryo synaptogenesis. *Nature* **361,** 350–353.
26. Broadie, K., and Bate, M. (1993). Activity-dependent development of the neuromuscular synapse during *Drosophila* embryogenesis. *Neuron* **11,** 607–619.
27. Broadie, K., Bellen, H. J., DiAntonio, A., Littleon, J. T., and Schwarz, T. L. (1994). The absence of synaptotagmin disrupts excitation-secretion coupling during synaptic transmission. *Proc. Natl. Acad. Sci. USA* **91,** 10727–10731.
28. Broadie, K., Prokop, A., Bellen, H. J., O'Kane, C. J., Schulze, K. L., and Sweeny, S. T. (1995). Syntaxin and synaptobrevin function downstream of vesicle docking in *Drosophila. Neuron* **15,** 663–673.
29. Schulze, K. L, Broadie, K., Perin, M. S., and Bellen, H. J. (1995). Genetic and electrophysiological studies of Drosophila syntaxin-1A demonstrate its role in non-neuronal secretion and its essential role in neurotransmitter release. *Cell* **80,** 311–320.
30. Sweeney, S .T., Broadie, K., Keane, J., Niemann, H., and O'Kane, C. J. (1995). Targeted expression of tetanus toxin light chain in *Drosophila* specifically eliminates synaptic transmission and causes behavioural defects. *Neuron* **14,** 341–351.
31. Prokop, A., Landgraf, M., Rushton, E., Broadie, K., and Bate, M. (1996). Presynaptic development at the *Drosophila* neuromuscular junction: Assembly and localization of presynaptic active zones. *Neuron* **17,** 617–626.
32. Broadie, K. S. (1994). Synaptogenesis in *Drosophila:* Coupling genetics and electrophysiology. *J. Physiol. (Paris)* **88,** 123–139.
33. Broadie, K., and Bate, M. (1995). The Drosophila NMJ: A genetic model system for synapse formation and function. *Sem. Dev. Biol.* **6,** 221–231.
34. Yoshihara, M., Rheuben, M. B., and Kidokoro, Y. (1997). Transition from growth cone to functional nerve terminal in *Drosophila* embryos. *J. Neurosci.* **17,** 8408–8426.
35. Broadie, K., Sink, H., Van Vactor, D., Fambrough, D., Whitington, P. M., Bate, M., and Goodman, C. S. (1993). From growth cone to synapse: The life history of the RP3 motor neuron. *Development Supple.* 227–238.
36. Broadie, K., Rushton, E., Skoulakis, E. M. C., and Davis, R. L. (1997). Leonardo, a *Drosophila* 14-3-3 protein involved in learning, regulates presynaptic function. *Neuron* **19,** 391–402.
37. Jan, L. Y., and Jan, Y. N. (1976). Properties of the larval neuromuscular junction in *Drosophila melanogaster. J. Physiol.* (*Lond.*) **262,** 189–214.
38. Broadie, K., Skaer, H., and Bate, M. (1992). Whole-embryo culture of *Drosophila:* Development of embryonic tissues in vitro. *Roux's Arch. Dev. Biol.* **201,** 364–375.
39. Keshishian, H., Broadie, K., Chiba, A., and Bate, M. (1996). The Drosophila neuromuscular junction: A model system for studying synaptic development and function. *Annu. Rev. Neurosci.* **19,** 545–575.

ULTRASTRUCTURAL CORRELATES OF NEUROMUSCULAR JUNCTION DEVELOPMENT

Mary B. Rheuben,* Motojiro Yoshihara,† and Yoshiaki Kidokoro†

*Department of Anatomy, College of Veterinary Medicine, Michigan State University, East Lansing, Michigan 48824, and †Institute for Behavioral Sciences, Gunma University School of Medicine, 3–39–22 Showa-machi, Maebashi 371-8511, Japan

I. Introduction
II. Features and Development of the Junctional Aggregate
A. Growth Cone Stage
B. Prevaricosity Stage
C. Varicosity Stage
III. Differentiation of Postsynaptic Specializations
A. Subsynaptic Reticulum
B. Postsynaptic Densities: Adhesion and Receptor Localization
IV. Characterization and Development of Presynaptic Specializations
A. Vesicle Types
B. Presynaptic Dense Bodies
References

I. Introduction

The larval and late embryonic neuromuscular junctions found on the abdominal muscles of *Drosophila* are a very useful experimental model system for examining the normal developmental processes underlying synapse formation and for examining the effects of lethal or semi-lethal mutations producing defects in proteins important to synaptic function. However, because some of the defects may be subtle, involving only quantitative changes in the sizes or number of pre- and postsynaptic organelles, it has been important to characterize the normal ultrastructure of the different types of nerve terminals that innervate these muscle fibers and to characterize the morphological responses that the muscle fiber makes to the presence of each of them in turn.

The nerves that innervate these fibers fall into three general classes: types I, II, and III, which are characterized more fully by S. Gramates and

0074-7742/99 $30.00

V. Budnik in this volume. Briefly, type I neurons use glutamate as a primary neurotransmitter to form functional excitatory neuromuscular junctions of two structural subtypes: Ib with larger varicosities of 2–5 μm and Is with smaller varicosities of 1–3 μm. On muscle fibers 6 and 7 it is known that the two axon types are also physiologically different, with Is having larger synaptic potentials than Ib.[1–4] Other fibers typically have two distinguishably different subtypes of type I terminal present, but they do not always correspond precisely in size to those found on 6 and 7, and the characteristics of their synaptic potentials have not been studied. If *Drosophila* muscles are similar to those of other insects in this regard, it can be expected that there will be some subtle physiological differences in the properties of these individual excitatory motor neurons, depending on the functions of their muscle fibers.

The functions of type II terminals are not fully understood. Action potentials in type II axons have not yet been associated with an electrically observable (either excitatory or inhibitory) postsynaptic response from the muscle fiber. Type II axons are derived from only two neurons per body segment that innervate nearly all the body wall fibers and that are predominantly octopaminergic.[5] Their terminals have varicosities smaller than those of type I terminals. The varicosities contain more dense cored vesicles than do type I terminals, with these vesicles being ovoid or irregularly shaped rather than spherical.[3] Type III terminals from muscle fiber 12 (which are insulin immunoreactive) and two other terminal types seen on muscle fiber 8 (one of which is immunoreactive to leucokinin) contain differing types of dense cored vesicles. They also have not been associated with an electrical response from the muscle fiber.[6,7] As will be elaborated later, both type II and Type III axons have structures suitable for neuromodulatory roles, as well as those that might give rise to other functions.

Each of the abdominal muscle fibers is innervated by a unique combination of these subtypes of motor and neuromodulatory neurons. The secondary motor nerve branch containing their axons typically contacts each muscle fiber at a particular spot, and the type I terminals then give rise to a fairly consistent number and placement of branches that spread out from this point on each muscle fiber. Other terminal types diverge in a less consistent pattern.[1] This region, referred to as the nerve entry point, is thus interesting from the point of view that it is the target for the ingrowing axons during development, as well as the point from which their terminals must disassociate from each other and form a specific branch pattern on the muscle fiber. The nerve entry point is also the location at which the glial sheath covering the motor nerve terminates and, at least in some muscles such as 1 and 9, the site where PT cells (persistent *twist* cells) are found.[8–11] Nuclei of glial sheath cells typically lie along the outer part of the nerve a few microns away from the nerve entry point itself. The group of terminals from the various motor and neuromodulatory neurons that innervate each fiber is referred to as the junctional aggregate.

II. Features and Development of the Junctional Aggregate

The multineuronal nature of larval junctional aggregates prompts several questions about their development. First, the arrival of innervating axons to their target muscle may not be synchronous, so the levels of maturity of individual junctions in the aggregate could differ. This question is difficult to resolve precisely. Light microscope methods involving single cell injections or those using cell surface markers cannot easily make distinctions between individual growth cones because of their structural complexity and diaphanous nature. Second, the specific developmental processes and interactions between the different terminal types and their target muscles may differ for each of the types of neuron. Some of these questions have been addressed, but additional work is needed. To begin we will describe the steps involved in forming the junctional aggregate stage by stage.

A. Growth Cone Stage

During the early stages of junction formation, 13 hr after egg laying (AEL) to 16 or 16.5 hr AEL, the distal ends of the axons contacting the muscle fibers have the shapes of broad flat growth cones. The central planar region of the growth cone might reach 5 μm in diameter, and the lengths of individual filopodial extensions might reach up to 20 μm (Fig. 1).[12,13] Filopodia are labile, extending and retracting over almost the entire surface of the muscle fiber over a time course of minutes.[13] The cytoplasm of the growth cone is relatively unspecialized and vacuolated, and for the most part it seems to be loosely associated with the muscle fibers over which it is passing or intending to be associated. Muscle fibers have thin, patchy basal laminae, but growth cones have little or no ultrastructurally visible basal lamina.

Muscle fibers at 13–16 hr AEL still have filopodial extensions and are interconnected via a few cytoplasmic bridges that are evident in scanning electron microscope (SEM) preparations. In sectioned material some of these bridges appeared to have true cytoplasmic coupling whereas others seemed to be two separate processes that adhered to each other at the

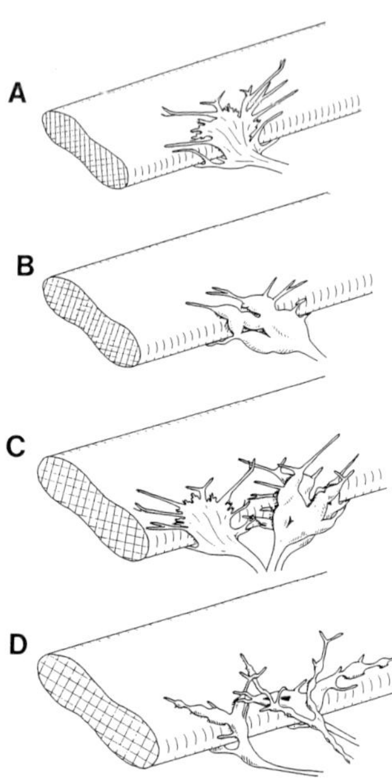

FIG. 1. Development of the junctional aggregate. (A) Growth cone stage. During the period from 13 to 16 hr AEL, the growth cone "explores" the surface of the muscle fiber. Late during this period, it often appears as though more than one growth cone are present and overlapping; for simplicity's sake only one is illustrated here. The growth cones are thin and flat with very long filopodia. (B) Prevaricosity stage, single terminal. After 16 hr AEL for ventral muscles and 16.5 hr AEL for more dorsal muscles, the central region of a growth cone at the nerve entry point enlarges. Distinct branches form. (C) Junctional aggregate at the prevaricosity stage. During the prevaricosity stage, multiple axons and their developing terminals are present. They are generally layered one on top of the other at the nerve entry point, but are shown diverging here for clarity. Occasionally different terminals at the same aggregate seem to be of differing degrees of development. (D) Varicosity stage. After 17 hr AEL, distinct varicosities form from the generalized swelling of the prevaricosity. Different terminal types may still adhere to one another quite closely as they diverge from the nerve entry point, with the overall branch plan conforming to that typical of the muscle fiber. Reprinted with permission from Yoshihara *et al.*[12] Copyright Society for Neuroscience.

tips. Some bridges may give rise to residual electrical coupling, which is substantial in earlier stages.[14,15] It is typical for one of these muscle bridges to lie beneath the nerve entry point, and they may connect adjacent fibers in different combinations such as muscle fibers 1 and 9, 1 and 2, 2 and 3, or 2 and 10 (Fig. 2B).

Late during the growth cone period more than one axonal profile can usually be seen at the nerve entry point of a muscle fiber adhering to or

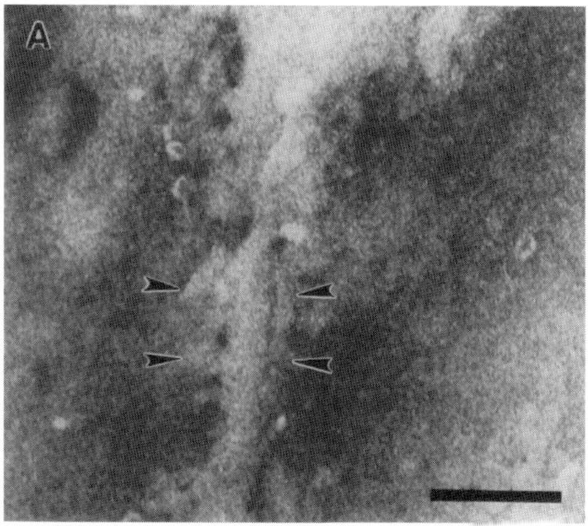

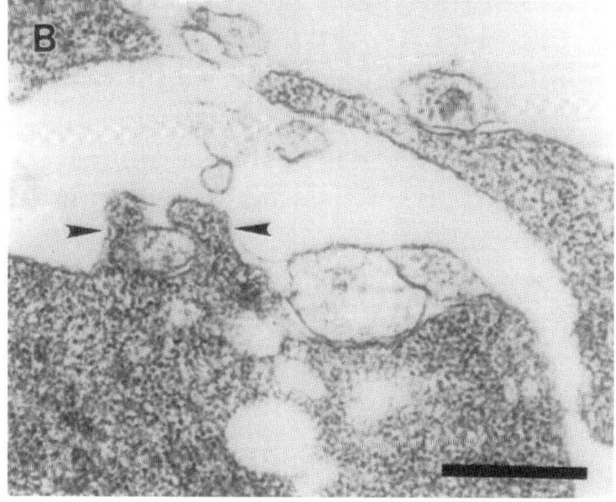

FIG. 2. Relationships of the developing nerve terminal to the muscle fiber. The muscle fiber responds to the presence of the terminal branches on its surface by forming short protuberances that look like ruffles (arrowheads, both A and B). The surface of the nerve terminal also has irregular projections. (A) Scanning electron micrograph, muscle fiber 12, 19 hr AEL. (B) the nerve entry points of muscle fibers 2 and 10 are shown at 16 hrs AEL. Portions of a muscle bridge between muscle fiber 2 (upper right) and muscle fiber 3 (upper left) are shown, with 10 being below. The completeness of the bridge was established in subsequent sections. Note the small neurites that are associated with it. Scale bars: 0.5 μm.

following a muscle bridge. The most common arrangement is for one axonal profile to be in direct contact with the fiber and the second axonal profile to be adhering to the first, rather than to the muscle fiber (Ref. 12 and unpublished data). At this stage, we cannot exclude the possibility that these may be branches from the same parent axon, but in long series that did not seem to be the case. In confocal and SEM overviews there is increasing complexity of the growth cone regions throughout the 13–16 hr AEL period. Together this suggests that more than one of the innervating axons has reached the muscle fiber with its growth cone, but the identities or types of these terminals and their precise times of arrival have not been determined for most muscles. In the case of muscle fibers 1 and 9, the group of terminals would include the axon from the aCC neuron, perhaps one from one of the U neurons, one from an unidentified motor neuron, and a branch from one of the two type II neurons. In the case of muscle fibers 6 and 7, the group would be expected to include axons from RP3 and 6/7b, but branches from type II neurons are usually lacking.[1,12,13,16]

B. Prevaricosity Stage

During the next phase of development, which is referred to as the prevaricosity stage (16 to 17 hr AEL), the planar region of each growth cone, just at the point at which the axon meets the muscle fiber, enlarges quite remarkably. It forms a structure that is larger than the typical larval varicosity and can be quite irregular in shape, tubular, roughly triangular, or polygonal (Figs. 3A and 3B). This process is short in duration and takes place earlier in the ventral muscles than in the dorsal muscles. Sections through the nerve entry point at the prevaricosity stage illustrate that multiple axons are present lying one on top of another, that they are beginning to form discrete branches rather than a sheet, that their terminals are beginning to differ from each other morphologically, particularly with regard to vesicle types, and that the structures associated with synapse formation, both nerve–nerve and nerve–muscle, are beginning to appear.[12]

Among the nerve entry points (on muscle fibers 1 and 9) that have been examined by serial sections, the deepest prevaricosity in the layers of terminals is often the largest and it consistently forms presynaptic dense bodies and clear vesicle clusters typical of type I neurons in apposition to the edge of the muscle fiber. Early divisions of the prevaricosity are laid out on the muscle surface in a way that appears to presage the distribution of mature type I terminals. Most remaining filopodia, and the prevaricosity itself, lie beneath a thickening basal lamina surrounding the muscle fiber. Thinner, stick-like terminals are seen, and because some axonal profiles have higher proportions of irregularly shaped, dense-cored vesicles, type II terminals are also probably present. Differentiation of the mature larval junctional features is well under way by this time.

C. Varicosity Stage

Next, rounded varicosities form, from 17 hr AEL through hatching. The first type I varicosities are derived from the prevaricosity by constriction

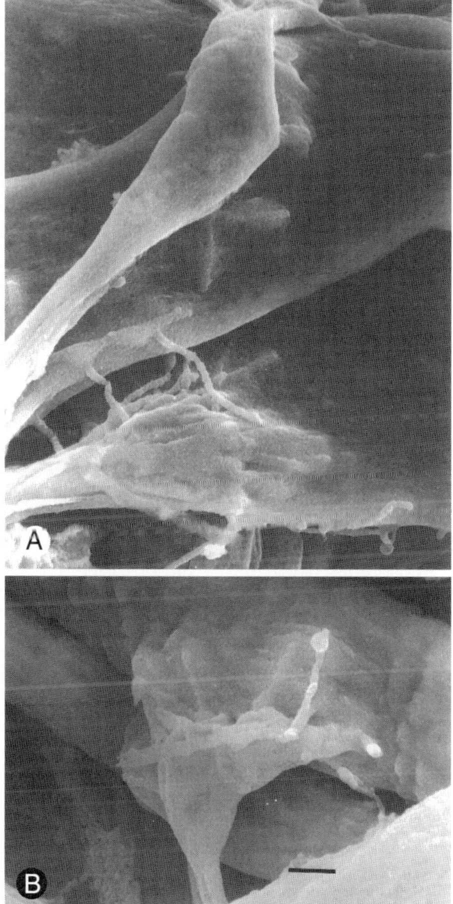

FIG. 3. Prevaricosity stage, 16.75 hr AEL. During the prevaricosity stage the number and length of exploring filopodia decrease. Some are embedded within the basal lamina while others are above it. In A the intersegmental nerve is shown crossing muscle fiber 2 (dorsal is toward the top, anterior to the left) and an early prevaricosity is shown on muscle fiber 10. In this junctional aggregate it looks as though two growth cones are overlapping, and not much thickening has taken place. In B, which is taken from the same animal and the same abdominal segment, A3, a well-formed prevaricosity is shown on a more ventral muscle, muscle fiber 24. Compare with Fig. 1. Note the multiple axons in the entering nerve trunk and the rounded swellings of the terminal. Scanning electron micrographs: nerve membranes were visualized with α-HRP followed with osmium-thiocarbohydrazide treatment. Scale bar: 1 μm.

and are largest at the base of the junction (Fig. 4).[12,14] Additional smaller varicosities are seen distally. During this time period, several axon terminals remain in intimate contact with each other as they extend away from the

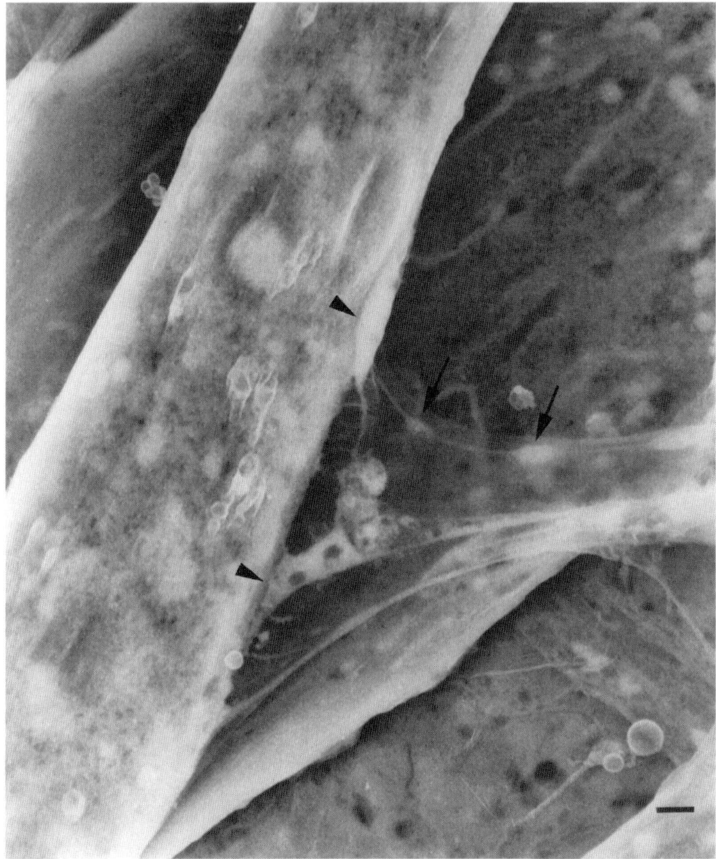

FIG. 4. Varicosity stage. The nerve entry point of muscle fiber 1 is illustrated at 19 hr AEL. Three axons are shown approaching the muscle fiber. Because of tension on the nerve during dissection the terminal branches are pulled away from the muscle fiber, and the developing varicosities are shown (arrowheads). The lowest axon passes beneath the fiber and its terminal is not visible. The middle axon branches within the nerve entry point and forms the two elongated varicosities indicated. Both of them appear to be in the process of dividing into two smaller varicosities; the central regions of type Ib boutons often give rise to a "hollow" appearance due to the lack of vesicles and other osmiophilic structures in the center. Note the degree to which the terminal adheres to the muscle fiber membrane (patches just above and below the lower arrowhead). The uppermost axon in the nerve entry point is thinner than the other two and contains varicosities along the region within the nerve trunk typical of type II axons (arrows). Spherical structures adjacent to the lower varicosity in the nerve entry point may be persistent *twist* cells. Scale bar: 1 μm. Reprinted with permission from Yoshihara *et al.*[12] Copyright Society for Neuroscience.

nerve entry point, with one lying on top of another in some regions and, less commonly, both lying side by side making contact with the muscle fiber.

There is also evidence for the formation of varicosities by type II axons and terminals during the varicosity stage. *Drosophila* type II axons have varicosities all along the length of the motor nerve as well as in the region associated with the muscle fiber surface.[5] In this regard they are similar to other neurosecretory axons of insects. At 19 hr AEL one of the three axons could often be seen forming small varicosities along the nerve trunk prior to the nerve entry point of muscle fibers 1 and 9 (Fig. 4). Because discrete enlargements are seen only on the smaller axon, with the two larger axons having uniform diameters and connecting to large varicosities on the muscle fiber, it is assumed that the small axonal varicosities belong to the type II axon. In addition, type II varicosities have been found on the muscle surface by hatching. This suggests that both type I and type II axons are beginning to form characteristic varicosities during the period 17 hr AEL through hatching.

The appearance of the developing junctions examined suggests the following scenario for further investigation. The two to four axons that innervate any given muscle fiber may not arrive completely synchronously. The work with the aCC axon and its role in pioneering the intersegmental nerve (ISN) suggests that synchrony is unlikely, but also that axons that follow the pioneer may be as little as 15 min behind it.[17] This difference may be insignificant given the 3-hr duration of the growth cone period during which considerable filopodial extension and retraction occur. However, it does seem to be true, at least for muscle fibers 1 and 9, that the first axon to form synaptic relationships with a muscle fiber is consistently one of the type I motor neurons, quite possibly the Ib for each fiber. This could arise if the Ib is always the first to arrive. Alternatively, it may be that the timing of arrival is less important than the relative adhesiveness of the growth cones in establishing this intimate relationship to the muscle fiber. As the branch pattern is established by a type I terminal, the affinity of the other terminals for this "pioneer" may continue for some time so that their initial branches follow the same course. Subsequently, as this affinity declines, they extend branches away from and beyond it.

Available evidence suggests that all of the different types of axons are present in the nerve entry point by the varicosity stage, and the process of initiating a junctional aggregate and determining multineuronal innervation appears to be complete prior to hatching, with further growth adding to the branches and increasing and enlarging varicosities. Additional work is needed with specific cellular markers to identify the individual axon terminals during these early developmental stages when morphological criteria are vague and to see how consistent their developmental patterns are. It will be important to explore not just nerve–muscle target interactions, but also nerve–nerve interactions. Of these stages, the prevaricosity stage, with its remarkable terminal swelling, is very important because synaptic transmission is initiated at this time.

III. Differentiation of Postsynaptic Specializations

The muscle fiber responds to the presence of each of the different types of motor and neuromodulatory neurons in quite different ways. Three structural features are important in this response: the subsynaptic reticulum (SSR), which consists of thin folded extensions of the muscle fiber, the patches of postsynaptic glutamate receptors and the associated membranous and cytoplasmic proteins that occur on the tops of the SSR folds, and the various specializations that provide the mechanical means of assuring the attachment and alignment of the terminal with the muscle fiber. These are features common to many synapses, with the subsynaptic reticulum being both analogous to the postsynaptic folds of the vertebrate neuromuscular junction or to dendritic spine, and to the terminal Schwann cells that normally cover a junction. However, as in vertebrate neuromuscular junctions, the degree to which each of these facets is developed in *Drosophila* varies with the terminal type, muscle fiber type, and age.

A. Subsynaptic Reticulum

The SSR ranges from being absent to spectacularly well developed in the various junctions formed by the different types of nerves innervating *Drosophila* muscle fibers. Third instar type Ib boutons have the deepest layers of folded subsynaptic reticulum, whereas type Is boutons have less.[2,3] Type II and III terminals are often embedded in the edge of a field of SSR belonging to a type I terminal by third instar, but there they appear to have only one or two layers of muscle fiber folds of their own,[3] and it is less clear that they induce the formation of any SSR when on the surface of the fiber alone. (See Fig. 1 in Chap. 5 for micrographs of the third instar subsynaptic reticulum.) The leucokinin immunoreactive terminals on muscle fiber 8 lie in shallow gutters.[7] Oddly the neuromuscular junctions from adult flies also lack an SSR, with both junctions found on cervical muscles and those on the fibrillar dorsal longitudinal muscles having been examined.[18,19]

The precise function of the elaborate SSR associated with type I terminals is not yet completely understood, but there is clear evidence that it houses higher densities of certain functionally important proteins than adjacent regions of the muscle membrane. These include, in addition to the glutamate receptors, some of the antigens recognized by α-HRP,[20–22] fasciclin II (an adhesion molecule that is also recognized by α-HRP),[23] the Shaker potassium channel, the protein recognized by antibodies to the mammalian insulin receptor (dInsR), and the protein defective in the *discs large* (*dlg*) mutation.[6,20,24] Of these, α-HRP, DLG, the glutamate receptor, and the Shaker potassium channel have been reported not to occur at detectable levels near type II terminals. The functional implications of these important proteins are explored in this volume in reviews by K. Broadie, by S. Gramates and V. Budnik, and by F. Hannan and Y. Zhong.

The postsynaptic processes that make up the SSR have many divisions, narrow regions, and constrictions. Path length for cytoplasmic current flow or diffusion between the postsynaptic receptor patches and the main muscle fiber can reach several microns. In that regard they may be viewed as being functionally analogous to the thin necks of dendritic spines. Similar muscle structures are conserved through phylogeny, being found in other insects, crustaceans, and vertebrates, where they are developed to varying degrees depending on the motor neuron and muscle fiber types.[25–29]

The subsynaptic folds of the vertebrate neuromuscular junction are thought to play an important role in amplifying the synaptic signal and aiding it in reaching the threshold for propagation along the muscle fiber. This is a consequence of the combined features of a narrow shape of the fold, which gives rise to increased resistance, and a high density of voltage-gated sodium channels in the fold membrane.[30–33] In the presence of voltage-gated channels placed optimally to respond to quantal currents, such constrictions are predicted to increase the amplitude of the total voltage change produced in the muscle by synaptic activity.[34] No such density of voltage-gated channels has yet been mapped to the SSR in *Drosophila,* so a variety of functions, which relate both to its shape and to the populations of proteins within its membrane, can be postulated. These functions might also include those normally attributed to the glial cell processes that are lacking in dipterans, but which in other insects are involved in the uptake of glutamate after synaptic activity.[35]

The subsynaptic reticulum associated with the type I fibers is formed gradually as development proceeds. Initial synaptic contacts of the growth cones and prevaricosities with the muscle fiber are on smooth surfaces, but subsequently during late embryonic development both short arm-like extensions of the muscle fiber and invaginations or holes have been reported nearby.[12,23,36] In some cases there are so many short extensions of the muscle toward the neurite that the muscle membrane looks ruffled (Figs. 2A and 2B). Some of these bear postsynaptic specializations. By hatching, there are typically only one or two folded thin layers that represent early stages of the SSR. The length of SSR membrane as exposed in a cross section increases exponentially with time during subsequent larval development, paralleling the large increases in volume of the muscle fiber, which is growing concurrently.[37] Consequently, from an electrophysiological point of view, voltage control of the synaptic region can best be obtained up through the first instar, making the embryonic preparation particularly valuable for that kind of work.

It should be noted that the appearance of the SSR is very sensitive to the conditions of fixation. The relative amounts of extracellular space around the folds, the thickness of the folds, and the electron density of their cytoplasm are affected strongly (quite possibly by the osmolarity of the fixative).

B. Postsynaptic Densities: Adhesion and Receptor Localization

Using ordinary ultrastructural methods it is difficult to separate the intramembranous structures involved in forming a cohesive cluster of receptors from those involved in sticking the terminal to the muscle fiber. They are in part or entirely colocalized to the structure seen as an electron-dense thickening or decorated region of the sarcolemma, referred to as the postsynaptic density. For example, micrographs of adult fly flight and cervical muscles illustrate *en passant* synapses in which the nerve terminal membrane is separated from the muscle fiber membrane by basal lamina and varying amounts of extracellular space except at the postsynaptic density, where the closely apposed nerve terminal membrane and the muscle membrane jointly form a curved, stiffer-appearing structure.[18,19] The terminal and the muscle fiber seem to be firmly attached to each other only at this location, so both functions are colocalized. However, in mature larval type I junctions the SSR nearly completely surrounds the terminal. The tops of the irregularly arranged muscle folds contact the terminal, with short interruptions, around its entire circumference.[2,3] However, only some of these contact regions bear electron-dense postsynaptic specializations. Other surfaces of the SSR membrane might then subserve functions such as adhesion, and the SSR as a whole could then function in mechanically keeping the terminal in place on the fiber and in protecting its outer surface. This arrangement is quite different from that seen in *Manduca* and other insects, where almost all direct contacts between terminal and muscle bear electron-dense specializations and where other regions of the terminal are separated from the muscle folds of the SSR by glial fingers.[11,25,26]

The appearance and dimensions of the postsynaptic densities from type Is and Ib terminals are very similar, with both being oval or round electron-dense, stiff-appearing regions averaging about 0.3 μm^2 in area or 0.6 μm in diameter.[2] The distinctiveness of the postsynaptic density is enhanced in tannic acid preparations where it can be seen that the synaptic cleft contains material that differs from adjacent regions (unpublished observations). It is presumed that the distribution of glutamate receptors required for excitatory transmission is largely restricted to the postsynaptic density by analogy with other synapses, and this idea is supported by immunolabeling studies.[22]

Ultrastructurally detectable postsynaptic density formation occurs in the muscle fiber membrane during late embryonic development. During the growth cone period, the initial associations between terminal and the muscle fiber appear to be stabilized by focal contacts with an intervening thin layer of basal lamina or by contacts in which the two membranes are interacting more closely via projections from intramembranous proteins (unpublished observations).

During the prevaricosity stage, 16–17 hr AEL, synapse formation, which can be recognized by clusters of presynaptic vesicles and electron-dense material in the cleft, is accompanied simultaneously by increased electron density and by the formation of submembranous postsynaptic projections in the apposing muscle membrane.[12,23] The borders of this developing postsynaptic density are somewhat less distinct than those from third instar larvae, the densities themselves are smaller, and they occur within a long uninterrupted contact zone between nerve terminal membrane and sarcolemma. The average diameters of a mixed population of type I postsynaptic densities increased from 0.24 $\pm$ 0.09 μm (SD) at 17 hr AEL, to 0.32 $\pm$ 0.11 μm at 19 hr AEL, to 0.42 $\pm$ 0.16 μm at first instar, and remained at 0.42 $\pm$ 0.16 μm for second instar larvae (Fig. 5). These values represent minimum estimates of average diameters because of the likelihood that some of the measurements were from sections passing through the synapse away from the exact center. Type Ib and Is terminals are generally indistinguishable in single sections early in development; however, data from both types could be combined because the mature dimensions of Ib and Is postsynaptic densities are the same.

Saitoe *et al.*[21] reported that there is localization of immunoreactivity to *Drosophila* glutamate receptors (DGluRII) to the junctional region between 17 and 22 hr AEL, with a concurrent decline in the visibility of immunoreactive patches elsewhere on the sarcolemma. However, because the dimensions of the developing postsynaptic densities, 0.2 to 0.6 μm, are below the resolution possible with confocal microscopy, it has not yet been determined if receptors are confined exclusively to them at this stage.

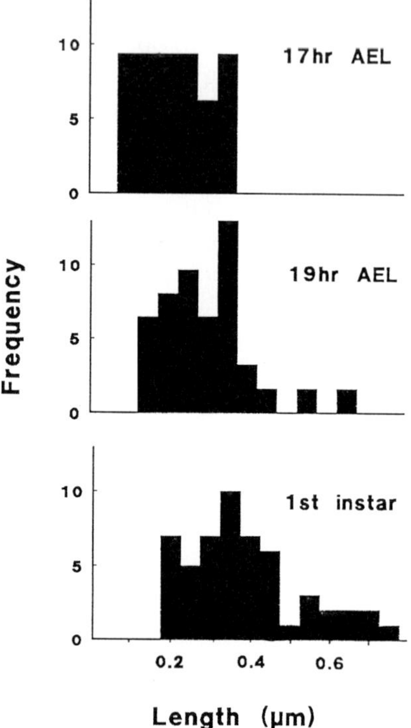

FIG. 5. Postsynaptic density diameters. The dimensions of the densities from type I terminals of three ages are plotted. Data are normalized to give the same total number. There is an increase in the frequency of full-sized (0.6 μm diameter) postsynaptic densities with age, but small diameter measurements continue to make up a preponderance of the sample. This arises from the likelihood of sampling the edge of a circular structure as well as from the continuing presence of immature densities as the terminal grows. Measurements were taken from all clearly defined densities encountered at these ages and were not from serial sections.

Data illustrated in Fig. 5 indicate that by 19 hr AEL a few synapses have reached the mature dimension of 0.6 μm in diameter. (In general, one would expect for all larval stages that a fraction of a randomly sampled population of synapses would consist of immature synapses that are being formed as the nerve terminal extends and the muscle fiber grows.) The approximately threefold increase in average calculated postsynaptic area between 17 hr AEL and hatching in this sample corresponds well to the average increase in miniature excitatory junctional current (mEJC) amplitude at these times.[38,39] The gradual increase in synapse area is similar to the situation found in developing neuromuscular junctions of *Manduca,* where freeze-fracture methods give good visualization of the postsynaptic specializations. In that instance the synapses increased in area and number of postsynaptic intramembranous particles as a function of time and of distance along the terminal from the nerve entry point.[40]

Dimensions of the mature larval postsynaptic densities of both type Is and Ib terminals are affected by mutations that reduce the amount of fasciclin II (FasII) produced.[41] FasII is a homophilic adhesion molecule that is present both in the presynaptic terminal and in the muscle membrane within the SSR.[23] In FasII hypomorphs, the areas of the postsynaptic densities are two to three times larger than in wild-type animals. Interestingly, the average miniature end plate potential amplitude is also increased slightly in hypomorphs, from 0.8 to 1.0 mV. The average number of presynaptic dense bodies per synapse increases from less than one to 1.9–2.6.[41] In hypomorphs, a greater proportion of the SSR folds in apposition to the terminal circumference are engaged in forming synapses, but not all of them. One possible interpretation is that there is an interplay between the adhesive functions of the SSR and those related to formation of the synapse, and reduction of FasII allowed expansion of the proportion of fold membrane devoted to postsynaptic density. Synaptic strength of the terminal as a whole is unaffected in these animals, with the total number of presynaptic densities throughout the terminal remaining about the same.

Muscle specializations associated with type II terminals differ markedly from those seen in type I terminals. The muscle forms structures that look similar to hemidesmosomes. They exhibit electron-dense material below the sarcolemma, and they can be found in apposition to the basal lamina associated with the nerve terminal (Fig. 6). These unusual structures were first noted for one of the dense cored vesicle containing terminals on muscle fiber 12.[2] They are more numerous along the outer edges of the type II varicosities and are not associated with any visible presynaptic specializations. Typical presynaptic structures in apposition to a muscle fiber have not been reported from surveys of third instar type II terminals, nor were structures comparable to type I postsynaptic densities found adjacent to them.[3]

Type II varicosities have been encountered in isolation on the surface of muscle fiber 1 at hatching. The terminals lie in slight depressions, with no adjacent fields of SSR to surround them. Therefore, the hemidesmosome-like structures (Figs. 6A and 6C) may function to stabilize the terminal in the absence of either glia or enveloping muscle folds. In addition, the observation helps confirm the association of these hemidesmosomes with type II terminals as type III terminals are not found on muscle fiber 1.

Insulin-immunoreactive type III terminals from muscle fiber 12, also referred to as DV terminals, with numerous large dense cored vesicles of differing internal densities, do form synaptoids with clusters of small

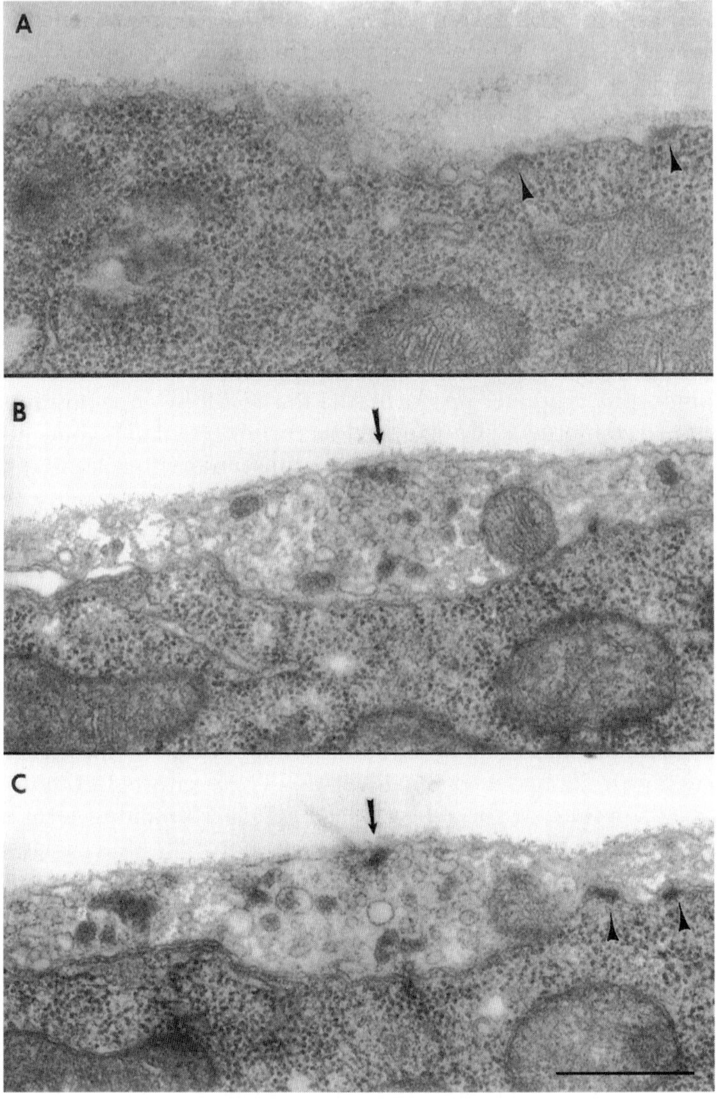

FIG. 6. Type II varicosity on muscle fiber 1, first instar. B and C are adjacent sections in the series from the center of the varicosity, and A is just at the edge of the varicosity. The dense cored vesicles are irregular in shape. Note the hemidesmosome-like structures near the edges of the varicosity (arrowheads). No pre- or postsynaptic specializations of the nature described for type I terminals have been found in this or any other type II varicosity in our sample. Instead, a putative dense body and a cluster of clear vesicles are occasionally found in apposition only to basal lamina on the outer surface (arrows), shown here from two adjacent sections. Scale bar: 0.5 μm.

33-nm clear cored vesicles in apposition to a more electron-dense patch of muscle cell membrane.[6] It is not known if these synapse-like structures relate to the release of insulin or to a cotransmitter such as glutamate. Little is known about the embryonic development of type III terminals or the specific properties of their postsynaptic densities. Insulin-like immunoreactivity and expression of *Drosophila* insulin receptors (dInsR) were not observed until late first to early second instars,[6] so the maturation of the type III terminals may be delayed.

IV. Characterization and Development of Presynaptic Specializations

Four structural features of the presynaptic terminal vary quantitatively and qualitatively among the different neuron types in *Drosophila* in parallel to the variations reported among functionally different terminal types from other species. These include the size and number of clear cored vesicles; size, number, and appearance of dense cored vesicles; number of mitochondria; and the number and shape of the presynaptic dense bodies that make up the active zones. These organelles interest neurophysiologists because jointly they may be involved in producing the diverse capabilities of any given synapse. For morphologists there are a number of challenges. The dimensions and relative numbers of clear cored vesicles are sensitive to fixation and physiological state, as is the appearance of the dark cores of dense cored vesicles, so comparisons and identifications must be made with some caution. In addition, very small presynaptic dense bodies may have dimensions comparable to section thickness so that their appearance is also variable depending on the plane of section.[42] Some of these difficulties can be overcome by combining light microscope immunocytochemistry with transmission electron microscopy or by making serial sections.

A. Vesicle Types

Mature type Is and Ib terminals differ subtly from each other in their presynaptic morphology. Aside from differences in varicosity size, they also differ in the relative sizes of their clear cored synaptic vesicles, with type Ib (= CV) terminals having a high density of consistently sized 44-nm vesicles and type Is (= CVo) terminals having a lower density of vesicles and vesicles of a greater range of sizes. Both Type Ib and Is terminals have occasional dense cored vesicle, with Is being more likely to contain them.[2,3] These characteristic differences in vesicle populations are to some degree recognizable in the developing terminals from 17 hr AEL onward in our studies, but often individual terminals cannot be assigned to either Ib or Is with confidence.

Type II and III terminals both have a higher proportion of dense cored vesicles than type I terminals, although both contain clear cored vesicles as well. The dense cored vesicles of type II terminals are elliptical, e.g., measuring 75 by 190 nm, whereas those of type III terminals are more spherical, can vary in electron density, and range from 73 to 108 nm in diameter.[3] Serial sections through a type II terminal just at hatching indicate that 16% of the vesicles present in the center of the terminal are dense cored and the remainder are clear cored (unpublished results).

B. Presynaptic Dense Bodies

The presynaptic dense bodies of type I terminals in *Drosophila* are similar in shape to those described for other dipterans,[2,3,10,11] but differ from those seen in frog, crustacean, rodent, or moth. Actually, the greatest species-specific variation in a presynaptic feature seems to occur in the shapes of the dense bodies. In *Drosophila,* the presynaptic dense body has three or four diverging electron-dense branches or "prongs" that lie immediately next to the terminal membrane. A perpendicular electron-dense layer on top of these branches seems to form a flat roof, giving rise to the "T-bar" appearance seen when only one of the prongs is in the plane of section. This unique structure can be clearly visualized from the micrographs in Atwood *et al.*[2] In *Manduca,* for example, the presynaptic active zone is a single narrow bar about 0.2 μm in length,[25,26] and in crayfish a synapse may contain several dense bodies that seem to be post-like in shape.[42,43]

The large type Ib boutons have more dense bodies per varicosity and for the terminal as a whole, but produce smaller evoked synaptic currents.[2,4] These morphological and physiological differences are similar to those associated with the tonic and phasic crustacean motor neurons or slow and fast motor neurons in *Manduca.*[26,43] It would appear for *Drosophila* as well as these other species that differing quantal contents generated by these different types of motor neurons cannot be accounted for simply by the numbers of presynaptic dense bodies present. It would also appear that morphologically indistinguishable presynaptic active zones can be associated with physiologically different properties. One factor that may contribute to the differing properties is the number of mitochondria present within the terminal. Type Ib boutons have more mitochondria than type Is. The tonic and phasic properties of crayfish motor neurons, which include differing capabilities of facilitation and susceptibility to depression, have been associated with the oxidative metabolism of their respective mitochondrial populations.[44]

In embryonic type I terminals from the prevaricosity stage, 16–17 hr AEL, most active zones observed in serial section appear to already have the multipronged shape. They are formed in apposition to the edge of the muscle fiber at the nerve entry point. Those active zones that did not have this shape consisted of a very small tuft of electron-dense material surrounded by three or four clear cored vesicles in apposition to a small patch of postsynaptic thickening. They sometimes occurred in the same prevaricosity as larger, more clearly formed dense bodies. So aside from the likelihood that the appearance of some of these is due to the plane of section, it is also possible that the initial (possibly functional) immature form of the dense body is a simple tuft of electron dense material. The number of presynaptic dense bodies per varicosity in 19-hr animals is relatively low (e.g., six and eight in each of two varicosities that were 1.0×1.7 and 1.5×2.1 μm in diameter, respectively, muscle fibers 1 and 9) (Ref. 12 and additional unpublished results).

Synaptotagmin, a protein associated with synaptic vesicles, can be detected by immunocytochemical means in small amounts in the growth cone and as discrete clusters of material in the prevaricosity[12,45]; its appearance and distribution within the terminal parallels both the appearance of presynaptic dense bodies and the clusters of vesicles in the prevaricosity stage as well as the varicosity stage that follows.

Type II terminals have not yet been extensively characterized by serial sections in mature larvae. Jia *et al.*[3] reported that no presynaptic dense bodies were identified in the MV terminals (which correspond to type II's) studied on muscle fiber 12, so they may be uncommon. Actually, type II neurites may be difficult to recognize in between varicosities, being reduced to simple profiles containing a few microtubules and mitochondria.

However, in examining the junctional aggregates formed on muscle fibers 1 and 9 during development, three examples have been encountered where a Type II terminal, identified by the type of dense cored vesicles present, the hemidesmosomal attachments to the muscle fiber, and the complete absence of folds of SSR, formed a synaptoid structure on its outer surface, facing away from the muscle fiber (Figs. 6B, and 6C). In one of these varicosities, the dense body persisted through two consecutive serial sections and consisted of several closely spaced tufts of electron-dense material. These could either be interpreted to be part of a pronged structure or as a cluster of separate structures; distinguishing would require a lucky *en face* view. The dense body was surrounded by irregularly sized and shaped clear cored vesicles.

These synaptoids are remarkably similar to those observed in other insects in what are presumed to be neurosecretory or neurohemal terminals because they do not have a consistent postsynaptic target (reviewed in Ref. 46). In proctolin-containing endings on the visceral muscles of locust and grasshopper, synaptoids were formed both in apposition to the muscle fiber and away from it.[47] Presumed neuromodulatory axons form synapses in apposition to the glial cells surrounding the motor nerve in the immediate vicinity of the muscle fibers in *Manduca.*[48] Given that no synaptic electrical responses from the muscle fiber have been attributed to type II terminals yet, given their octopaminergic nature, and given the similar structures, it is very likely that type II terminals subserve a neurohemal or neuromodulatory role, but additional functions are not ruled out.

Axons that appear to belong to type II neurons on the (relatively weak) basis of their vesicle types have been identified within the nerve entry point on muscle fiber 9 during the prevaricosity and varicosity stages. Some of these form synaptoids in apposition to the type I terminal that is concurrently forming presynaptic densities in apposition to the muscle fiber.[12] If these are directly functioning endings, this might be the basis for a nonneurohemal action, perhaps inhibitory. Alternatively, it could be a transient developmental phenomenon, since such synapses have not yet been noted in third instar animals. Prokop *et al.,*[49] sampling all peripheral nerves in embryos just at the hatching stage, found that up to 11% of the active zones encountered are neurohemal, nerve–nerve, or neuroglial, which suggests that nonneuromuscular synapses are a relatively common phenomenon. However, their sample population probably included synapses formed by the motor nerves that contain "real" motor axons such as type I's and the accompanying type II's and III's, as well as sections from the transverse nerve, which would be more likely to include neuromodulatory axons. (For a review of insect neurosecretory systems, see Ref. 50.)

In mutants in which muscle fiber development is perturbed, the relative numbers of neuroneural, neuroglial, and neurohemal type synapses greatly increase compared to wild type.[49] These synapses occurred in varicosities or enlargements along the lengths of axons when normal muscle was missing. It is not clear what fraction of these synaptoids occur in type I axons and which occur in neuromodulatory type axons. In the flightless grasshopper, neuroglial synaptoids are commonly present in the nerve to the dorsal longitudinal muscle at the stage when the muscle is greatly reduced or absent.[51] This nerve also presumably contains both motor axons and a subpopulation of neurosecretory axons. Two hypotheses can be suggested: either that motor neurons respond to the absence of muscle by forming synaptoids with anything that is handy, including basal lamina or glial cells, or that peripheral neuromodulatory axons, which normally produce such structures, produce them in much greater numbers in the absence of a muscle target.

Consequently, one can conclude on morphological grounds that these neuromodulatory type terminals may be capable of performing several different functions, some of which may be important to normal development of the muscle fiber, and further study is greatly needed.

Acknowledgments

The previously unpublished work from the authors that was included was supported by grants from the Ministry of Education, Science, Sports, and Culture to Yoshiaki Kidokoro and Motojiro Yoshihara, grants from the Brain Science Foundation and Kato Memorial Science Foundation to M. Y., a grant from the Mitsubishi Foundation to Y. K., an All-University Research Initiation Grant, Genetic Research Grant, Companion Animal Fund Grant from Michigan State University, and grants from the Narishige Neurosciences Foundation and the Yamada Science Foundation to Mary B. Rheuben. We gratefully acknowledge the technical assistance of Dawn M. Autio and the art work by Jessica Hoane.

References

1. Johansen, J., Halpern, M. E., Johansen, K. M., and Keshishian, H. (1989). Stereotypic morphology of glutamatergic synapses on identified muscle cells of *Drosophila* larvae. *J. Neurosci.* **9,** 710–725.
2. Atwood, H. L., Govind, C. K., and Wu, C.-F. (1993). Differential ultrastructure of synaptic terminals on ventral longitudinal abdominal muscles in *Drosophila* larvae. *J. Neurobiol.* **24,** 1008–1024.
3. Jia, X., Gorczyca, M., and Budnik, V. (1993). Ultrastructure of neuromuscular junctions in *Drosophila:* Comparison of wild type and mutants with increased excitability. *J. Neurobiol.* **24,** 1025–1044.
4. Kurdyak, P., Atwood, H. L., Stewart, B. A., and Wu, C.-F. (1994). Differential physiology and morphology of motor axons to ventral longitudinal muscles in larval *Drosophila. J. Comp. Neurol.* **350,** 463–472.
5. Monastirioti, M., Gorczyca, M., Rapus, J., Eckert, M., White, K., and Budnik, V. (1995). Octopamine immunoreactivity in the fruit fly *Drosophila melanogaster. J. Comp. Neurol.* **35,** 275–287.
6. Gorczyca, M., Augart, C., and Budnik, V. (1993). Insulin-like receptors and insulin-like peptide are localized at neuromuscular junctions in *Drosophila. J. Neurosci.* **13,** 3692–3704.
7. Cantera, R., and Nässel, D. R. (1992). Segmental peptidergic innervation of abdominal targets in larval and adult dipteran insects revealed with an antiserum against leucokinin I. *Cell Tissue Res.* **269,** 459–471.
8. Auld, V. J., Fetter, R. D., Broadie, K., and Goodman, C. S. (1995). Gliotactin, a novel transmembrane protein on peripheral glia, is required to form the blood-nerve barrier in *Drosophila. Cell* **81,** 757–767.

9. Bate, M., Rushton, E., and Currie, D. A. (1991). Cells with persistent *twist* expression are the embryonic precursors of adult muscles in *Drosophila. Development* **113,** 79–89.
10. Osborne, M. P. (1967). The fine structure of neuromuscular junctions in the segmental muscles in the blowfly larva. *J. Insect Physiol.* **13,** 827–833.
11. Osborne, M. P. (1995). The ultrastructure of nerve-muscle synapses. *In* "Insect Muscle" (P. N. R. Usherwood, ed.), pp. 151–205. Academic Press, London.
12. Yoshihara, M., Rheuben, M. B., and Kidokoro, Y. (1997). Transition from growth cone to functional motor nerve terminal in *Drosophila* embryos. *J. Neurosci.* **17,** 8408–8426.
13. Keshishian, H., Chiba, A., Chang, T. N., Halfon, M. S., Harkins, E. W., Jarecki, J., Wang, L., Anderson, M. D., Cash, S., Halpern, M. E., and Johansen, J. (1993). Cellular mechanisms governing synaptic development in *Drosophila melanogaster. J. Neurobiol* **24,** 757–787.
14. Broadie, K. S., and Bate, M. (1993). Development of the embryonic neuromuscular synapse of *Drosophila melanogaster. J. Neurosci.* **13,** 144–166.
15. Gho, M. (1994). Voltage-clamp analysis of gap junctions between embryonic muscles in *Drosophila. J. Physiol.* **481,** 371–383.
16. Sink, H., and Whitington, P. M. (1991). Location and connectivity of abdominal motoneurons in the embryo and larva of *Drosophila melanogaster. J. Neurobiol.* **22,** 298–311.
17. Jacobs, J. R., and Goodman, C. S. (1989). Embryonic development of axon pathways in the *Drosophila* CNS. II. Behavior of pioneer growth cones. *J. Neurosci.* **9,** 2412–2422.
18. Koenig, J. H., Saito, K., and Ikeda, K. (1983). Reversible control of synaptic transmission in a single gene mutant of *Drosophila melanogaster. J. Cell Biol.* **96,** 1517–1522.
19. Kosaka, T., and Ikeda, K. (1983). Possible temperature-dependent blockage of synaptic vesicle recycling induced by a single gene mutation in *Drosophila. J. Neurobiol.* **14,** 207–225.
20. Lahey, T., Gorczyca, M., Jia, X.-X. and Budnik, V. (1994). The *Drosophila* tumor suppresser gene *dlg* is required for normal synaptic bouton structure. *Neuron* **13,** 823–835.
21. Saitoe, M., Tanaka, S., Takata, K., and Kidokoro, Y. (1997). Neural activity affects distribution of glutamate receptors during neuromuscular junction formation in *Drosophila* embryos. *Dev. Biol.* **184,** 48–60.
22. Petersen, S. A., Fetter, R. D., Noordermeer, J. N., Goodman, C. S., and DiAntonio, A. (1997). Genetic analysis of glutamate receptors in *Drosophila* reveals a retrograde signal regulating presynaptic transmitter release. *Neuron* **19,** 1237–1248.
23. Schuster, C. M., Davis, G. W., Fetter, R. D., and Goodman, C. S. (1996). Genetic dissection of structural and functional components of synaptic plasticity. I. Fasciclin II controls synaptic stabilization and growth. *Neuron* **17,** 641–654.
24. Tejedor, F. J., Bokhari, A., Rogero, O., Gorczyka, M., Zhang, J., Kim, E., Sheng, M., and Budnik, V. (1997). Essential role for *dlg* in synaptic clustering of Shaker K^+ channels in vivo. *J. Neurosci.* **17,** 152–159.
25. Rheuben, M. B., and Reese, T. S. (1978). Three-dimensional structure and membrane specializations of moth excitatory neuromuscular synapse. *J. Ultrastruct. Res.* **65,** 95–111.
26. Rheuben, M. B. (1985). Quantitative comparison of the structural features of slow and fast neuromuscular junctions in *Manduca. J. Neurosci.* **5,** 1704–1716.
27. Pearce, J., Govind, C. K., and Shivers, R. R. (1986). Intramembranous organization of lobster excitatory neuromuscular synapses. *J. Neurocytol.* **15,** 241–252.
28. Atwood, H. L., and Wojtowicz, J. M. (1986). Short-term and long-term plasticity and physiological differentiation of crustacean motor synapses. *Int. Rev. Neurobiol.* **28,** 275–362.
29. Fertuck, H. C., and Salpeter, M. M. (1974). Localization of acetylcholine receptor by ^{125}I-labeled alpha-bungarotoxin binding at mouse motor endplates. *Proc. Natl. Acad. Sci. USA* **71,** 1376–1378.

30. Couteaux, R., and Spacek, J. (1988). Specializations of subsynaptic cytoplasms: Comparison of axospinous synapses and neuromuscular junctions. *In* "Cellular and Molecular Basis of Synaptic Transmission." (H. Zimmermann, ed.), pp. 25–50. Springer-Verlag, Berlin.
31. Vautrin, J., and Mambrini, J. (1989). Synaptic current between neuromuscular junction folds. *J. Theor. Biol.* **140,** 479–498.
32. Boudier, J.-L., Le Treut, T., and Jover, E. (1992). Autoradiographic localization of voltage-dependent sodium channels on the mouse neuromuscular junction using ^{125}I-alpha scorpion toxin. II. Sodium channel distribution on postsynaptic membranes. *J. Neurosci.* **12,** 454–466.
33. Flucher, B. E., and Daniels, M. P. (1989). Distribution of sodium channels and ankyrin in neuromuscular junctions is complementary to that of acetylcholine receptors and the 43kd protein. *Neuron* **3,** 163–175.
34. Baer, S. M., and Rinzel, J. (1991). Propagation of dendritic spikes mediated by excitable spines: A continuum theory. *J. Neurophysiol.* **65,** 874–890.
35. Faeder, I., and Salpeter, M. (1970). Glutamate uptake by a stimulated insect nerve muscle preparation. *J. Cell Biol.* **46,** 300–307.
36. Broadie, K. S., Prokop, A., Bellen, H. J., O'Kane, C. J., Schulze, K. L., and Sweeney, S. T. (1995). Syntaxin and synaptobrevin function downstream of vesicle docking in *Drosophila*. *Neuron* **15,** 663–673.
37. Guan, B., Hartmann, B., Kho, Y.-H., Gorczyca, M., and Budnik, V. (1996). The *Drosophila* tumor suppressor gene, *dlg*, is involved in structural plasticity at a glutamatergic synapse. *Curr. Biol.* **6,** 695–706.
38. Kidokoro, Y., and Nishikawa, K. (1994). Miniature endplate currents at the newly formed neuromuscular junction in *Drosophila* embryos and larvae. *Neurosci. Res.* **19,** 143–154.
39. Nishikawa, K., and Kidokoro, Y. (1995). Junctional and extrajunctional glutamate receptor channels in *Drosophila* embryos and larvae. *J. Neurosci.* **15,** 7905–7915.
40. Rheuben, M. B., and Kammer, A. E. (1981). Membrane structures and physiology of an immature synapse. *J. Neurocytol.* **10,** 557–575.
41. Stewart, B. A., Schuster, C. M., Goodman, C. S., and Atwood, H. L. (1996). Homeostasis of synaptic transmission in *Drosophila* with genetically altered nerve terminal morphology. *J. Neurosci.* **16,** 3877–3886.
42. Atwood, H. L., and Cooper, R. L. (1996). Assessing ultrastructure of crustacean and insect neuromuscular junctions. *J. Neurosci. Methods* **69,** 51–58.
43. Atwood, H. L., and Cooper, R. L. (1995). Functional and structural parallels in Crustacean and *Drosophila* neuromuscular systems. *Am. Zool.* **35,** 556–565.
44. Nguyen, P. V., Marin, L., and Atwood, H. L. (1997). Synaptic physiology and mitochondrial function in crayfish tonic and phasic motor neurons. *J. Neurophysiol* **78,** 281–294.
45. Littleton, J. T., Bellen, H. J., and Perin, M. S. (1993). Expression of synaptotagmin in *Drosophila* reveals transport and localization of synaptic vesicles to the synapse. *Development* **118,** 1077–1088.
46. Rheuben, M. B. (1995). Specific associations of neurosecretory or neuromodulatory axons with insect skeletal muscles. *Am. Zool.* **35,** 566–577.
47. Klemm, N., Hustert, R., Cantera, R., and Nässel, D. R. (1986). Neurons reactive to antibodies against serotonin in the stomatogastric nervous system and in the alimentary canal of locust and crickets (Orthoptera, Insecta). *Neuroscience* **17,** 247–261.
48. Rheuben, M. B., and Kammer, A. E. (1983). Mechanisms influencing the amplitude and time course of the excitatory junction potential. *In* "The Physiology of Excitable Cells" (A. D. Grinnell and W. Moody, eds.), pp. 393–409. Alan Liss, New York.
49. Prokop, A., Landgraf, M., Rushton, E., Broadie, K., and Bate, M. (1996). Presynaptic development at the *Drosophila* neuromuscular junction: Assembly and localization of presynaptic active zones. *Neuron* **17,** 617–626.

50. Raabe, M. (1989). "Recent Developments in Insect Neurohormones." Plenum Press, New York.

51. Arbas, E. A., and Tolbert, L. P. (1986). Presynaptic terminals persist following degeneration of "flight" muscle during development of a flightless grasshopper. *J. Neurobiol.* **17,** 627–636.

ASSEMBLY AND MATURATION OF THE *DROSOPHILA* LARVAL NEUROMUSCULAR JUNCTION

L. Sian Gramates and Vivian Budnik

Department of Biology and Molecular and Cellular Biology Program, University of Massachusetts, Amherst, Massachusetts 01003

I. Introduction
II. A Bit of Anatomy
III. A Whirlwind Tour of Development
IV. Mechanisms of Synapse Assembly
   A. That Postsynaptic Apparatus: How Does It Get There Anyway?
   B. DLG and Synaptic Protein Clustering
   C. DLG and Synaptic Function
   D. Glutamate Receptor Clustering by PDZ-Containing Proteins
V. Structural Plasticity at the Neuromuscular Junction
   A. The Ever-Changing Fly Neuromuscular Junction
   B. DLG and Structural Plasticity during Neuromuscular Junction Development
   C. Electrical Activity and Presynaptic Plasticity
   D. Role of Cell Adhesion in Synapse Maintenance and Structural Plasticity
VI. Concluding Remarks and Future Directions
   References

I. Introduction

Synaptic transmission between neurons and their targets is crucially dependent on the precise arrangement of proteins at the synapse and on the spatial organization of the pre- and postsynaptic apparatus. For example, the organization of neurotransmitter receptors into high density clusters, which are directly apposed to the presynaptic release machinery, enables the generation of local changes in membrane potential that are sufficient to trigger or to prevent action potentials in the postsynaptic cell. Several events required for functional synaptic transmission occur immediately after the first interactions between the neuron and its postsynaptic partner. However, many of the processes involved in the formation of a mature synapse happen well after synaptogenesis, during a period referred to as synapse maturation. Moreover, it has become clear that, once formed, synapses are not static structures but rather dynamic entities that change depending on a number of factors such as modifications in target size or

0074-7742/99 $30.00

electrical activity. A long-standing question is how synapses are assembled and mature and how the synaptic cytoskeleton changes during development and plasticity. In recent years the *Drosophila* larval neuromuscular junction has emerged as a powerful model system to examine the physiological significance of molecules involved in synapse development and function (reviewed in Refs. 1 and 2). This review attempts to recount some of the major advances in our understanding of synapse development and maturation using the fruit fly neuromuscular junction as a model system.

II. A Bit of Anatomy

The wild-type *Drosophila* larva has an elaborate but well-characterized body wall musculature, organized in a segmentally repeated pattern of multinucleated muscle cells (Appendix).[3] Each abdominal hemisegment has 30 uniquely identifiable muscle fibers, each with a characteristic morphology, body wall insertion sites, and pattern of innervation by multiple specific motor neurons. The muscle pattern is precisely repeated in abdominal segments 2–7, with some variation in the first and last abdominal segments. These 30 muscle fibers are innervated by approximately 40 motor neurons, each with its own characteristic set of synaptic contacts on particular target muscles, complement of neurotransmitters, degree of terminal branching, and bouton morphology (Figs. 1 and 2).[1,2] All the motor neurons express glutamate as an excitatory neurotransmitter.[4–7] However, subsets also express cotransmitters, such as octopamine[8] and peptide neurotransmitters, including proctolin,[9] leukokinin I-like peptide,[10] pituitary adenylyl cyclase activating peptide (PACAP-like peptide),[11] and insulin-like peptide.[12]

The motor neuron endings branch daintily upon the surface of the target muscle fiber (Fig. 1). These endings are composed of a series of varicosities, or synaptic boutons, connected by axonal processes.[5] The axon endings fall into three morphologically distinct classes (Fig. 2) (reviewed in Ref. 2).

Type I axon endings typically project onto one or sometimes two muscle fibers and innervate all body wall muscles. Boutons appear roughly rounded in shape and extend only a short distance from the branch point area—the region of the muscle cell first contacted by axons emerging from the nerve (Figs. 1 and 2).[5] A relatively thick axonal process connects type I boutons. The presynaptic aspect of these boutons is intimately enveloped by a baroquely layered structure formed by the muscle junctional membrane, the subsynaptic reticulum (SSR) (Fig. 2A).[13,14] The SSR corresponds to the junctional muscle membrane where many molecules required for neurotransmission and neurotransmitter uptake are clustered. Type I boutons

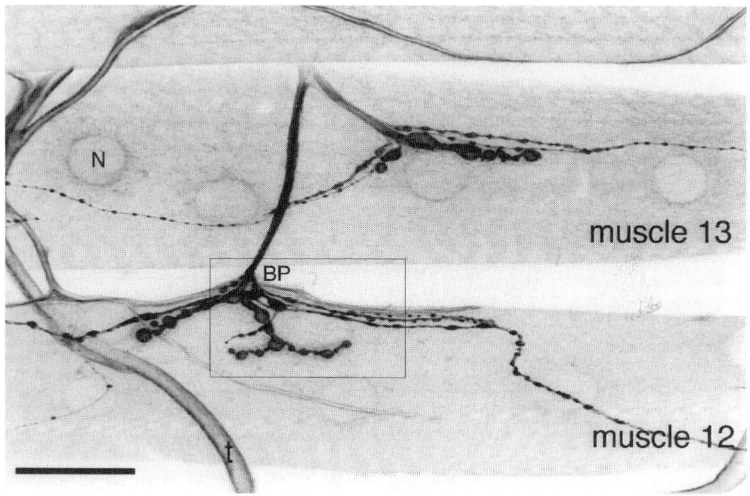

FIG. 1. Neuromuscular junctions in muscle fibers 12 and 13 in a third instar larval preparation labeled with anti-HRP antibodies. For identification of bouton types see Fig. 2D. The inset corresponds to the branch point region (BP), which is shown at a higher magnification in Fig. 2. N, nucleus; t, trachea. Bar: 20 μm.

are nearly filled with small round clear vesicles that contain glutamate[5,13,14]; a subset may also contain peptide neurotransmitters such as PACAP-like peptide.[11,13,14] Active zones, or putative neurotransmitter release sites, of type I boutons are characterized in cross section by electron-dense T bars composed of a stalk surmounted by a perpendicular bar.[13,14] Vesicle density is higher in the vicinity of T bars than in the rest of the bouton, and the membrane under the T bars is sometimes marked by putative exocytotic pits.[14] Type I boutons are further subdivided into type Ib and type Is boutons, respectively, big and small (Figs. 2A and 2E). Type Is boutons are smaller than type Ib, extend further out from the branch point, have a considerably less extensive SSR, and contain a few large clear and small dense core vesicles as well as small clear vesicles (Figs. 1 and 2E).[13,14] Type Ib synapses evoke smaller excitatory junctional potentials (EJPs) than type Is synapses.[15] In addition, these two types of axonal endings differ in their short-term plasticity properties; short-term facilitation is more pronounced at type Ib boutons. It has been suggested that type Ib and type Is endings are, respectively, analogous to the crustacean tonic and phasic nerve endings at the neuromuscular junction.[15]

Type II axons innervate multiple muscle fibers. Two motor neurons per segment give rise to these endings, which usually innervate all but eight muscles per hemisegment.[8,16] Type II boutons, which are the smallest and

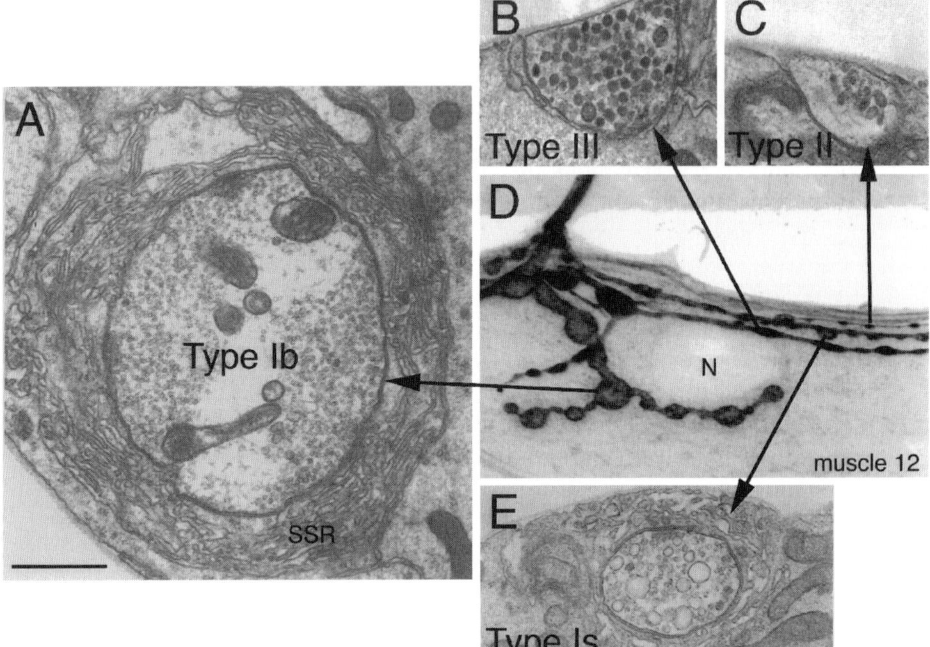

FIG. 2. Correlation of bouton morphology at the light and electron microscope level. (A) Type Ib bouton examined in cross section with TEM. arrow, active zone; SSR, subsynaptic reticulum. (B) Type III bouton examined in cross section with TEM. (C) Type II bouton examined in cross section with TEM. (D) A high magnification view of the boxed area is shown in Fig. 1, corresponding to the branch point area of muscle 12. N, nucleus. (E) Type Is bouton examined in cross section with TEM. Bar: 0.9 μm in A–C and E and 10 μm in D.

most numerous, are connected by thin axonal processes and extend nearly the entire length of the muscle surface, further than any of the other types (Figs. 1 and 2C).[5] They are localized to superficial grooves in the surface of the muscle fiber, with little or no surrounding SSR.[14] Within type II boutons are both clear round vesicles that contain glutamate and elliptical dense core vesicles believed to contain octopamine.[14] Evoked EJPs are significantly larger on bath application of 0.1–1 μM octopamine in wild-type larva (L. Griffith, personal communication), yet their amplitude is also higher in larvae unable to make tyrosine-β-hydroxylase, the enzyme that converts tyramine to octopamine.[17]

That just two octopamine-containing motor neurons innervate most muscles in every segment suggests that octopamine may modulate muscle activity in a global fashion. Release of octopamine by type II endings may influence an ‘‘excitation state’’ in the entire body wall musculature. This is in contrast to the case of type I motor neurons, each of which innervates one to a few muscle fibers. This different innervation modality by type I motor neurons may allow a fine control of the activity of each muscle cell.

Type III axon endings innervate only muscle 12 (and sometimes muscle 13) in segments A2 to A5.[12] The boutons are elongated and intermediate in size between type I and type II boutons (Figs. 1 and 2B). The endings extend more distally along the muscle fiber than type I endings, usually running adjacent to the even longer type II endings.[12] Similarly to type II endings, they have a superficial localization on the muscle cell surface and almost completely lack SSR (Fig. 2B).[14] Type III boutons are filled with large round vesicles of a varying degree of electron density, with a few small translucent vesicles,[14] and express insulin-like peptide.[12] The putative synaptic release sites are quite unlike those in type I boutons, consisting of electron-dense thickenings of the presynaptic membrane, surrounded by clusters of translucent vesicles.[14] The presence of a peptidergic phenotype, combined with the fact that so few muscle fibers are innervated by these endings, raises the possibility that type III boutons may subserve a neurohemal function, releasing modulatory substances to the entire body wall cavity.

Although only three major types of endings have been described in detail so far, it is possible that other bouton types and modulatory substances exist. Most of the ultrastructural studies have centered only on muscles 12, 13, 6, and 7, and recently in muscles 1 and 9.[18] Investigation of other neurotransmitter systems combined with EM studies in other muscles may reveal the presence of other terminals not yet described.

III. A Whirlwind Tour of Development

In the *Drosophila* larva, synapse formation is a dynamic collaboration between motor neuron and target muscle that begins late in embryogenesis and continues throughout the entire period of larval development. This review only briefly recounts the main aspects of initial synapse formation, as this is an important basis for our review of synapse maturation. For a more extensive review of synaptogenesis, the reader is referred to reviews in this volume by A. Chiba, by K. Broadie, and by M. Rheuben *et al.*

Synaptogenesis starts at about 13 hours after egg laying (AEL), as motor neuron growth cones begin to make contact with their target muscles. At this point, the motor neurons have begun expressing glutamate, and the muscle fibers express a few widely distributed glutamate receptors.[6,19] Each growth cone extends a profusion of filopodial processes, which actively explore muscle fibers in the vicinity of its target muscle.[6,18,20–25] As the motor neurons find their appropriate target muscles, about 14.5 hr AEL, they form embryonic synapses and the superfluous contacts are pruned back. Glutamate receptors in the target muscle cluster at the site of innervation, a process that depends on presynaptic contact[6] (see next section and K. Broadie of this volume for a more detailed discussion). In its earliest form, the embryonic neuromuscular junction resembles a growth cone[18] (see M. Rheuben *et al* of this volume). However, as the synapse matures, the axon ending branches and develops boutons. The muscle cells undergo a second wave of glutamate receptor synthesis at 16.5 hr AEL; these new receptors cluster at the synapses.[6,26] When the first instar larva hatches the synapse is relatively rudimentary; only a few to several boutons exist[12] and in type I boutons the SSR is not yet developed.[27] In contrast to the mature third instar type I bouton, at this stage the bouton lies in a shallow depression in the postsynaptic membrane and is not enveloped by SSR.[27,28]

Many changes in neuromuscular junction structure occur during the larval period, as the larval muscle fibers increase their volume by more than 150-fold.[27] The presynaptic arbor continues to expand during the larval stages, developing more extensive and elaborate branching, and increasing the number and size of synaptic boutons.[12,29] Type III endings begin extending onto muscle fiber 12 by the end of the first instar larval stage.[12] At the postsynaptic muscle the SSR begins to increase in surface area, developing infoldings at mid-first instar, completely surrounding the type I bouton by the late second instar, and becoming fully elaborated by the third larval instar.[27,28]

Development of muscle fibers and motor neurons is separable prior to synaptic contact. A motor neuron will grow to its target area, even if its target muscle has been ablated, and will often form synapses with remaining neighbor muscles.[30,31] In *twist* mutant embryos, which completely lack body wall muscles, motor neurons find their target regions and form morphologically normal presynaptic active zones.[32] Likewise, a normal pattern of body wall muscles will develop in the absence of motor neurons. Moreover, cell surface molecules, such as fasciclin III, localize to muscle synaptic domains prior to innervation, even if innervation is genetically ablated or delayed, as in the case of *prospero* mutants.[33] However, later stages of neuromuscular development require interactive participation of both the motor neuron and the muscle fiber. Neither presynaptic expansion of the axon arbor nor postsynaptic expansion of the junctional membrane will proceed without the participation of both partners.[1,12,27] Likewise, glutamate receptors remain homogeneously scattered on uninnervated muscle fibers, and the second wave of glutamate receptor synthesis never occurs.[26,34] Conversely, in the absence of muscle cells, active zones remain scattered in bouton-like structures along the axon and normal boutons fail to form.[32]

IV. Mechanisms of Synapse Assembly

A. That Postsynaptic Apparatus: How Does It Get There Anyway?

A highly precise organization of proteins in the pre- and postsynaptic apparatus is crucial to the efficient transmission of electrical signals from neuron to target.[35] The means by which those proteins are organized during development is the object of vigorous inquiry. Studies of the clustering of nicotinic acetylcholine receptors (AChRs) at the vertebrate neuromuscular junction[36] and of the aggregation of glycine receptors in the vertebrate central nervous system[37] have shed some light on the subject. Release of the nerve-derived extracellular matrix molecule agrin initiates clustering of AChRs within a few hr of the motor neuron's first contact with muscle.[36] Evidence suggests that rapsyn, a protein that colocalizes with AChR clusters, is directly involved in organizing AChRs into clusters, which are anchored to the underlying cytoskeleton at synaptic regions.[38] Similarly, the protein gephyrin is necessary for the clustering of glycine receptors at inhibitory synapses in the vertebrate central nervous system.[37]

Just as AChRs cluster at the vertebrate neuromuscular junction, so do glutamate receptors cluster around synaptic boutons of the *Drosophila* neuromuscular junction.[6,26] Two glutamate receptors, GluRIIa and GluRIIb, encoded by different genes, are expressed at the larval body wall muscles.[39] Both GluRIIa and GluRIIb appear to be clustered at type I synaptic boutons during larval stages. At least one of these two receptors, GluRIIa, is expressed before the onset of synaptogenesis.[19] Similarly, immunoreactivity to GluRII is detectable as early as 10.5 hr AEL in spheroidal mesodermal cells prior to myoblast fusion.[26] However, it is not clear if both GluR types are present at this stage, as the ability of the antibody to distinguish between the two GluR types is not known.

At 14 hr AEL, just before the motor neurons find their correct target muscles, GluRs appear to be in clusters associated with muscle fiber nuclei.[26] This observation suggests that the early GluR immunoreactivity may correspond to the association of the receptors with the perinuclear membranes during trafficking. Alternatively, GluR observed in the early stages may correspond to an immature inactive state of the receptor that accumulates near the site of synthesis.

By 17 hr AEL a functional neuromuscular junction is observed.[6] At this stage GluRs are still present in extrajunctional clusters, but these clusters are smaller and more numerous, perhaps as a result of a breaking down of the large clusters or of the second wave of GluR synthesis. These smaller clusters are not associated with muscle nuclei. By the end of embryogenesis at 22 hr AEL, these extrajunctional clusters have disappeared,[26] and GluRs accumulate at the junctions.[6,26] The blockade of neural activity during synaptogenesis by genetic[34] or pharmacological[40] means prevents the junctional clustering and second wave of synthesis of glutamate receptors and affects the dispersal of extrajunctional clusters. Thus, neural activity appears to have an important role in the clustering and upregulation of receptors during synaptogenesis.

B. DLG and Synaptic Protein Clustering

Many of the events and actors involved in synapse assembly are conserved across vertebrate and invertebrate species. However, clustering proteins homologous to the vertebrate proteins rapsyn, gephyrin, and agrin have yet to be identified in *Drosophila.* Nonetheless, a recently identified family of proteins, the membrane-associated guanylate kinases (MAGUKs), appears to be involved in clustering function at glutamatergic synapses (Fig. 3).[41–43] Proteins in the MAGUK family are composed of a number of modular domains involved in protein–protein interactions: three PDZ repeats (named for the first three proteins in which they were identified, PSD95/SAP90, DLG, and ZO1),[44] a src homology 3 (SH3) domain,[43] a HOOK domain,[45] and a guanylate kinase-like (GUK) domain.[43] Several

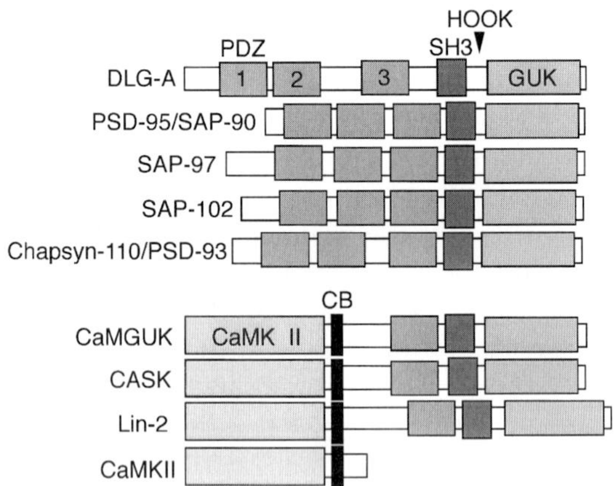

Fig. 3. Protein organization of the MAGUK family of synaptic proteins (top) showing the *Drosophila* DLG-A and vertebrate PSD-95/SAP-90, SAP-97, SAP-102, and Chapsyn-110/PSD-93. (Bottom) Organization of the MAGUK related family of synaptic proteins, which includes the *Drosophila* CaMGUK, the vertebrate CASK, and the *C. elegans* Lin-2. The *Drosophila* CaMKII is also shown for comparison. CB, calmodulin-binding domain.

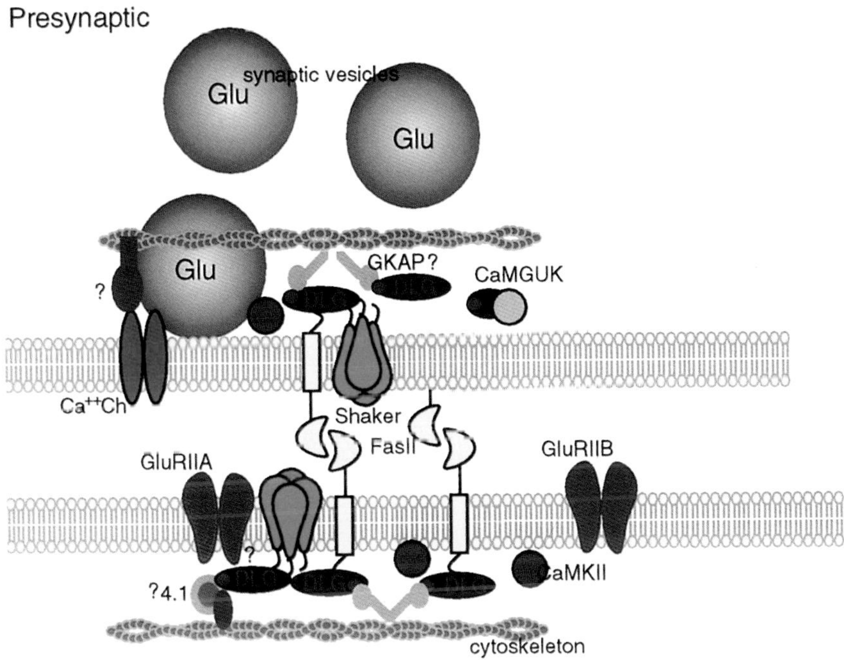

FIG. 4. Model of the interactions between synaptic proteins at the fly neuromuscular junction.

mammalian MAGUKs with synaptic expression have been isolated, including PSD95/SAP90, SAP97, SAP102, and chapsyn-110/PSD93.[42] Some of these proteins have been demonstrated by *in vitro* studies to interact directly with both NMDA receptors[44,46,47] and Shaker-type K^+ channels.[48] Studies in heterologous cells demonstrate that coexpression of PSD95/SAP90 with either of those molecules in COS7 cells results in the formation of NMDA receptor[46,49] or Shaker-type K^+ channel clusters.[48] This interaction has been shown to be mediated via binding between the PDZ domains of PSD95/SAP90 and and the carboxyl-terminal X-T/S-X-V/I (tSXV) motif in clustered molecules.[44,48,50]

The tumor suppressor gene *discs-large* (*dlg*) encodes a *Drosophila* MAGUK with a 60% identity at the amino acid level to vertebrate brain MAGUKs.[43] DLG protein is expressed at synaptic junctions[51] and epithelial septate junctions.[52] Mutations in *dlg* lead to a loss of cell polarity in epithelial tissues and to the formation of extensive tumors in the central nervous system and the imaginal discs.[52] DLG is expressed at the neuromuscular junction in a developmentally dynamic pattern at glutamatergic type I boutons, appearing first pre- and then postsynaptically.[27,51] In *dlg* mutants, the postsynaptic SSR is markedly less extensive and complex, with a reduction in the folding of the subsynaptic membranes.[53,54] At the presynaptic site, boutons have a bloated appearance[54] and an increase in the number of active zones.[55]

Immunocytochemical studies show Shaker K^+ channels to be clustered around type I boutons, colocalizing with DLG (Fig. 4, see color insert).[56] Like vertebrate Shaker-type channels, they have a carboxyl-terminal tSXV motif (ETDV).[57,58] This motif has been demonstrated to interact with PDZ1 and PDZ2 of DLG using the yeast two-hybrid assay.[56] The first evidence that the proposed clustering function of MAGUK was physiologically significant in the intact animal was obtained at the fly larval neuromuscular junction.[56] These studies provided compelling evidence that the MAGUK DLG is essential for the clustering of Shaker channels at type I boutons.[56] Severe hypomorphic *dlg* mutants, or *Shaker* (*Sh*) mutants with a carboxyl tail deletion, prevent the clustering of Shaker channels at type I synapses. Other *dlg* mutants, in which only the region encompassing PDZ1-PDZ2 is functionally present, are able to cluster Shaker channels, but not target them to the synaptic boutons, so that they exist ectopically in the muscle. Thus, the clustering and targeting of Shaker channels, while intimately related, are separable processes, both of which require DLG, but only the former requiring the PDZ domains, with the region of DLG required for targeting as yet unidentified.[56]

A possible candidate for the targeting function is the HOOK domain, which has been shown to be required for the targeting of DLG to septate junctions in epithelial cells (Fig. 3).[45] It has also been determined, using transgenic flies carrying an epitope-tagged form of DLG in which the HOOK domain was deleted, that the tagged DLG fails to localize to type I synaptic boutons, just as it fails to localize to epithelial junctions (M. Gorczyca and V. Budnik, unpublished results). These observations suggest that the HOOK domain may serve to anchor DLG to the synaptic cytoskeleton and may represent a DLG domain required for targeting Shaker channels. The HOOK domains of the MAGUKs hDLG and p55 have been shown to bind certain members of the protein 4.1 family, which interact with glycophorin, linking membrane proteins to the underlying spectrin/actin cytoskeleton.[59–62] Coracle and Expanded, *Drosophila* homologs of protein 4.1, colocalize with DLG at septate junctions[63] and appear to depend on DLG for correct localization.[45,64] However, a direct interaction between these proteins and DLG has not been demonstrated,[65] and Coracle is not expressed at the *Drosophila* neuromuscular junction (V. Budnik, unpublished data). It is possible that an alternative *Drosophila* protein 4.1 family member, such as dMoesin or dMerlin,[65] could be involved in targeting DLG to the neuromuscular junction, although none has yet been identified as either being appropriately localized or having an interaction with DLG.

Several researchers have suggested that SH3 domains have a role in subcellular targeting,[66–68] although DLG molecules lacking the SH3 domain are correctly targeted to the septate junction.[45] Similarly, it has been determined that absence of the SH3 domain does not prevent the association of DLG with type I synapses (M. Gorczyca and V. Budnik, unpublished results).

Other domains of DLG could also have a potential role in its localization. The SH3 and GUK domains of PSD95/SAP90, a synaptically expressed vertebrate MAGUK, have been shown to interact with a set of newly identified proteins of as yet unknown function, called alternatively GKAP[69,70] or SAPAPs.[71] These proteins could possibly be involved with targeting DLG or with linking DLG with the underlying cytoskeleton or internal second messenger pathways; however, no such interaction has yet been demonstrated. Likewise, the PDZ3 region of PSD95/SAP90 interacts with the novel protein CRIPT, which redistributes PSD95/SAP90 to the reticular network when both proteins are transfected into COS7 cells.[72] The PDZ3 region of PSD95/SAP90 has also been shown to bind to neuroligin,[73] a transmembrane protein with an extracellular domain that binds tightly to the extracellular domain of β-neurexin.[74,75] Neurexins are a large family of proteins encoded by several genes, with each gene giving rise to multiple alternatively spliced neurexin forms.[76,77] This family of proteins has been speculated to function in cell–cell recognition in the mammalian brain.[78] In *Drosophila*, a single neurexin gene has been isolated, and it appears to be involved in the formation of septate junctions in epithelial and glial cells.[79] Interestingly, the intracellular carboxy-terminal tail of the mammalian β-neurexin binds to CasK, a novel MAGUK related protein containing an N-terminal similar to CaM kinase in addition to PDZ, SH3, and GUK domains (Fig. 3).[80] In *Drosophila,* a CasK homolog, CaMGUK, has been identified (Figs. 3 and 4).[81] Flies mutant for *caki,* the gene coding for CaMGUK, have behavioral abnormalities, including altered walking speed.[82] However, its expression at the neuromuscular junction has not yet been detected, and it is not known if it has a role in synaptic development.

Another *Drosophila* synaptic protein that is localized at type I boutons is the product of the gene *still life* (*sif*). This protein contains a domain related to PDZ domains and produces structural alterations at the neuromuscular junction.[83] As is the case with CaMGUK, its binding partners have not been identified.

C. DLG and Synaptic Function

Voltage clamp analysis of synaptic currents at the body wall muscles of *dlg* mutants reveals a doubling in the amplitude of nerve-evoked currents over that of wild type.[53] This is not accompanied by any significant change in the size of miniature junctional currents, suggesting that the defect in synaptic transmission is presynaptic rather than postsynaptic. This change in amplitude can be rescued by using a GAL4 enhancer trap line to drive DLG in only the presynaptic motor neuron. In contrast, driving DLG in only the postsynaptic muscle fiber fails to rescue the mutant phenotype. This result is consistent with DLG affecting the amount of transmitter released by the presynaptic cell, rather than from a change in receptor density or distribution in the postsynaptic cell.[53]

The mechanism by which synaptic efficacy is enhanced in *dlg* mutants has not yet been conclusively determined. A persuasive model is that Shaker channels fail to be appropriately targeted to the presynaptic terminals so that repolarization of action potentials is retarded by a reduced efflux of K^+ ions, resulting in a prolonged period of neurotransmitter release. The observation that Shaker channels are not localized to synaptic boutons in *dlg* mutants lends support to this model,[56] as does analysis of I_A-deficient *Sh* mutants, which have a dramatically enhanced level of neurotransmitter release.[84] An alternative explanation is that other presynaptic properties of type I synaptic boutons are altered in mutants. Consistent with this idea is the observation that in severe hypomorphic *dlg* alleles the number of active zones is increased three- to fourfold.[55]

D. Glutamate Receptor Clustering by PDZ-Containing Proteins

At the mammalian central synapses GluRs are clustered by several types of PDZ-containing proteins. As described earlier, NMDA-type GluRs are clustered by members of the MAGUK family. However, other GluR types appear to interact with other PDZ-containing proteins such as GRIP[85] and Homer.[86] For example, the AMPA receptor subunit GluR2 interacts with the PDZ4-6 domains of GRIP, a protein containing seven PDZ domains.[85] A metabotropic GluR, mGluR5, interacts with the single PDZ domain of Homer.[86] A *Drosophila* Homer homolog has been cloned (U. Thomas and P. Worley, personal communication), but its functional role, as well as its binding partners, is not known.

In *Drosophila,* GluRs only partially colocalize with DLG. Moreover, fly GluRs lack any typical PDZ-binding motif. However, immunoprecipitation of body wall muscle extracts with anti-DLG antibodies results in coimmunoprecipitation of GluRs, and the distribution of GluRs at body wall muscles is abnormal in a *dlg* mutant allele (G. Popova, M. Packard, and V. Budnik, unpublished results). Further analysis will be required to determine if these observations are the result of a direct or indirect interaction with DLG.

V. Structural Plasticity at the Neuromuscular Junction

A. The Ever-Changing Fly Neuromuscular Junction

As described in the previous section, the larval neuromuscular junction is a dynamic system. New synaptic boutons are added throughout larval development.[12,29] Individual boutons are not static in nature; they become larger and develop more active zones per bouton.[27,28] This expansion parallels the exponential growth of larval muscles, which have a 150-fold increase in volume between hatching and pupariation; the size of the presynaptic surface increases to match the increased size of the growing postsynaptic target.[27]

Uninnervated muscles grow and develop normally; thus, it can be proposed that a retrograde signal transmitted from the muscle to the presynaptic cell is involved in the neuromuscular junction's ability to keep pace with the growing muscle.[32,39] Some support for the existence of such a retrograde signal is given by studies of embryos mutant for *myocyte-enhancing factor 2* (*mef-2*). In *mef-2* mutant embryos, myoblasts fail to fuse and do not differentiate into muscles. These unfused myoblasts are successfully contacted by motor neurons but target recognition is not followed by proper localization of presynaptic active zones at neuromuscular contact sites.[32] Another line of evidence indicating the existence of a retrograde signal are studies in which one of the two GluR genes has been mutated. These studies show that in null GluRIIA mutants the size of miniature EJCs is dramatically decreased, consistent with a reduction in the number of receptors in the muscle. However, the amplitude of junctional currents is similar to wild-type animals. Recordings of single synaptic boutons revealed that in these mutants the size of junctional currents per bouton is dramatically increased, possibly as a mechanism to compensate the decrease in GluR number.[39] These experiments provide clear evidence that the postsynaptic cell signals and influences the structure and physiology of the presynaptic cell. The identity of the putative retrograde signal remains a mystery. However, the cell adhesion molecule Fasciclin II (FasII) may be an attractive candidate, as FasII is localized at both pre- and postsynaptic sites, and changes in FasII levels elicit dramatic changes in NMJ structure and physiology.[28,55,87,88]

On the postsynaptic side of the neuromuscular junction, the growing muscle undergoes changes more complex than mere increase in size. The muscle surface at the site of synaptic contact elaborates from a simple unfolded membrane apposed to the presynaptic bouton to the ornate infoldings and convolutions of the mature SSR.[14,27] This increase in surface area may allow the accumulation of an increasing number of molecules required for synaptic transmission, which could then lead to an amplification of the synaptic response of increasingly larger muscles.

The dynamic nature of the neuromuscular junction can be demonstrated by the results of manipulating the expression levels of cell adhesion molecules (CAMs) in a muscle-specific manner. For example, altering the levels of FasII[89] or connectin[90] during embryogenesis can cause synaptic contacts to be delayed or induced to form on inappropriate muscle fibers. However, if these alterations are transient and the levels of the CAMs return to normal, the errors in synaptic connectivity are partially corrected, suggesting that neuromuscular junctions continually form and retract in response to external signals after synapse maturation.

B. DLG and Structural Plasticity during Neuromuscular Junction Development

Normal expansion of the SSR fails to occur in *dlg* mutants, so that the SSR of third instar larvae is underdeveloped.[51] This phenotype can be partially rescued by driving DLG expression in mutant muscle fibers. Further, overexpression of DLG protein in normal muscle fibers leads to an overdeveloped SSR.[53] These data point to a correlation between levels of DLG expression in the postsynaptic cell and the degree to which the SSR expands. However, the postsynaptic role of DLG does not tell the entire story of its influence on SSR expansion; there is also a presynaptic contribution. Selective expression of DLG in only the presynaptic motor neuron is sufficient to substantially rescue the reduced SSR phenotype in a mutant muscle cell.[53] Intriguingly, driving presynaptic expression of the vertebrate synaptically expressed MAGUKs PSD95/SAP90 and SAP102 also rescues at least some of the synaptic defects of *dlg* mutants, demonstrating an *in vivo* functional similarity to their *Drosophila* homolog.[54]

C. Electrical Activity and Presynaptic Plasticity

Axon pathfinding and target recognition, both of which happen before glutamate-induced currents can be detected in muscle fibers, appear to be independent of electrical activity.[91] Electrical activity is, in contrast, intimately involved in the refinement of synaptic connections. Experiments using current-blocking toxins and synaptic activity mutants have shown that reducing the level of presynaptic, though not postsynaptic, activity during synaptogenesis and early larval development promotes collateral sprouting of axon terminals.[92] Similarly, denervation of a muscle fiber leads to collateral sprouting of axons innervating neighboring muscles.[93] This sprouting leads to the formation of inappropriate connections, suggesting that electrical activity has a developmental role in preventing the consolidation of inappropriate connections so that precise patterns of connectivity are maintained.

Electrical activity levels also have a role in the larval expansion of the presynaptic apparatus.[94,95] Mutants affecting K^+ channels, such as *ether-a-go-go* (*eag*) and *Sh* double mutants, have an increased number of type II axonal branches and type I and type II boutons. Mutant larvae show a hyperexcitable phenotype,[84] with an increased level of neurotransmitter release and in frequency of endogenous activity in the body wall.[95] The synaptic boutons of these mutants have fewer vesicles and more active zones, suggesting that an abnormally large number of vesicles is released with each nerve impulse.[14] The opposite phenotype is seen in *no action potential* (*nap*) mutants, which have a decreased expression of Na^+ channels.[95,96]

The activity-dependent overgrowth of the neuromuscular junction may be mediated via the activation of a cAMP-dependent second messenger pathway, possibly through accumulation of Ca^{2+} in the presynaptic terminals.[97] Synaptic overgrowth similar to that seen in K^+ channel mutants is also observed in the learning mutant *dunce* (*dnc*).[97] *dnc* encodes phosphodiesterase II, thus mutations in this locus result in increased levels of cAMP. In contrast, mutations in *rutabaga,* the gene encoding a Ca^{2+}/calmodulin-dependent adenylate cyclase,[98] result in increased levels of cAMP and rescue the neuromuscular junction phenotype of both *dnc* and the K^+ channel hyperexcitable mutants.[97] An enzyme involved in the activity-dependent Ca^{2+} increase in neurons, Ca^{2+}/calmodulin-dependent protein kinase II (CaM kinase II), has been implicated in mechanisms of behavioral and physiological plasticity.[99,100,101] CaM kinase II is highly enriched in *Drosophila* motor endings and muscles.[101] Transgenic flies expressing *alanine CaM kinase II inhibitory peptide* (*ala*), which inhibits CaM kinase II by competing for its catalytic site, have an increased lower-order branching of type I motor endings, as well as defects in synaptic efficacy resembling those of *dnc* mutants,[101] which have an increased higher-order branching of type II endings.[97] The accumulation of activity-dependent Ca^{2+} in the nerve terminals could serve to activate both Ca^{2+}/calmodulin-dependent adenylate cyclase and CaM kinase II, thus invoking two pathways that could modulate different aspects of synapse structure and function.[101]

D. Role of Cell Adhesion in Synapse Maintenance and Structural Plasticity

Another component that appears to have a role in the regulation of both the size of presynaptic endings and the amount of neurotransmitter released by those endings is cell adhesion molecules. One of the best characterized of *Drosophila* synaptic CAMs is FasII, a homophilic member of the immunoglobulin superfamily related to the vertebrate neural CAM (NCAM) and to the *Aplysia* CAM (apCAM).[91,102]

The expression pattern of FasII in the *Drosophila* neuromuscular system is consistent with a molecule homophilically influencing a cell adhesion event. It is expressed at high levels in motor neuron axons and growth cones and at low levels at the surface of muscle fibers during the period of growth cone exploration.[28] FasII levels begin to fall in both muscles and axons at the time of synaptic contact and continue to sharply decrease throughout larval development. By the third larval instar, the remaining FasII is localized to all synaptic terminals of the neuromuscular junction. At type I terminals it is found in both the boutons and in the axonal processes connecting them, as well as in a population of vesicles in the center of many of the boutons.[28] Postsynaptically, FasII is localized to the SSR, where it has been reported to be expressed in a gradient, with the highest concentration nearest the boutons.[28]

FasII appears to have a role in the stabilization of synapses. Null mutants have relatively normal synapse formation, but the boutons retract during larval development, and null larvae die during the first instar,[28] suggesting an essential function in synapse maintenance. Both the lethality and the synapse elimination phenotype can be rescued by driving expression of a FasII transgene both pre- and postsynaptically; neither pre- nor postsynaptic expression alone is sufficient.[55] That FasII is required for synapse stabilization is also demonstrated in transgenic flies that express differential levels of FasII in two adjacent muscle fibers innervated by the same motor neurons. In these flies the formation of synaptic boutons is biased toward the muscle cell containing enhanced FasII levels.[87]

In severe *fasII* hypomorphs expressing less than 10% of normal levels of FasII, synapses are formed and are maintained during larval stages. These neuromuscular junctions have a decreased number of boutons, but the overall synaptic strength is similar to the wild type. The lack of change in synaptic strength despite a substantial decrease in bouton number is due to an increase in bouton size, number of active zones, and quantal content in each isolated bouton.[103] Similarly, synaptic strength is maintained when FasII is differentially expressed in two adjacent muscle fibers; the expansion of the axon arbor on the muscle with increased FasII expression is accompanied by a corresponding reduction of the arbor in the neighboring muscle.[87] These observations suggest that there is a mechanism to compensate for a change in the number of boutons, maintaining overall synaptic activity. This compensation mechanism occurs at the level of the nerve terminal and appears to require a retrograde signal from the muscle—the same motor neuron could generate both a nerve terminal with fewer boutons and increased synaptic strength per bouton in a muscle containing lower levels of FasII and a nerve terminal with an enhanced number of boutons and decreased synaptic strength per bouton in a muscle cell that overexpresses FasII.[87]

A different result is obtained in mutants in which FasII is expressed at 50% of the wild-type levels. These mutants have an increase in synaptic sprouting and bouton number, but synaptic strength is maintained as in severe hypomorphs due to a decrease in quantal content per bouton.[104] The enhanced number of boutons in mild *fasII* hypomorphs is reminiscent of the phenotype seen in *dnc* mutants and *eag-Sh* double mutants.[95,97,104] Intriguingly, *eag Sh* and *dnc* mutants have a decrease in the level of expression of FasII at the neuromuscular junction, suggesting that electrical activity and levels of calcium in the terminals regulate the accumulation of FasII.[104] Driving transgenic FasII expression rescues the excess sprouting phenotype, demonstrating that the decrease in FasII expression is both necessary and sufficient to induce excess presynaptic sprouting.[104] While FasII affects structural plasticity, the cAMP-dependent transcription factor CREB appears to regulate the changes in synaptic strength[105] (see review by F. Hannan and Y. Zhong in this volume for a discussion).

Intriguingly, a transmembrane form of FasII has a tSXV motif (NSAV), raising the possibility that DLG might be involved in the localization of FasII to type I synapses (Fig. 4). Several lines of evidence support this possibility. Strong *fasII* and *dlg* hypomorphs have similar effects on presynaptic ultrastructure, increasing the number of active zones per bouton. In addition, double mutants have a nonadditive phenotype, suggesting that *dlg* and *fasII* are likely to be involved in a common developmental pathway.[55] Immunocytochemistry reveals that DLG and FasII colocalize at the bouton borders of type I synapses. In *dlg* mutants, FasII is abnormally localized and becomes widely and diffusely dispersed in a broad region around the boutons.[55] This phenotype is partially rescued by pre- or postsynaptic expression of transgenic DLG or SAP97 and is fully rescued by simultaneous pre- and postsynaptic DLG or SAP97 expression.[55]

When DLG antiserum is used to immunoprecipitate larval body wall muscle extracts, FasII is coimmunoprecipitated with DLG from wild type, but not *dlg* mutant, protein extracts. This result suggests that DLG and FasII exist in a biochemical complex at the body wall muscles. Direct interaction between FasII and DLG has been demonstrated in yeast 2-hybrid, heterologous COS7 cells, and ELISA assays.[55] As in the case of Shaker K^+ channels, this interaction is mediated by binding between PDZ1-2 and tSXV in the intracellular tail of the transmembrane FasII isoform. Transgenic chimeric FasII constructs, in which the extracellular domain involved in homophilic interactions has been replaced with the CD8 extracellular domain, are targeted to type I synapses only when the tSXV motif is intact.[88] This direct interaction provides a mechanism by which the FasII-mediated homophilic adhesion can be coupled to the underlying cytoskeleton and intracellular signaling pathways both pre- and post synaptically.

Intriguingly, FasII binds to the same PDZ domains as Shaker K+ channels.[55,56] Recent evidence points to multimerization of synaptic MAGUKs as a mechanism of clustering.[106] Multimerization of DLG could allow for coaggregation of two or more molecules even if they had competing specificities for the same PDZ domains. Alternatively, Shaker channels and FasII may be able to bind independently to either PDZ1 or PDZ2 domains and be ''rafted'' together via the bivalency of DLG monomers. This idea is supported by triple transfection experiments in heterologous cells. When *Sh, dlg,* and *fasII* were triply transfected into COS7 cells, and COS7 cell extracts immunoprecipitated with anti-Shaker antibodies, both DLG and FasII coimmunoprecipitated with Shaker, suggesting that the three molecules could establish a ternary complex. Similarly, visualization of triply transfected COS7 cells revealed the formation of large clusters, in which DLG, FasII, and Shaker were exactly colocalized. These interactions were mediated by binding between Shaker and DLG, and between FasII and

DLG, because double transfection of *Sh* and *fasII* alone did not result in the organization of these molecules into clusters. In addition, mutations of the *fasII* and *Sh* PDZ-binding consensus sequences prevented the formation of the ternary complex.

The ability of DLG to organize Shaker channels and FasII into the same cluster raises the possibility that MAGUKs could cocluster any set of membrane proteins that contain the appropriate C-terminal PDZ-binding motif. Thus, DLG could serve as a scaffold for the recruitment of many classes of proteins localized at the neuromuscular junction, such as ligand-gated ion channels, seven-transmembrane receptors, and neurotransmitter transporters, as long as these membrane proteins terminate intracellularly with a compatible tSXV sequence. It is conceivable that DLG/PSD-95 family MAGUKs may directly organize many classes of membrane proteins into heteroclusters at synaptic junctions. It has been observed that different PDZ domains preferentially bind to different C-terminal motifs.[107, 108] The various recognition specificities of the PDZ domains in the protein directing clustering should determine the identities of the proteins recruited into the cluster; thus each distinct MAGUK should to be able to organize its own specific complement of proteins into a distinct junctional complex.

VI. Concluding Remarks and Future Directions

Studies in the last several years have demonstrated that the body wall muscle neuromuscular junction is an excellent model system to unravel the mechanisms by which synapses are formed during development and are modified during plasticity. The variety of genetic, molecular, anatomical, and electrophysiological techniques in this preparation makes it unique, allowing many alternative sets of approaches to be undertaken to study the question of synapse development. Perhaps more important is the realization that there is a great degree of evolutionary conservation in major aspects of synapse development and molecular participants of this process. This allows careful extrapolations to be made, and the intact fly can be used to test the roles of mammalian proteins for which a functional characterization is difficult.

A particularly exciting area that exemplifies this evolutionary conservation is the mechanism of clustering ion channels. The studies reviewed here demonstrate that similar sets of proteins subserve this function in both mammals and flies. While the cell biological and biochemical approaches in mammals permitted the initial characterization of this family as clustering proteins *in vitro,* their physiological role was demonstrated by the use of genetics in flies. Similarly, cell adhesion molecules have long been implicated in synaptic plasticity. However, a dissection of the pathways likely to be involved in the regulation of structural plasticity and changes in synaptic strength was obtained in flies through a genetic approach.

While the feasibility of an intimate collaboration between the approaches in a genetic system such as the fruit fly and more complex systems such as the mammalian brain has began to be realized, many of the questions posed at the beginning of this review are just starting to be resolved. We anticipate that over the next several years a host of new synaptic proteins will be discovered and their interactions determined and that they will be placed in the network of events that orchestrate the dynamics of synapse development.

Acknowledgments

We thank the members of the Budnik lab, Michael Gorczyca, Evgenya Popova, Young-Ho Koh, and Mary Packard, for their comments on this manuscript and for their patience with us while we were assembling this book. We also thank Michael Gorczyca for help in obtaining the confocal images in the figures. Work in the Budnik lab is supported by NIH Grants RO1-NS30072, RO1-NS37061, and KO4-NS01786

References

1. Keshishian, H., Broadie, K., Chiba, A., and Bate, N. (1996). The *Drosophila* neuromuscular junction: A model system for studying synaptic development and function. *Annu. Rev. Neurosci.* **19,** 545–575.
2. Budnik, V. (1996). Synapse assembly and structural plasticity at *Drosophila* neuromuscular junctions. *Curr. Opin. Neurobiol.* **6,** 858–867.
3. Crossley, C. A. (1977). The morphology and development of the *Drosophila* muscular system. *In* "Genetics and Biology of Drosophila" (M. Ashburner and T. R. F. Wright eds.), Vol 2b, pp. 499–560. Academic Press, New York.
4. Jan, J. Y., and Jan, Y. H. (1976). L-glutamate as an excitatory transmitter at the *Drosophila* larval neuromuscular junction. *J. Physiol.* **262,** 215–236.
5. Johansen, J. Halpern, M. E., Johansen, K. M. and Keshishian, H. (1989). Stereotypic morphology of glutamatergic synapses on identified muscle cells of *Drosophila* larvae. *J. Neurosci.* **9,** 710–725.
6. Broadie, K., and Bate, M. (1993). Development of the embryonic neuromuscular synapse of *Drosophila melanogaster. J. Neurosci.* **13,** 144–166.
7. Broadie, K., and Bate, M. (1993). Development of larval muscle properties in the embryonic myotubes of *Drosophila melanogaster. J. Neurosci.* **13,** 167–180.
8. Monastriani, M., Gorczyca, M., Rapus, J., Eckert, M., White, K., and Budnik, V. (1995). Octopamine immunoreactivity in the fruit fly *Drosophila melanogaster. J. Comp. Neurol.* **356,** 275–287.

9. Anderson, M. S., Halpern, M. E., and Keshishian, H. (1988). Identification of the neuropeptide transmitter proctolin in *Drosophila* larvae: Characterization of fiber-specific neuromuscular endings. *J. Neurosci.* **8,** 242–255.
10. Cantera, R., and Nassel, D. R. (1992). Segmental peptidergic innervation of abdominal targets in larval and adult dipteran insects revealed with an antiserum against leucokinin I. *Cell Tissue Res.* **269,** 459–471.
11. Zhong, Y., Pena, L. (1995). A novel synaptic transmission mediated by a PACAP-like neuropeptide in *Drosophila. Neuron* **15,** 2354–2366.
12. Gorczyca, M., Augart, C., and Budnik, V. Insulin-like receptor and insulin-like peptide are localized at neuromuscular junctions in *Drosophila. J. Neurosci.* **13,** 3692–3704.
13. Atwood, H. L., Govind, C. K., and Wu, C.-F. (1993). Differential ultrastructure of synaptic terminals on ventral longitudinal abdominal muscles in *Drosophila* larvae. *J. Neurobiol* **24,** 1009–1024.
14. Jia, X.-X., Gorczyca, M., and Budnik, V. (1993). Ultrastructure of neuromuscular junctions in *Drosophila:* Comparison of wild type and mutants with increased excitability. *J. Neurobiol.* **24,** 1025–1044.
15. Kurdyak, P., Atwood, H. L., Steward, B. A., and Wu, C.-F. (1994). Differential physiology and morphology of motor axons to ventral longitudinal muscles in larval *Drosophila. J. Comp. Neurol.* **350,** 463–472.
16. Budnik, JV., and Gorczyca, M. (1992). SSB, an antigen that selectively labels morphologically distinct synaptic boutons at the *Drosophila* larval neuromuscular junction. *J. Neurobiol.* **23,** 1054–1066.
17. Coleman, M. J., White, K., and Griffith, L. C. (1977). Synaptic transmission in octopamine deficient *Drosophila. Soc. Neurosci.* **23,** 978. [Abstract]
18. Yoshihara, M., Rheuben, M. B., and Kidokoro, Y. (1997). Transition from growth cone to functional motor nerve terminal in *Drosophila* embryos. *J. Neurosci.* **17,** 8408–8426.
19. Currie, D. A., Truman, J. R., and Burden, S. J. (1995). *Drosophila* glutamate receptor RNA expression in embryonic and larval muscle fibers. *Dev. Dyn.* **203,** 311–316.
20. Halpern, M. E., Chiba, A., Johansen, J., and Keshishian, H. (1991). Growth cone behavior underlying the development of stereotypic synaptic connections in *Drosophila* embryos. *J. Neurosci.* **11,** 3227–3238.
21. Johansen, J., Halpern, M. E., and Keshishian, H. (1989). Axonal guidance and the development of muscle fiber-specific innervation of *Drosophila* embryos. *J. Neurosci.* **9,** 4318–4332.
22. Sink, H., and Whitington, P. W. (1991). Pathfinding in the central nervous system and periphery by identified embryonic *Drosophila* motor axons. *Development* **112,** 307–316.
23. Sink, H., and Whitington, P. W. (1991). Location and connectivity of abdominal motor neurons in the embryo and larva of *Drosophila melanogaster. J. Neurobiol.* **22,** 298–311.
24. Van Vactor, D., Sink, H., Fambrough, D., Tsoo, R., and Goodman, C. S. (1993). Genes that control neuromuscular specificity in *Drosophila. Cell* **73,** 1137–1153.
25. Sink, H., and Whitington, P. W. (1991). Early ablation of target muscles modulates the arborisation pattern of an identified embryonic *Drosophila* motor axon. *Development* **113,** 701–707.
26. Saitoe, M., Tanaka, S., Takata, K., and Kidokoro, Y. (1997). Neural activity affects distribution of glutamates receptors during neuromuscular junction formation in *Drosophila* embryos. *Dev. Biol.* **184,** 48–60.
27. Guan, B., Hartmann, B., Koh, Y.-H., Gorczyca, M., and Budnik, V. (1996). The *Drosophila* tumor suppressor gene. *dlg,* is involved in structural plasticity at the glutamergic synapse. *Curr. Biol.* **6,** 695–706.

28. Schuster, C. M., Davis, G. W., Fetter, R. D., and Goodman, C. S. (1996). Genetic dissection of structural and functional components of synaptic plasticity. I. Fasciclin II controls synaptic stabilization and growth. *Neuron* **17,** 641–654.
29. Keshishian, H., Chiba, A., Chang, T. N., Halfon, M. S., Harkins, E. W., Jarecki, J., Wang, L., Anderson, M., Cash, S., and Halpern, M. E. (1993). Cellular mechanisms governing synaptic development in *Drosophila melanogaster. J. Neurobiol.* **24,** 757–787.
30. Cash, S., Chiba, A., and Keshishian, H. (1992). Alternate neuromuscular target selection following the loss of single muscle fibers in *Drosophila. J. Neurosci.* **12,** 2051–2064.
31. Chiba, A., Hing, H., Cash, S., and Keshishian, H. (1993). Growth cone choices of *Drosophila* motor neurons in response to muscle fiber mismatch. *J. Neurosci.* **13,** 714–732.
32. Prokop, A., Landgraf, M., Rushton, E., Broadie, K., and Bate, M. (1996). Presynaptic development at the *Drosophila* neuromuscular junction: Assembly and localization of presynaptic active zones. *Neuron* **17,** 617–626.
33. Broadie, K., and Bate, M. (1993). Muscle development is independent of innervation during *Drosophila* embryogenesis. *Development* **119,** 533–543.
34. Broadie, K., and Bate, M. (1993). Innervation directs receptor synthesis and localization in *Drosophila* embryo synaptogenesis. *Nature* **361,** 350–353.
35. Hall, Z. W., and Sanes, J. R. (1993). Synaptic structure and development: The neuromuscular synapse. *Neuron* **10,** Suppl. 99–121.
36. Fallon, J. R., and Hall, Z. W. (1994). Building synapses: Agrin and dystroglycan stick together. *Trends Neurosci.* **17,** 469–473.
37. Khuse, J., Betz, H., and Kirsch, J. (1995). The inhibitory glycine receptor: Architecture, synaptic localization and molecular pathology of a postsynaptic ion-channel complex. *Curr. Opin. Neurobiol.* **5,** 318–323.
38. Apel, E. D., and Merlie, J. P. (1995). The assembly of the postsynaptic apparatus. *Curr. Opin. Neurobiol.* **5,** 62–67.
39. Petersen, S. A., Fetter, R. D., Noordermeer, J. N., Goodman, C. S., and DiAntonio, A. (1997). Genetic analysis of glutamate receptors in *Drosophila* reveals a retrograde signal regulating presynaptic transmitter release. *Neuron* **19,** 1237–1247.
40. Broadie, K., and Bate, M. (1993). Activity-dependent development of the neuromuscular synapse during *Drosophila* embryogenesis. *Neuron* **11,** 607–619.
41. Gomperts, S. N. (1996). Clustering membrane proteins: It's all coming together with the PSD-95/SAP90 family. *Cell* **84,** 659–662.
42. Garner, C. C., and Kindler, S. (1996). Synaptic proteins and the assembly of synaptic junctions. *Trends Cell Biol.* **6,** 429–433.
43. Woods, D. F., and Bryant, P. J. (1993). ZO1, DlgA and PSD-95/SAP90: Homologous proteins in tight, septate and synaptic cell junctions. *Mech. Dev.* **44,** 85–89.
44. Kornau, H.-C., Schenker, L. T., Kennedy, M. B., and Seeburg, P. H. (1995). Domain interaction between NMDA receptor subunits and the postsynaptic density protein PSD-95. *Science* **269,** 1737–1740.
45. Hough, C. D., Woods, D. F., Park, S., and Bryant, P. J. (1997). Organizing a functional junctional complex required specific domains of the *Drosophila* MAGUK Discs large. *Genes Dev.* **11,** 3242–3253.
46. Kim, E., Cho, K.-O., Rothschild, A., and Sheng, M. (1996). Heteromultimerization and NMDA receptor-clustering activity of chapsyn-110, a member of the PSD-95 family of proteins. *Neuron* **17,** 103–113.
47. Müller, B. M., Kistner, U., Kindler, S., Chung, W. J., Kuhlendahl, S., Fenster, S. D., Lau, L. F., Veh, R. W., Huganir, R. L., Gundelfinger, E. D., and Garner, C. C. (1996). SAP102, a novel postsynaptic protein that interacts with NMDA receptor complexes in vivo. *Neuron* **17,** 255–265.

48. Kim, E., Niethammer, M., Rothschild, A., Jan, Y. N., and Sheng, M. (1995). Clustering of Shaker-type K^+ channels by interaction with a family of membrane-associated guanylate kinases. *Nature* **378,** 85–88.
49. Niethammer, M., Kim, E., and Sheng, M. (1996). Interaction between the C terminus of NMDA receptor subunits and multiple members of the PSD-95 family of membrane-associated guanylate kinases. *J. Neurosci.* **16,** 2157–2163.
50. Doyle, D. A., Lee, A., Lewis, J., Kim, E., Sheng, M., and Mackinnon, R. (1996). Crystal structures of a complexed and peptide-free membrane protein-binding domain: Molecular basis of peptide recognition by PDZ. *Cell* **85,** 1067–1076.
51. Lahey, T., Gorczyca, M., Jia, X.-X., and Budnik, V. (1994). The *Drosophila* tumor suppressor gene *dlg* is required for normal synaptic bouton structure. *Neuron* **13,** 823–835.
52. Woods, D. F., and Bryant, P. J. (1991). The discs-large tumor suppressor gene of *Drosophila* encodes a guanylate kinase homolog localized at septate junctions. *Cell* **66,** 451–464.
53. Budnik, V., Koh, Y.-H., Guan, B., Hartmann, B., Hough, C., Woods, D., and Gorczyca, M. (1996). Regulation of synapse structure and function by the *Drosophila* tumor suppressor gene *dlg. Neuron* **17,** 627–640.
54. Thomas, U., Phannavong, B., Muller, B., Garner, C. C., and Gundelfinger, E. D. (1997). Functional expression of rat synapse-associated proteins SAP97 and SAP102 in *Drosophila dlg-1* mutants: Effects on tumor suppression and synaptic bouton structure. *Mech. Dev.* **62,** 161–174.
55. Thomas, U., Kim, E., Kuhlendahl, S., Koh, Y. H., Gundelfinger, E. D., Sheng, M., Garner, C., and Budnik, V. (1997). Synaptic clustering of the cell adhesion molecule Fasciclin II by Discs-large and its role in the regulation of presynaptic structure. *Neuron* **19,** 787–799.
56. Tejedor, F., Bokhari, A., Rogero, O., Gorczyca, M., Zhang, J., Kim, E., Sheng, M., and Budnik, V. (1997). Essential role for *dlg* in synaptic clustering of Shaker K^+ channels in vivo. *J. Neurosci.* **17,** 152–159.
57. Pongs, O., Kecskemethyl, N., Muller, R., Krah-Jentgens, I., Baumann, A., Kiltz, H. H., Canal, I., Llamazares, S., and Ferrus, A. (1988). *Shaker* encodes a family of putative potassium channel proteins in the nervous system of *Drosophila. EMBO J.* **7,** 1087–1096
58. Schwartz, T. L., Tempel, B. L., Papazian, D. M., Jan, Y. N., and Jan, L. Y. (1988). Multiple potassium-channel components are produced by alternative splicing at the *Shaker* locus in *Drosophila. Nature* **331,** 137–142.
59. Lue, R. A., Marfatia, S. M., Branton, D., and Chishti, A. H. (1994). Cloning and characterization of hDlg: The human homologue of the *Drosophila* discs large tumor suppressor binds to protein 4.1. *Proc. Natl. Acad. Sci. USA* **91,** 9818–9822.
60. Lue, R. A., Brandin, E., Chan, E. P., and Branton, D. (1996). Two independent domains of hDlg are sufficient for subcellular targeting: The PDZ1-2 conformational unit and an alternatively spliced domain. *J. Cell Biol.* **135,** 1125–1137.
61. Marfatia, S. M., Lue, R. A., Branton, D., and Chishti, A. H. (1994). In vitro binding studies suggest a membrane-associated complex between erythroid p55, protein 4.1, and glycophorin C. *J. Biol. Chem.* **269,** 8631–8634.
62. Marfatia, S. M., Cabral, J. H., Lin, L., Hough, C., Bryant, P. J., Stolz, L., and Chishti, A. H. (1996). Modular organization of the PDZ domains in the human discs-large protein suggests a mechanism for coupling PDZ domain-binding proteins to ATP and the membrane cytoskeleton. *J. Cell Biol.* **135,** 753–766.
63. Fehon, R. G., Dawson, I. A., and Artavanis-Tsakonas, S. (1994). A *Drosophila* homologue of membrane-skeleton protein 4.1 is associated with septate junctions and is encoded by the *coracle* gene. *Development* **120,** 545–557.
64. Woods, D. F., Hough, C., Peel, D., Callaini, G., and Bryant, P. J. (1996). Dlg protein is required for junction structure, cell polarity, and proliferation control in *Drosophila* epithelia. *J. Cell Biol.* **134,** 1469–1482.

65. Fehon, R. G., LaJeunesse, D., Lamb, R., McCartney, B. M., Schweizer, L., and Ward, R. E. (1997). Functional studies of the protein 4.1 family of junctional proteins in *Drosophila. Soc. Gen. Physiol. Ser.* **52,** 149–159.
66. Bar-Sagi, D., Rotin, D., Batzer, A., Mandiyan, V., and Schlessinger, J. (1993). SH3 domains direct cellular localization of signalling molecules. *Cell* **74,** 83–91.
67. Schlessinger, J. (1994). SH2/SH3 signaling proteins. *Curr. Opin. Genet. Dev.* **4,** 25–30.
68. Cohen, G. B., Ren, R., and and Baltimore, D. (1995). Modular binding domains on signal transduction proteins. *Cell* **80,** 237–248.
69. Kim, E., Naisbitt, S., and Hsueh, Y. P. (1996). GKAP, a novel synaptic protein that interacts with the guanylate kinase-like domain of the PSD-95/SAP90 family of channel clustering molecules. *J. Cell Biol.* **136,** 669–678.
70. Naisbitt, S., Kim, E., Weinberg, R., Rao, A., Yang, F. C., Craig, A. M., and Sheng, M. (1997). Characterization of guanylate kinase-associated protein, a postsynaptic density protein at excitatory synapses that interacts directly with Postsynaptic Density-95/Synapse-Associated Protein 90. *J. Neurosci.* **17,** 5687–5696.
71. Takeucki, M., Hata, Y., Hirao, K., Toyoda, A., Irie, M., and Takai, Y. (1997). SAPAPs: A family of PSD-95/SAP90-associated proteins localized at postsynaptic density. *J. Biol. Chem.* **272,** 11943–11951.
72. Niethammer, M., Kim, E., Pechan, P., and Sheng, M. (1997). CRIPT, a novel protein interacting with PDZ domains of PSD-95 family proteins. *Soc. Neurosci.* **23,** 1466. [Abstract]
73. Irie, M., Hata, Y., Takeuchi, M., *et al.* (1997). Binding of neuroligins to PSD-95. *Science* **277,** 1511–1515.
74. Ichtchenko, K., Hata, Y., and Nguyen, T. (1995). Neuroligin 1: A splice site-specific ligand for b-neurexins. *Cell* **84,** 435–443.
75. Ichtchenko, K., Nguyen, T., and Sudhof, T. C. (1996). Structures, alternative splicing, and neurexin binding of multiple neuroligins. *J. Biol. Chem.* **271,** 2676–2682.
76. Ushkaryov, Y. A., and Sudhof, T. C. (1993). Neurexin IIIa: Extensive alternative splicing generates membrane-bound and soluble forms. *Proc. Natl. Acad. Sci. USA* **90,** 6410–6414.
77. Ullrich, B., Ushkaryov, Y. A., and Sudhof, T. C. (1995). Cartography of neurexins: More than 1000 isoforms generated by alternative splicing and expressed in distinct subsets of neurons. *Neuron* **14,** 497–507.
78. Missler, M., and Sudhof, T. C. (1998). Neurexins: Three genes and 1001 products. *Trends Genet.* **14,** 20–26.
79. Baumgartner, S., Littleton, T. J., Broadie, K., Bhat, M. A., Harbecke, R., Lengyel, J. A., Chiquet-Ehrismann, R., Prokop, A., and Bellen, H. J. (1996). A *Drosophila* neurexin is required for septate junction and blood-nerve barrier formation and function. *Cell* **87,** 1–20.
80. Hata, Y., Butz, S., and Sudhof, T. C. (1996). CASK: A novel *dlg*/PSD95 homolog with an N-terminal calmodulin-dependent protein kinase domain identified by interaction with neurexins. *J. Neurosci.* **16,** 2488–2494.
81. Dimitratos, S. D., Woods, D. F., and Byrant, P. J. (1997). Camguk, Lin-2, and CASK: Novel membrane-associated guanylate kinase homologs that also contain CaM kinase domains. *Mech. Dev.* **63,** 127–130.
82. Martin, J.-R., and Ollo, R. (1996). A new *Drosophila* Ca^{2+}/calmodulin-dependent protein kinase (Caki) is localized in the central nervous system and implicated in walking speed. *EMBO J.* **15,** 1865–1876.
83. Sone, M., Hishino, M., Suzuki, E., Kuroda, S., Kaibuchi, K., Nakagoshi, H., Saigo, K., Nabeshimo, Y., and Hama, C. (1997). Still life, a protein in synaptic terminals of *Drosophila* homologous to GDP-GTP exchangers. *Science* **275,** 543–547.

84. Ganetzky, B., and Wu, C.-F. (1983). Neurogenetic analysis of potassium currents in *Drosophila. J. Neurogenet.* **1,** 17–28.
85. Dong, H., O'Brien, R. J., Fung, E. T., Lanahan, A. A., Worley, P. F., and Huganir, R. L. (1997). GRIP: A synaptic PDZ domain-containing protein that interacts with AMPA receptors. *Nature* **386,** 279–284.
86. Brakeman, P. R., Lanahan, A. A., O'Brien, R. J., Roche, K., Barnes, C. A., Huganir, R. L., and Worley, P. F. (1997). Homer: A protein that selectively binds metabotropic glutamate receptors. *Nature* **386,** 284–288.
87. Davis, G. W., Schuster, C. M., and Goodman, C. S. (1997). Genetic analysis of the mechanisms controlling target selection: Target-derived Fasciclin II regulates the pattern of synapse formation. *Neuron* **19,** 561–573.
88. Zito, K., Fetter, R. D., Goodman, C. S., and Isacoff, E. Y. (1997). Synaptic clustering of Fasciclin II and Shaker: Essential targeting sequences and role of Dlg. *Neuron* **19,** 1007–1016.
89. Lin, D. M., and Goodman, C. S. (1994). Ectopic and increased expression of Fasciclin II alters motor neuron growth cone guidance. *Neuron* **13,** 507–523.
90. Nose, A., Takeichi, M., and Goodman, C. S. (1994). Ectopic expression of connectin reveals a repulsive function during growth cone guidance and synapse formation. *Neuron* **13,** 525–539.
91. Goodman, C. S., and Schatz, C. J. (1993). Developmental mechanisms that generate precise patterns of neuronal connectivity. *Neuron* **10,** 77–98.
92. Jarecki, J. and Keshishian, H. (1995). Role of neural activity during synaptogenesis in *Drosophila. J. Neurosci.* **15,** 8177–8190.
93. Halfon, M. S., Hashimoto, C., and Keshishian, H. (1995). The *Drosophila Toll* gene functions zygotically and is necessary for proper motor neuron and muscle development. *Dev. Biol.* **161,** 151–167.
94. Ganetzky, B., and Wu, C.-F. (1993). Neurogenetic analysis of potassium currents in *Drosophila:* Synergistic effects on neuromuscular transmission in double mutants. *J. Neurogenet.* **1,** 17–28.
95. Budnik, V., Zhong, Y., and Wu, C.-F. (1990). Morphological plasticity of motor axons in *Drosophila* mutants with altered excitability. *J. Neurosci.* **10,** 3754–3768.
96. Kernan, M. J., Kuroda, M. I., Kreber, R., Baker, B. S., and Ganetzky, B. (1991). *nap*^{ts}, a mutation affecting sodium channel activity in *Drosophila,* is an allele of *mele,* a regulator of X chromosome transcription. *Cell* **66,** 949–959.
97. Zhong, Y., Budnik, V., and Wu, C.-F. (1992). Synaptic plasticity in *Drosophila* memory and hyperexcitable mutants: Role of cAMP cascade. *J. Neurosci.* **12,** 644–651.
98. Davis, R. L. (1996). Physiology and biochemistry of *Drosophila* learning mutants. *Physiol. Rev.* 299–317.
99. Malenka, R. C., Kauer, J. A., Perkel, D. J., Mauk, M. D., Kelly, P. T., Nicoll, R. A., and Warham, M. N. (1989). An essential role for postsynaptic calmodulin and protein kinase activity in long-term potentiation. *Nature* **340,** 554–557.
100. Griffith, L. C., Verselis, L. M., Aitken, K. M., Kynacou, C. P., Danho, W., and Greenspan, R. J. (1993). Inhibition of calcium/calmodulin-dependent protein kinase in *Drosophila* disrupts behavioral plasticity. *Neuron* **10,** 501–509.
101. Wang, J., Renger, J. J., Griffith, L. C., Greenspan, R. J., and Wu, C. F. (1994). Concomitant alteration of physiological and developmental plasticity in *Drosophila* CaM kinase II-inhibited synapses. *Neuron* **13,** 1373–1384.
102. Grenningloh, G., Rehm, E. J., and Goodman, C. S. (1991). Genetic analysis of growth cone guidance in Drosophila: Fasciclin II functions as a neuronal recognition molecule. *Cell* **67,** 45–57.

103. Stewart, B. A., Schuster, C. M., Goodman, C. S., and Atwood, H. L. (1996). Homeostasis of synaptic transmission in *Drosophila* with genetically altered nerve terminal morphology. *J. Neurosci.* **16,** 3877–3886.
104. Schuster, C. M., Davis, G. W., Fetter, R. D. and Goodman, C. S. (1996). Genetic dissection of structural and functional components of synaptic plasticity. II. Fasciclin II controls presynaptic structural plasticity. *Neuron* **17,** 655–667.
105. Davis, G. W., Schuster, C. M., and Goodman, C. S. (1996). Genetic dissection of structural and functional components of synaptic plasticity. III. CREB is necessary for presynaptic functional plasticity. *Neuron* **17,** 669–679.
106. Hsueh, Y., Kim, E., and Sheng, M. (1997). Disulfide-linked head-to-head multimerization in the mechanism of ion channel clustering by PSD95. *Neuron* **18,** 803–814.
107. Sheng, M. (1996). PDZs and receptor/channel clustering: Rounding up the latest suspects. *Neuron* **17,** 575–578.
108. Songyang, Z., Fanning, A. S., Fu, C., Xu, J., Marfatia, S. M., Chishti, A. H., Crompton, A., Chan, A. C., Anderson, J. M., and Cantley, L. C. (1997). Recognition of unique carboxyl-terminal motifs by distinct PDZ domains. *Science* **275,** 73–77.

SECOND MESSENGER SYSTEMS UNDERLYING PLASTICITY AT THE NEUROMUSCULAR JUNCTION

Frances Hannan and Yi Zhong

Cold Spring Harbor Laboratory, Cold Spring Harbor, New York 11724

I. Introduction
II. Short-Term Plasticity
III. Neuromodulation
   A. PACAP
   B. Octopamine
   C. Other Neuropeptides
IV. Structural Plasticity
   A. Hyperexcitable Mutants
   B. Signal Transduction Mutants
   C. Cell Adhesion Mutants
V. Long-Term Functional Plasticity
   References

I. Introduction

The *Drosophila* neuromuscular junction provides an elegant experimental model to study signal transduction pathways that mediate synaptic plasticity because of its accessibility to electrophysiological techniques, morphological analyses, and pharmacological and genetic manipulations at identifiable synapses. Much of what is known concerning synaptic transmission and plasticity in *Drosophila* learning and memory mutants has been analyzed using the larval neuromuscular junction preparation, since the *Drosophila* central nervous system (CNS) is relatively inaccessible to electrophysiological recordings.

Although the neuromuscular junction itself is not involved in learning and memory, these studies have yielded profound insights into the mechanisms underlying both functional and structural plasticity. Many of the second messenger molecules found at the larval neuromuscular junction are involved in learning and memory in both invertebrates and vertebrates: studies at the neuromuscular junction thus have intriguing implications for mechanisms governing behavioral plasticity.

0074-7742/99 $30.00

Previous reviews have focused on the development of the neuromuscular system and the presynaptic molecules whose activity is triggered by the arrival of an action potential in the presynaptic terminal. This review examines the role that second messenger systems play in both structural and functional plasticity at the *Drosophila* neuromuscular junction.

II. Short-Term Plasticity

Glutamate is the major neurotransmitter at the *Drosophila* neuromuscular junction.[1,2] It is found at motor endings innervating all larval body wall muscles, in both type I and type II synaptic boutons.[3] Both ionotropic and metabotropic glutamate receptors have been identified in insect muscles (reviewed in Ref. 4). Clustering of cation selective ionotropic receptors on the postsynaptic side of the neuromuscular junction is induced during development of the neuromuscular junction in response to presynaptic electrical activity[5,6] in a manner analogous to the clustering of nicotinic acetylcholine receptors at the vertebrate neuromuscular junction (reviewed in Ref. 7).

Three different ionotropic glutamate receptors (GluR) and one metabotropic GluR have been identified by cloning in *Drosophila*.[8,9] Of the cloned receptors, only the DGluRII ionotropic receptor is expressed in muscle.[10] DGluRII mRNA is expressed uniformly throughout all muscle fibers,[11] and extrajunctional clusters of receptor proteins become localized to junctional sites in response to electrical activity at the developing synapses.[12] It is likely that DGluRII is the major receptor for glutamate, but there may be other GluR subunits involved since DGluRII is insensitive to argiotoxin when expressed in *Xenopus* oocytes.[10] At least three pharmacologically distinct ionotropic glutamate receptors have been observed at the locust neuromuscular junction.[4]

Application of glutamate or direct stimulation of the motorneuron results in an immediate fast depolarization in the larval body wall muscle, caused by the direct opening of the cation channels at the neuromuscular junction (Fig. 1).[1,2] The evoked response exhibits four major types of short-term plasticity, including facilitation, augmentation, posttetanic potentiation (PTP), and depression (Fig. 2).[13,14] Facilitation and augmentation are enhanced responses following single stimuli. Augmentation lasts several seconds longer than facilitation. PTP lasting 1–10 min can be induced by a brief high frequency (5–10 Hz) tetanic stimulation of the motor neuron. Classically, these phenomena are presumed to result mainly from a buildup of residual Ca^{2+} in the presynaptic terminal, leading to enhanced transmitter release.[13] PTP may also involve the reversible phosphorylation of synaptic molecules[15] and accumulation of Na^+ in the nerve terminal.[14] Continued stimulation results in depression, most probably due to depletion of the pool of releasable synaptic vesicles containing transmitter, although receptor desensitization may also be involved.[13,16]

Mutations that affect the cAMP signaling pathway, such as *dunce (dnc)* and *rutabaga (rut),* show impaired short-term facilitation (STF) and completely abolish PTP (Fig. 2).[17] The *dunce* mutation also enhances the size of excitatory junctional currents (EJCs) and affects the Ca^{2+} dependence of transmitter release.[17] The *rut* gene encodes a Ca^{2+}/sensitive adenylyl cyclase[18] and *dnc* encodes a cAMP phosphodiesterase.[19] The *dnc* and *rut* mutants were originally isolated on the basis of their defects in olfactory learning and memory.[20,21] A mouse type I adenylyl cyclase knockout also shows altered behavior and long-term potentiation (LTP).[22] Although *dnc* and *rut* mutations have opposite biochemical effects (*rut* reduces cAMP levels whereas *dnc* increases them), both cause deficits in behavioral plasticity and short-term synaptic plasticity, suggesting that cAMP homeostasis is important for neural plasticity. Genetic manipulations disrupting the activity of molecules upstream and downstream of *rut* and *dnc,* including Gsα and cAMP-dependent protein kinase A (PKA), also affect learning and memory in *Drosophila.*[23–25]

Facilitation, augmentation, and PTP are severely affected in transgenic flies expressing peptide inhibitors (*ala1* and *ala2*) of Ca^{2+}/calmodulin-dependent protein kinase (CaMKII).[26] In *ala* transformants, amplitudes of EJCs and spontaneous activity are also enhanced, and the Ca^{2+} dependence of transmitter release is altered.[26,27] Overexpression of these *ala* peptides has been shown to affect the learning of courtship behaviors in *Drosophila.*[28] The CaMKII molecule has also been strongly implicated in learning and memory in mammals, as knockout mice show deficits in hippocampal LTP and spatial learning tasks,[29,30] and targeted expression of an activated CaMKII transgene in the hippocampus also affects LTP and spatial memory formation.[31,32]

Both augmentation and PTP are disrupted in the *leonardo* learning and memory mutant, whereas short-term facilitation is strengthened.[33] The *leonardo* gene encodes a 14-3-3 protein[34] that affects a wide range of signaling pathways principally by interacting with kinases. The *leonardo* mutant also reduces basal synaptic transmission fidelity and fatigue resistance.[33] Another learning and memory mutant that blocks PTP is *linotte* (M. Saitoe and T. Tully, personal communication).

Mutations affecting neuronal excitability also cause defects in short-term plasticity. For example, the *Shaker* mutation disrupts a voltage-gated K^+ channel,[35] enhances facilitation, and increases EJCs.[36] Mutations in *Shaker*

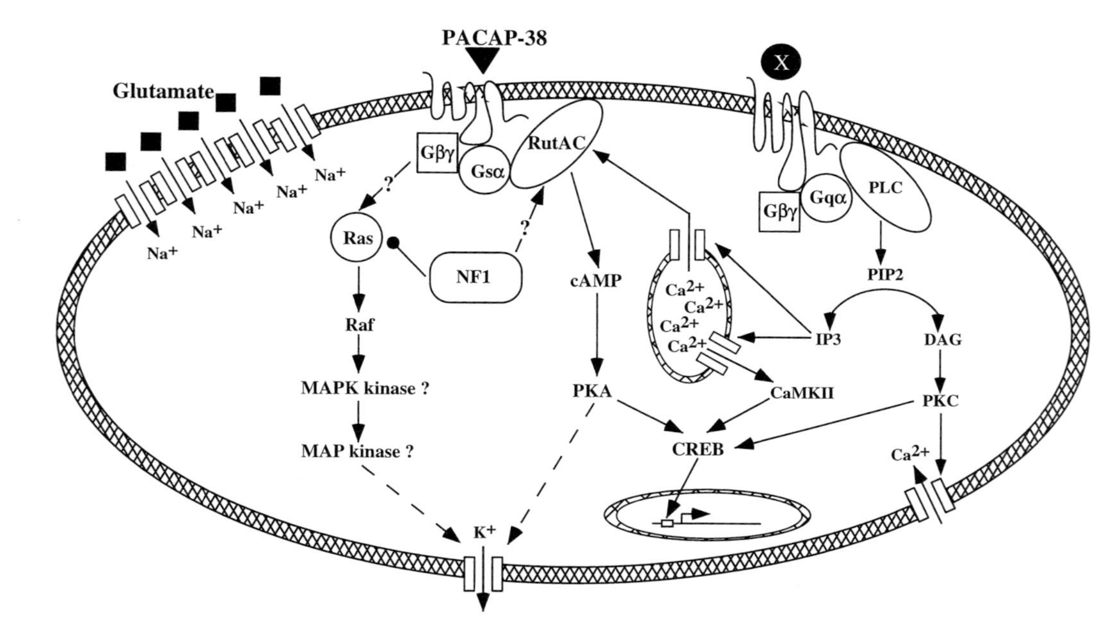

Glutamate
PACAP-38
X
Na+
Na+
Na+
Na+
Na+
Gβγ
Gsα
RutAC
Gβγ
Gqα
PLC
?
?
Ras
NF1
cAMP
PIP2
Ca2+
Ca2+
Ca2+
Ca2+
IP3
DAG
Raf
MAPK kinase ?
PKA
CaMKII
PKC
MAP kinase ?
CREB
Ca2+
K+

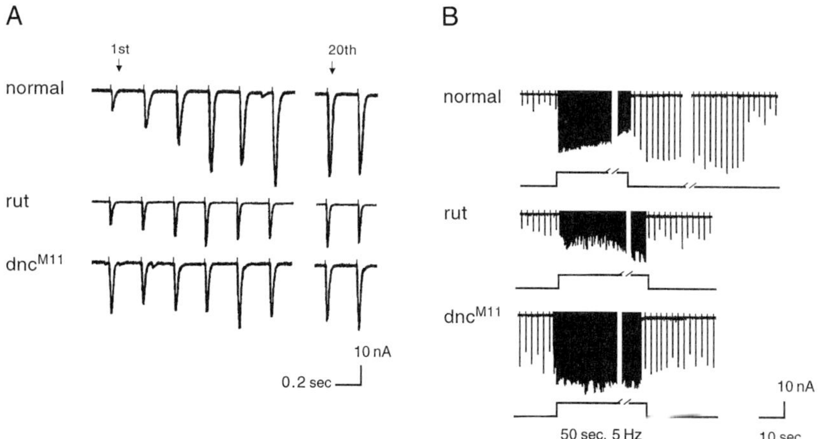

FIG. 2. Short-term plasticity at the larval neuromuscular junction in wild-type and mutant *Drosophila*. (A) Facilitation is reduced in *rut* and *dnc* mutants. (B) Posttetanic potentiation is abolished in *rut* and *dnc* mutants. Adapted from Zhong and Wu.[17]

also affect olfactory learning and courtship conditioning.[37,38] An allele of the *bang-sensitive* mutant *bas*^{MW1} also disrupts augmentation; however, STF is normal.[14]

In addition to functional effects at the neuromuscular junction, both hyperexcitable mutants and signal transduction mutants can affect the morphology of the neuromuscular junction (discussed later).

FIG. 1. Schematic representation of receptor-linked second messenger systems at the *Drosophila* larval neuromuscular junction. (A) Release of glutamate from the presynaptic terminal triggers an influx of sodium through ionotropic glutamate receptors. (B) Stimulation of the PACAP receptor activates the *rut*-encoded adenylyl cyclase (Rut) via the Gsα subunit of the heterotrimeric G-protein, thereby causing an increase in cAMP levels leading to an increased activity of protein kinase A (PKA). Activation of the Ras/Raf/MAPK pathway is thought to be mediated by the (G$\beta\gamma$) subunits of the G-protein complex. The simultaneous activation of both pathways leads to the modulation of voltage-gated K^+ channel activity. (C) Binding of an unidentified ligand (X) to its receptor causes dissociation of the Gqα-like subunit from the heterotrimeric G-protein complex, thereby stimulating phospholipase C (PLC) to cleave phosphatidylinositol bisphosphate (PIP2) into inositol trisphosphate (IP3) and diacylglycerol (DAG). IP3 initiates the release of Ca^{2+} from internal stores, which may activate Ca^{2+}/calmodulin-sensitive molecules, including Rut and Ca^{2+}/calmodulin-dependent kinase (CaMKII). DAG activates protein kinase C (PKC), which may modulate the activity of dihydropyridine-sensitive Ca^{2+} channels. All three kinases (PKA, CaMKII, and PKC) can also phosphorylate the cAMP response element binding protein (CREB), which may then translocate to the nucleus and activate the transcription of target genes.

III. Neuromodulation

A. PACAP

Pituitary adenylyl cyclase-activating polypeptide 38 (PACAP38) is a vertebrate neuropeptide that induces both an immediate depolarization and a delayed enhancement of voltage-gated K^+ channel activity when applied to *Drosophila* larval body wall muscles (Fig. 3).[39] The effect of PACAP38 can be mimicked by high frequency stimulation of the motor nerve fibers, which is presumed to result in release of an endogenous *Drosophila* PACAP38-like peptide.[39] This peptide is probably unrelated to the product(s) of the *amnesiac* locus[40] since no response is seen when *amnesiac* peptides are applied to the *Drosophila* neuromuscular junction (Y. Zhong, unpublished data). PACAP38 immunoreactivity is observed at the neuromuscular junctions of almost all body wall muscles but, unlike glutamate, is seen exclusively in type I synaptic boutons.[39] Three types of PACAP receptors have been identified in vertebrates, which differ in their affinity for PACAP and the types of G-protein that can be stimulated following receptor activation (e.g., Refs. 41–43). The high-affinity type I PACAP receptor couples to both adenylyl cyclase and phospholipase C.[41] No receptor responsive to PACAP38 has yet been cloned in *Drosophila* or any other invertebrate. PACAP receptors at the *Drosophila* neuromuscular junction may not be restricted to junctional sites, as significant extrajunctional responses are also observed.[39]

Studies of the postsynaptic second messenger pathways activated by PACAP38 at the *Drosophila* larval neuromuscular junction have revealed complex and unexpected biochemistries (Fig. 1).[44] Binding of PACAP38 to its G-protein-coupled receptor leads to the simultaneous activation of

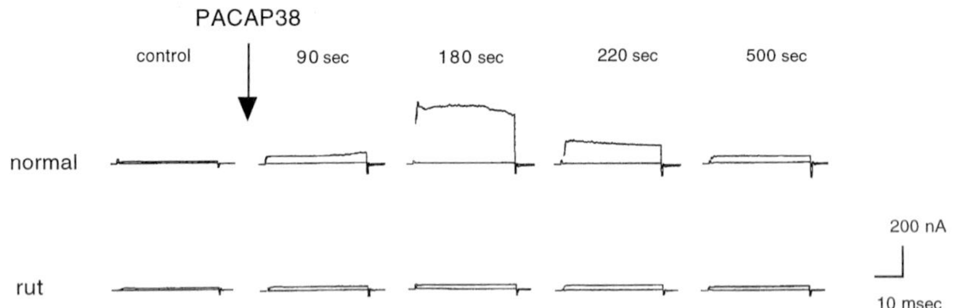

FIG. 3. PACAP38 response at the larval neuromuscular junction in wild-type and mutant *Drosophila*. Application of PACAP38 results in the delayed modulation of voltage-gated K^+ channel activity. This response is abolished in *rut* mutants. Adapted from Zhong.[44]

both Ras/Raf/MAPK and cAMP pathways. Stimulation of the *rut*-encoded adenylyl cyclase (Rut) via the Gsα subunit of the heterotrimeric G-protein causes an increase in cAMP levels, leading to increased activity of PKA. Activation of the Ras/Raf/MAPK pathway is thought to be mediated by the G$\beta\gamma$ subunits of the G-protein complex. The final link(s) between these pathways and the activation of voltage-gated K^+ channels is unclear at present. Mutations in *ras, raf,* and *rut* block the response to PACAP38 (Fig. 3), and activation of either pathway alone is insufficient to enhance K^+ channel activity, suggesting that coactivation of both pathways is required.[44]

It has been shown that a mutation that affects the *Drosophila* homolog of type 1 neurofibromatosis (NF1) also blocks the PACAP response at the larval neuromuscular junction.[45] The human NF1 disease causes benign and malignant tumors of the peripheral nervous system, as well as learning deficits in roughly half of affected individuals.[46,47] The *NF1* gene encodes a large protein containing a domain with Ras GTPase-activating protein (RasGAP) activity.[46,48] It has been assumed that the phenotypes of *NF1* mutations result from enhanced Ras activity, as GAPs normally act by stimulating the intrinsic rate of GTP hydrolysis by Ras, converting the active Ras-GTP form into inactive Ras-GDP.[46,49] Genetic analysis of the NF1 function at the larval neuromuscular junction has, however, revealed an unsuspected role for the NF1 protein. It has been shown that the defective PACAP response in *Drosophila NF1* mutants can be rescued by the application of cAMP and is not mimicked by manipulations that increase levels of active Ras.[45] Furthermore, the small body size defect of *Drosophila NF1* mutants can be rescued by overexpression of an activated form of PKA.[48] Localization and activity of Rut appear normal in *NF1* mutant flies,[45] suggesting that NF1 somehow regulates the activity of Rut in addition to any effect it may have on the activity of Ras (Fig. 1). Preliminary results suggest that the *NF1* mutation also affects olfactory learning in *Drosophila* (H.-F. Guo and Y. Zhong, unpublished data), which is consistent with an effect on the cAMP pathway. A mouse knockout of NF1 also displays impaired spatial learning, suggesting a role in hippocampal function.[50]

B. Octopamine

The biogenic amine octopamine is a major modulator of muscle activity in many invertebrates (reviewed in Refs. 51–54). Octopamine immunoreactivity is also distributed widely in *Drosophila* larval neuromuscular junctions; however, in contrast to PACAP38 it is restricted to type II synaptic boutons and is never or rarely seen at neuromuscular junctions in muscles 3–7, 25, and 28.[55] At least four types of octopamine receptors can be distinguished pharmacologically, differing in both G-protein coupling and tissue distribution.[51–54] The cloned *Drosophila* octopamine/tyramine receptor (OcTyR) most closely resembles the type I class of locust octopamine receptors, which couple to both adenylyl cyclase and phospholipase C, thus affecting both cAMP and Ca^{2+} levels.[56,57] Expression of the OcTyR receptor mRNA is restricted to the CNS and is not observed in any muscles in *Drosophila*.[58]

Flies carrying a null mutation in the tyramine-β-hydroxylase (Tβh) gene are unable to synthesize octopamine and show deficits in egg-laying activity.[59] Synaptic strength is also increased in these octopamine-deficient *Drosophila,* as excitatory junctional potential (EJP) amplitudes are significantly higher in Tβh null larvae than wild-type larvae, possibly to compensate for a lack of neurally released octopamine in the mutants.[60]

C. Other Neuropeptides

Several other neuropeptides with a much more restricted distribution are found at the *Drosophila* larval neuromuscular junction. Proctolin immunoreactivity is seen mainly in boutons on muscles 4, 12, and 13 plus a few other muscles[61]; leucokinin immunoreactivity is restricted to muscle 8 boutons[62]; and insulin immunoreactivity is found only at muscle 12.[63] In contrast, a *Drosophila* insulin receptor appears to be located at synaptic regions in all body wall muscles.[63] It has been proposed that neuropeptide-containing boutons form a third class of synapse, designated type III.[55] It is not clear whether these peptides serve as neuromodulators at the neuromuscular junction or as circulating hormones excreted into the hemolymph.

Several peptide products of the *Drosophila FMRFamide* gene[64] have excitatory effects on larval body wall muscles, enhancing both nerve-stimulated and spontaneous contractions (R. S. Hewes, E. C. Snowdeal, and P. H. Taghert, personal communication). These peptides probably function primarily as neurohormones, since the *FMRFamide* gene is expressed strongly in several neurosecretory cells in the larval thoracic ganglion,[65,66] but a small number of motor neurons may also weakly express the *FMRFamide* gene (R. Benveniste and P. H. Taghert, personal communication).

Other roles for peptides or biogenic amines at the larval neuromuscular junction remain to be discovered. For example, it is likely that G-protein-coupled receptors linked to phospholipase C will play some role at the *Drosophila* larval neuromuscular junction (Fig. 1), since the phospholipase C-mediated pathway modulates the activity of dihydropyridine-sensitive Ca^{2+} channels in larval body wall muscles.[67] Binding of the unidentified ligand to such a receptor would cause dissociation of a Gqα-like subunit from the heterotrimeric G-protein complex, thereby stimulating phospholipase C to cleave phosphatidylinositol bisphosphate (PIP2) into inositol trisphosphate (IP3) and diacylglycerol (DAG). IP3 initiates the release of Ca^{2+} from internal stores and DAG activates protein kinase C (PKC), leading to a variety of cellular responses. Alterations in Ca^{2+} levels may in turn activate Ca^{2+}/calmodulin-sensitive molecules, including Rut and CaMKII.

IV. Structural Plasticity

During larval development the body wall muscles increase in size more than 150-fold.[68] Not surprisingly, a parallel increase in the size of motor neurons and the number of boutons is also observed.[69] The synaptic boutons observed in embryos do not begin to become distinguishable as type I or type II endings until late stage 17 just prior to hatching.[70] The complexity of the subsynaptic reticulum (SSR) underlying type I boutons is also developmentally regulated.[71] The *discs-large (dlg)* gene product is important for the proper development of this postsynaptic structure[71–73] and for the clustering of channels and cell adhesion molecules at type I synapses[74–76] as discussed earlier by K. Broadie.

This capacity for developmental change at the larval neuromuscular junction underscores a latent potential for dramatic changes in structure caused by several classes of mutations that alter the activity of ion channels, second messengers, and cell adhesion molecules. Long-term changes in synaptic function that contribute to learning and memory, including LTP in the mammalian hippocampus and long-term facilitation (LTF) in *Aplysia,* are also accompanied by structural changes (reviewed in Ref. 77). The molecular mechanisms underlying developmental rearrangements of synaptic connections also closely parallel the mechanisms involved in learning and memory (reviewed in Ref. 78). An understanding of the mechanisms underlying structural and functional plasticity at the *Drosophila* neuromuscular junction may lead to insights regarding the cellular processes involved in learning and memory.

Figure 4 shows some examples of activity-dependent structural plasticity. Table 1 summarizes the phenotypes of three classes of mutations affecting morphology and function at the *Drosophila* larval neuromuscular junction: (1) hyperexcitable mutants, (2) signal transduction mutants, and (3) cell adhesion mutants. The effects of these mutations have allowed us to elucidate the molecular basis of activity-dependent synaptic plasticity. This section reviews the effects of these mutations and combinations thereof and

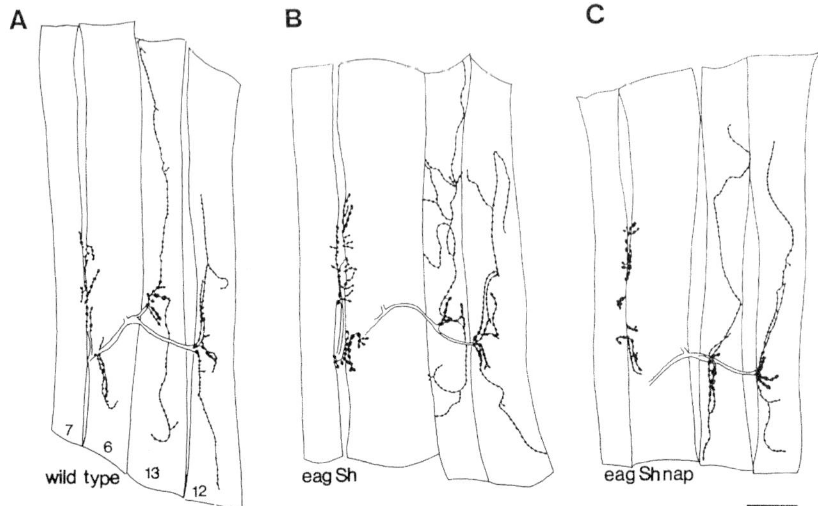

FIG. 4. Neuronal arborization on muscles 6, 7, 12, and 13 in wild-type and mutant *Drosophila*. (A) Wild-type arborization pattern. (B) The *eag Sh* hyperexcitable mutant shows increased branching and numbers of varicosities in type II terminals on muscles 12 and 13. (C) This effect is suppressed in the *eag Sh nap^{ts}* triple mutant. Reproduced with permission from Budnik *et al.*[86]

summarizes the current understanding of the correlation between structural and functional plasticity.

A. Hyperexcitable Mutants

Mutations that delay presynaptic repolarization either by decreased K^+ channel activity or by increased Na^+ channel activity result in a hyperexcitable phenotype. *Shaker (Sh), ether-a-go-go (eag), Hyperkinetic (Hk),* and *seizure (sei)* all encode K^+ channel subunits,[35,79–82] whereas *paralytic (para)* encodes a Na^+ channel.[83] The *no-action-potential temperature-sensitive (nap^{ts})* mutant reduces the number of functional Na^+ channels[84] whereas the *temperature-induced-paralysis E (tipE)* gene encodes a regulatory subunit that enhances the activity of the *para* Na^+ channel.[85]

When raised at room temperature, single mutations in *eag, Sh,* or *Hk* do not affect the morphology of the neuromuscular junction; however, double mutants *eag Sh* and *eag Hk* show an increase in both the number of axonal branches and the number of varicosities of type II processes.[86] A duplication of the wild-type *para* locus shows a similar but less extensive phenotype, either alone or in combination with the *Sh* mutation.[86] The *nap^{ts}* mutant shows a slight decrease in branching and is able to completely

TABLE I
MUTATIONS AFFECTING STRUCTURAL AND FUNCTIONAL PLASTICITY AT THE *Drosophila* LARVAL NEUROMUSCULAR JUNCTION[a]

Mutant(s)	Structure	Function
1. Hyperexcitable mutants[86]		
eag Sh, eag Hk	↑ ↑ Type II	↑ tonic activity
Dp para+, Dp para+ Sh	↑ Type II	—
nap^{ts}	↓ Type II	—
eag Sh nap^{ts}	Normal	Normal
eag Sh para	↑ Type II	—
Sh	Normal	↑ EJCs, ↑ STF
2. Signal transduction mutants[17,26,27,88,89,90]		
dnc	↑ Type II	↑ EJCs, ↓ STF, no PTP
rut	Normal	Normal EJCs, ↓ STF, no PTP
dnc rut	Normal	—
hs ala1, hs ala2	↑ ↑ Type I	↑ EJCs, ↓ STF, ↓ PTP
frq	↓ Type II	↑ firing
3. Cell adhesion mutants[69,99,101,102,108]		
fasI	↑ ↑ Type II	↓ EJCs, ↓ QC
Dp fasI+	↓ Type I & II	↑ EJCs, ↑ QC
10% fasII	↓ Type I	↑ EJPs
50% fasII	↑ ↑ Type I	Normal
4. Mutant combinations[88,102,108]		
dnc eag, dnc Sh, dnc eag Sh	↑ ↑ Type II	—
dnc rut Sh	Normal	—
eag Sh 100% fasII	Normal	—
dnc 100% fasII	Normal	—
dnc hs CREBb	↑ ↑ Type I	Normal
10% fasII hs CREBa	↓ Type I	Normal
50% fasII hs CREBa	↑ ↑ Type I	↑ QC
100% fasII hs CREBa	Normal	Normal

[a] See text for most mutant designations. Hypomorphic alleles *fasII^{e76}* and *fasII^{e86}* are designated *10% fasII* and *50% fasII,* respectively. The transgenic line *fasII^{e93}* expressing wild-type levels of FasII is designated *100% fasII.* Structural changes occur in type I or type II nerve terminals. Dp, duplication; hs, heat shock; EJC, excitatory junctional current; EJP, excitatory junctional potential; STF, short term facilitation; PTP, posttetanic potentiation; QC, quantal content.

reverse the phenotype of the *eag Sh* mutants; however, *para^{ts1}* has no effect on *eag Sh* morphology.[86] The normal rhythmic activity pattern seen in wild-type larvae is disrupted by periods of strong tonic activity in *eag Sh* mutants, which can be partially rescued by *nap^{ts}*.[86] Mutations in *nap, sei,* and *tipE,* which reduce excitability, also increase the number of ectopic connections on inappropriate muscle targets.[87] Nerve activity therefore plays a fundamental role in shaping neuronal arborizations at the *Drosophila* neuromuscular junction.

Mutations in *sh, eag,* and *nap^{ts}* also disrupt learning and memory in a courtship conditioning paradigm.[37] In addition, *Sh* and *nap^{ts}* show defects in olfactory learning.[38] Membrane excitability therefore is also important in learning and memory.

B. Signal Transduction Mutants

The structural changes induced by neural activity are probably mediated through the activation of various signal transduction pathways. Mutations affecting cAMP levels such as *dnc* and *rut* have dramatic effects on the function and morphology of the larval neuromuscular junction.[17,88] The numbers of branches and varicosities in type II processes are increased in *dnc* mutants, which have increased cAMP levels.[88] This effect can be suppressed in the *rut dnc* double mutant; however, the lowered levels of cAMP in the *rut* single mutant do not alter synaptic morphology.[88] Interestingly, the effects of *dnc* can be enhanced in double mutant combinations *dnc eag* and *dnc Sh,* although the triple mutant *dnc eag Sh* shows no additional effect over that seen in double mutants.[88] The *rut* mutation also suppresses the effect of the *dnc Sh* double mutant combination, since a *dnc rut Sh* triple mutant shows normal morphology.[88] These results suggest that enhanced neural activity increases cAMP synthesis by stimulating Rut, which in turn increases activity-dependent arborization.

Manipulating CaMKII activity also results in altered morphology at the larval neuromuscular junction.[26] Overexpression of the CaMKII peptide inhibitors *ala1* and *ala2* increases branching and varicosity number at type I synapses.[26] Furthermore, these effects seem to be limited to primary or secondary branches near the point of entry of the nerve to the muscle.[26] This is in direct contrast to the effect of the cAMP mutants and the hyperexcitable mutants which affect growth of type II synapses, particularly in the higher order branches.[86,88]

Interestingly, mutations in the Ca^{2+}-binding protein Frequenin, which also cause a hyperexcitable phenotype,[89] show reductions in both the length and the number of branches in type II motor neurons.[90] Thus levels of Ca^{2+} and/or the activity of Ca^{2+}-binding proteins may be important for plasticity at type II synapses as well as type I synapses.

C. Cell Adhesion Mutants

Signal transduction pathways activated by neural activity lead to the modulation of function or regulation of expression of downstream molecules such as cell adhesion molecules (CAMs), which can directly influence nerve terminal structure. CAMs play important roles in the guidance of motor neurons as they leave the CNS and find their way to the correct muscle targets in the periphery, as discussed previously (reviewed in Ref. 91 and 92). Mutations in individual CAMs have little effect on the stereotypic pattern of larval motor neurons, suggesting that there are multiple, overlapping cues involved in the guidance process.[91,92] A mouse neural CAM (NCAM) knockout affects spatial learning and LTP[93,94] and the *Aplysia* NCAM homolog is downregulated in response to stimuli that enhance LTF.[95]

In *Drosophila,* the *fasciclin I (fasI)* gene encodes a cell surface glycoprotein[96] that is expressed in all PNS neurons and some CNS neurons[97] and is capable of homophilic adhesion.[98] A *fasI* null mutant shows enhanced branching and numbers of varicosities in type II motor neurons, whereas overexpression or duplication of *fasI* shows a reduction in branching and varicosity numbers.[99] Interestingly, the overexpression of FasI affects both type I and type II motor neurons.[99] The *fasI* null mutation also reduces the amplitude of EJCs and the quantal content, whereas increased levels of FasI once again have the exact opposite effect.[99] Thus overall synaptic strength is balanced by reducing EJCs when the numbers of terminals are increased and by increasing EJCs when the terminal number is reduced. The involvement of FasI in mediating activity-dependent structural changes has not yet been examined.

The *fasciclin II (fasII)* gene encodes the *Drosophila* homolog of the vertebrate NCAM.[100] FasII is required both pre- and postsynaptically to stabilize synapses.[69] Mutations that reduce FasII expression to ~10% reduce branching and numbers of varicosities.[69,101] Reduction of FasII expression to ~50% has the opposite effect, including an increase in synaptic growth and concomitant alterations in synaptic function such as decreased quantal content.[69,102] The expression of FasII is reduced in *eag Sh* and *dnc* mutants, suggesting that the effect of these mutations on activity-dependent growth is mediated by a reduction in FasII levels.[102] This is further supported by the observation that an increased expression of FasII suppresses the increased synaptic growth seen in *eag Sh* or *dnc* mutants.[102]

Structural plasticity at the *Drosophila* larval neuromuscular junction therefore appears to be mediated by regulation of the levels of CAMs by second messenger systems that are stimulated in response to neural activity.

V. Long-Term Functional Plasticity

A presumption underlying our interest in the structural plasticity of nerve terminal arborization is that it provides a basis for long-term functional plasticity. More synapses formed should lead to enhanced synaptic transmission. This has been difficult to examine closely in any model system; however, results from studies of the *Drosophila* neuromuscular junction, although far from clear and conclusive, have provided some intriguing insights.

The relationship between structure and function is not as simple as is often assumed. For instance, a reduction in the number of varicosities at the motor nerve terminals induced by overexpression of FasI or reduced expression of FasII does not reduce synaptic transmission.[99,101,102] Conversely, an increase in the number of varicosities in *fasI* and *fasII* mutants also does not lead to increased transmitter release.[99,101,102] In *dnc* mutants, however, synaptic transmission is enhanced[17] and the number of varicosities is increased.[88]

In contrast, manipulation of the function of the cAMP response element binding (CREB) protein mainly affects function but not structure. CREB is a transcription factor that is phosphorylated by PKA and activates the transcription of many genes presumably important in synaptic plasticity.[103] Overexpression of a CREB repressor (CREBb) isoform in *Drosophila* specifically inhibits long-term memory formation, whereas expression of a CREB activator (CREBa) isoform dramatically enhances long-term memory formation.[104,105] CREB also affects LTF in *Aplysia*[106] and CREB knockout mice show impaired spatial learning.[107]

At the neuromuscular junction, expression of the CREBb repressor in a *dnc* background does not affect enhanced motor nerve terminal arborization but it does reduce the increase in synaptic transmission.[108] Conversely, expression of the CREBa activator in 10% FasII mutants does not affect reduced arborization but it also suppresses enhanced EJCs.[108] The CREBa activator does enhance synaptic transmission in the 50% FasII mutants where enhanced synaptic arborization has occurred, suggesting that structural growth is required in order to accommodate increased functionality.[108]

The influence of structure on function is likely to be multidimensional. For example, ultrastructural analyses of *fasII* mutants has revealed that the reduction in the number of varicosities may be compensated by an increase in both the size of synapses and the number of active zones formed.[101] Thus, the overall numbers of synaptic vesicles ready for release may not be changed. In addition to morphology, there are also nonstructural components, such as the number of receptors or proteins important for mediating release, that contribute to synaptic function.

Neuronal activity and the stimulation of signal transduction cascades lead to changes in structural and functional plasticity by regulating the activity of CAMs and CREB. These processes are clearly distinct, as FasI and FasII mainly affect structure while CREB appears to mainly affect function. Structural changes appear to be necessary, however, for achieving CREB-mediated increases in functional plasticity. We believe that further analysis of various mutants will allow us to determine the relationship between structural and functional change.

Acknowledgments

We thank Paul Taghert and co-workers and Tim Tully and Minoru Saitoe for communicating unpublished results. We also thank Jim DeZazzo and Minoru Saitoe for their comments on the manuscript. Frances Hannan is supported by a National Neurofibromatosis Foundation Young Investigator Award. Yi Zhong is supported by NIH Grant RO1-NS34779 and a Pew Scholarship.

References

1. Jan, L. Y., and Jan, Y. N. (1976). L-Glutamate as an excitatory transmitter at the *Drosophila* larval neuromuscular junction. *J. Physiol.* **262,** 215–236.
2. Broadie, K. S., and Bate, M. (1993). Development of the embryonic neuromuscular synapse of *Drosophila melanogaster. J. Neurosci.* **13,** 144–166.
3. Johansen, J., Halpern, M. E., Johansen, K. M., and Keshishian, H. (1989). Stereotypic morphology of glutamatergic synapses on identified muscle cells of *Drosophila* larvae. *J. Neurosci.* **9,** 710–725.
4. Usherwood, P. N. R. (1994). Insect glutamate receptors. *Adv. Insect Physiol.* **24,** 309–341.
5. Broadie, K., and Bate, M. (1993). Innervation directs receptor synthesis and localization in *Drosophila* embryo synaptogenesis. *Nature* **361,** 350–353.
6. Broadie, K., and Bate, M. (1993). Activity-dependent development of the neuromuscular synapse during *Drosophila* embryogenesis. *Neuron* **11,** 607–619.
7. Hall, Z. W., and Sanes, J. R. (1993). Synaptic structure and development: The neuromuscular junction. *Neuron* **10**(Suppl.), 99–121.
8. Betz, H., Schuster, C., Ultsch, A., and Schmitt, B. (1993). Molecular biology of ionotropic glutamate receptors in *Drosophila melanogaster. Trends Pharmacol. Sci.* **14,** 428–431.
9. Parmentier, M.-L., Pin, J.-P., Bockaert, J., and Grau, Y. (1996). Cloning and functional expression of a *Drosophila* metabotropic glutamate receptor expressed in the embryonic CNS. *J. Neurosci.* **16,** 6687–6694.
10. Schuster, C. M., Ultsch, A., Schloss, P., Cox, J. A., Schmitt, B., and Betz, H. (1991). Molecular cloning of an invertebrate glutamate receptor subunit expressed in *Drosophila* muscle. *Science* **254,** 112–114.
11. Currie, D. A., Truman, J. W., and Burden, S. J. (1995). *Drosophila* glutamate receptor RNA expression in embryonic and larval muscle fibers. *Dev. Dyn.* **203,** 311–316.
12. Saitoe, M., Tanaka, S., Takata, K., and Kidikoro, Y. (1997). Neural activity affects distribution of glutamate receptors during neuromuscular junction formation in *Drosophila* embryos. *Dev. Biol.* **184,** 48–60.
13. Zucker, R. S. (1989). Short-term synaptic plasticity. *Annu. Rev. Neurosci.* **12,** 13–31.

14. Jan, Y. N., and Jan, L. Y. (1978). Genetic dissection of short-term and long-term facilitation at the *Drosophila* neuromuscular junction. *Proc. Natl. Acad. Sci. USA* **75,** 515–519.
15. Llinas, R., McGuinness, T. L., Leonard, C. S., Sugimori, M., and Greengard, P. (1985). Intraterminal injection of synapsin I or calcium/calmodulin-dependent protein kinase II alters neurotransmitter release at the squid giant synapse. *Proc. Natl. Acad. Sci. USA* **82,** 3035–3039.
16. Adelsberger, H., Heckmann, M., and Dudel, J. (1997). The amplitude of quantal currents is reduced during short-term depression at neuromuscular synapses in *Drosophila. Neurosci. Lett.* **225,** 5–8.
17. Zhong, Y., and Wu, C.-F. (1991). Altered synaptic plasticity in *Drosophila* memory mutants with a defective cyclic AMP cascade. *Science* **251,** 198–201.
18. Levin, L. R., Han, P.-L., Hwang, P. M., Feinstein, P. G., Davis, R. L., and Reed, R. R. (1992). The *Drosophila* learning and memory gene *rutabaga* encodes a Ca^{2+}/calmodulin-responsive adenylyl cyclase. *Cell* **68,** 479–489.
19. Byers, D., Davis, R. L., Kiger, J. A., Jr. (1981). Defect in cyclic AMP phosphodiesterase due to the *dunce* mutation of learning in *Drosophila melanogaster. Nature* **289,** 79–81.
20. Livingstone, M. S., Sziber, P. P., and Quinn, W. G. (1984). Loss of calcium/calmodulin responsiveness in adenylate cyclase of *rutabaga,* a *Drosophila* learning mutant. *Cell* **37,** 205–215.
21. Dudai, Y., Jan, Y. N., Byers, D., Quinn, W. G., and Benzer, S. (1976). *dunce,* a mutant of *Drosophila* deficient in learning. *Proc. Natl. Acad. Sci. USA* **73,** 1684–1688.
22. Wu, Z.-L., Thomas, S. A., Villacres, E. C., Xia, Z., Simmons, M. L., Chavkin, C., Palmiter, R. D., and Storm, D. R. (1995). Altered behavior and long-term potentiation in type I adenylyl cyclase mutant mice. *Proc. Natl. Acad. Sci. USA* **92,** 220–224.
23. Connolly, J. B., Roberts, I. J. H., Armstrong, J. D., Kaiser, K., Forte, M., Tully, T., and O'Kane, C. J. (1996). Associative learning disrupted by impaired Gs signaling in *Drosophila* mushroom bodies. *Science* **274,** 2104–2107.
24. Drain, P., Folkers, E., and Quinn, W. G. (1991). cAMP-dependent protein kinase and the disruption of learning in transgenic flies. *Neuron* **6,** 71–82.
25. Skoulakis, E. M. C., Kalderon, D., and Davis, R. L. (1993). Preferential expression in mushroom bodies of the catalytic subunit of protein kinase A and its role in learning and memory. *Neuron* **11,** 197–208.
26. Wang, J., Renger, J. J., Griffith, L. C., Greenspan, R. J., and Wu, C.-F. (1994). Concomitant alterations of physiological and developmental plasticity in *Drosophila* CaM kinase II-inhibited synapses. *Neuron* **13,** 1373–1384.
27. Griffith, L. C., Wang, J., Zhong, Y., Wu, C.-F., and Greenspan, R. J. (1994). Calcium/calmodulin-dependent protein kinase II and potassium channel subunit Eag similarly affect plasticity in *Drosophila. Proc. Natl. Acad. Sci. USA* **91,** 10044–10048.
28. Griffith, L. C., Verselis, L. M., Aitken, K. M., Kyriacou, C. P., Danho, W., and Greenspan, R. J. (1993). Inhibition of calcium/calmodulin-dependent protein kinase in *Drosophila* disrupts behavioral plasticity. *Neuron* **10,** 501–509.
29. Silva, A. J., Stevens, C. F., and Tonegawa, Y. W. (1992). Deficient hippocampal long-term potentiation in α-calcium-calmodulin kinase II mutant mice. *Science* **257,** 201–206.
30. Silva, A. J., Paylor, R., Wehner, J. M., and Tonegawa, S. (1992). Impaired spatial learning in α-calcium-calmodulin kinase II mutant mice. *Science* **257,** 206–211.
31. Rotenberg, A., Mayford, M., Hawkins, R. D., Kandel, E. R., and Muller, R. U. (1996). Mice expressing activated CaMKII lack low frequency LTP and do not form stable place cells in the CA1 region of the hippocampus. *Cell* **87,** 1351–1361.
32. Mayford, M., Bach, M. E., Huang, Y.-Y., Wang, L., Hawkins, R. D., and Kandel, E. R. (1996). Control of memory formation through regulated expression of a CaMKII transgene. *Science* **274,** 1678–1683.

33. Broadie, K., Rushton, E., Skoulakis, E. M. C., and Davis, R. L. (1997). Leonardo, a *Drosophila* 14-3-3 protein involved in learning, regulates presynaptic function. *Neuron* **19,** 391–402.
34. Skoulakis, E. M. C., and Davis, R. L. (1996). Olfactory learning deficits in mutants for *leonardo,* a *Drosophila* gene encoding a 14-3-3 protein. *Neuron* **17,** 931–944.
35. Tempel, B. L., Papazian, D. M., Schwartz, T. L., Jan, Y. N., and Jan, L. Y. (1987). Sequence of a probable potassium channel component encoded at the *Shaker* locus of *Drosophila. Science* **237,** 770–775.
36. Jan, Y. N., Jan, L. Y., and Dennis, M. J. (1977). Two mutations of synaptic transmission in *Drosophila. Proc. R. Soc. Lond. B* **198,** 87–108.
37. Cowan, T. M., and Siegel, R. W. (1984). Mutational and pharmacological alterations of neuronal membrane function disrupt conditioning in *Drosophila. J. Neurogenet.* **1,** 333–344.
38. Cowan, T. M., and Siegel, R. W. (1986). *Drosophila* mutations that alter ionic conduction disrupt acquisition and retention of a conditioned odor avoidance response. *J. Neurogenet.* **3,** 187–201.
39. Zhong, Y., and Pena, L. A. (1995). A novel synaptic transmission mediated by a PACAP-like neuropeptide in *Drosophila. Neuron* **14,** 527–536.
40. Feany, M. B., and Quinn, W. G. (1995). A neuropeptide gene defined by the *Drosophila* memory mutant *amnesiac. Science* **268,** 869–873.
41. Spengler, D., Waeber, C., Pantaloni, C., Holsboer, F., Bockaert, J., Seeburg, P. H., and Journot, L. (1993). Differential signal transduction by five splice variants of the PACAP receptor. *Nature* **365,** 170–175.
42. Ishihara, T., Shigemoto, R., Mori, K., Takahashi, K., and Nagata, S. (1992). Functional expression and tissue distribution of a novel receptor for vasoactive intestinal polypeptide. *Neuron* **8,** 811–819.
43. Inagaki, N., Yoshida, H., Mizuta, M., Mizuno, N., Fujii, Y., Gonoi, T., Miyazaki, J. I., and Seino, S. (1994). Cloning and functional characterization of a third pituitary adenylate cyclase-activating polypeptide receptor subtype expressed in insulin-secreting cells. *Proc. Natl. Acad. Sci. USA* **91,** 2679–2683.
44. Zhong, Y. (1995). Mediation of PACAP-like neuropeptide transmission by coactivation of Ras/Raf and cAMP signal transduction pathways in *Drosophila. Nature* **375,** 588–592.
45. Guo, H.-F., The, I., Hannan, F., Bernards, A., and Zhong, Y. (1997). Requirement of *Drosophila* NF1 for activation of adenylyl cyclase by PACAP38-like neuropeptides. *Science* **276,** 795–798.
46. Gutmann, D. H., and Collins, F. S. (1993). The neurofibromatosis type 1 gene and its protein product, neurofibromin. *Neuron* **10,** 335–343.
47. North, K., Joy, P., Yuille, D., Cocks, N., Mobbs, E., Hutchins, P., McHugh, K., and deSilva, M. (1994). Specific learning disability in children with neurofibromatosis type 1: Significance of MRI abnormalities. *Neurology* **44,** 878–883.
48. The, I., Hannigan, G. E., Cowley, G. S., Reginald, S., Zhong, Y., Gusella, J. F., Hariharan, I., and Bernards, A. (1997). Rescue of a *Drosophila NF1* mutant phenotype by protein kinase A. *Science* **276,** 791–794.
49. McCormick, F. (1995). Ras signaling and NF1. *Curr. Opin. Biol.* **5,** 51–55.
50. Silva, A. J., Frankland, P. W., Marowitz, Z., Friedman, E., Lazlo, G., Cioffi, D., Jacks, T., and Bourtchuladze, R. (1997). A mouse model for the learning and memory deficits associated with neurofibromatosis type I. *Nature Genet.* **15,** 281–284.
51. Evans, P. D. (1980). Biogenic amines in the insect nervous system. *Adv. Insect. Physiol.* **15,** 317–473.

52. David, J.-C., and Coulon, J.-F. (1985). Octopamine in invertebrates and vertebrates: A review. *Prog. Neurobiol.* **24,** 141–185.
53. Evans, P. D. (1993). Molecular studies on insect octopamine receptors. *In* "Comparative Molecular Neurobiology" (Y. Pichon, ed.), pp. 286–296. Birkhauser, Basel.
54. Roeder, T. (1994). Biogenic amines and their receptors in insects. *Comp. Biochem. Physiol.* **107C,** 1–12.
55. Monastirioti, M., Gorczyca, M., Rapus, J., Eckert, M., White, K., and Budnik, V. (1995). Octopamine immunoreactivity in the fruit fly *Drosophila melanogaster. J. Comp. Neurol.* **356,** 275–287.
56. Arakawa, S., Gocayne, J. D., McCombie, W. R., Urquhart, D. A., Hall, L. M., Fraser, C. M., and Venter, J. C. (1990). Cloning, localization, and permanent expression of a *Drosophila* octopamine receptor. *Neuron* **4,** 343–354.
57. Saudou, F., Amlaiky, N., Plassat, J.-L., Borrelli, E., and Hen, R. (1990). Cloning and characterization of a *Drosophila* tyramine receptor. *EMBO J.* **9,** 3611–3617.
58. Hannan, F., and Hall, L. M. (1996). Temporal and spatial expression patterns of two G-protein coupled receptors in *Drosophila melanogaster. Invertebr. Neurosci.* **2,** 71–83.
59. Monastirioti, M., Linn, C. E. Jr., and White, K. (1996). Characterization of *Drosophila tyramine-β-hydroxylase* gene and isolation of mutant flies lacking octopamine. *J. Neurosci.* **16,** 3900–3911.
60. Coleman, M. J., White, K., and Griffith, L. C. (1997). Synaptic transmission in octopamine deficient *Drosophila. Soc. Neurosci.* **23,** 978. [Abstract]
61. Anderson, M. S., Halpern, M. E., and Keshishian, H. (1988). Identification of the neuropeptide transmitter proctolin in *Drosophila* larvae: Characterization of muscle fiber-specific neuromuscular endings. *J. Neurosci.* **8,** 242–255.
62. Cantera, R., and Nassel, D. R. (1992). Segmental peptidergic innervation of abdominal targets in larval and adult dipteran insects revealed with an antiserum against leucokinin I. *Cell Tissue Res.* **269,** 459–471.
63. Gorczyca, M., Augart, C., and Budnik, V. (1993). Insulin-like receptor and insulin-like peptide are localized at neuromuscular junctions in *Drosophila. J. Neurosci.* **13,** 3692–3704.
64. Schneider, L. E., and Taghert, P. H. (1988). Isolation and characterization of a *Drosophila* gene that encodes multiple neuropeptides related to Phe-Met-Arg-Phe-NH2 (FMRFamide). *Proc. Natl. Acad. Sci. USA* **85,** 1993–1997.
65. Schneider, L. E., O'Brien, M. A., and Taghert, P. H. (1991). In situ hybridization analysis of the *FMRFamide* neuropeptide gene in *Drosophila.* I. Restricted expression in embryonic and larval stages. *J. Comp. Neurol.* **304,** 608–622.
66. Schneider, L. E., Sun, E. T., Garland, D. J., and Taghert, P. H. (1993). An immunocytochemical study of the *FMRFamide* neuropeptide gene products in *Drosophila. J. Comp. Neurol.* **337,** 446–460.
67. Gu, G.-G., and Singh, S. (1997). Modulation of the dihydropyridine-sensitive calcium channels in *Drosophila* by a phospholipase C-mediated pathway. *J. Neurobiol.* **33,** 265–275.
68. Keshishian, H., Chiba, A., Chang, T. N., Halfon, M. S., Harkins, E. W., Jarecki, J., Wang, L., Anderson, M., Cash, S., Halpern, M. E., and Johansen, J. (1993). Cellular mechanisms governing synaptic development in *Drosophila melanogaster. J. Neurobiol.* **24,** 757–787.
69. Schuster, C. M., Davis, G. W., Fetter, R. D., and Goodman, C. S. (1996). Genetic dissection of structural and functional components of synaptic plasticity. I. Fasciclin II controls synaptic stabilization and growth. *Neuron* **17,** 641–654.
70. Johansen, J., Halpern, M. E., and Keshishian, H. (1989). Axonal guidance and the development of muscle fiber-specific innervation in *Drosophila* embryos. *J. Neurosci.* **9,** 4318–4332.

71. Guan, B., Hartmann, B., Kho, Y.-H., Gorczyca, M., and Budnik, V. (1996). The *Drosophila* tumor suppressor gene, *dlg*, is involved in structural plasticity at a glutamatergic synapse. *Curr. Biol.* **6,** 695–706.

72. Lahey, T., Gorczyca, M., Jia, X.-X., and Budnik, V. (1994). The *Drosophila* tumor suppressor gene *dlg* is required for normal synaptic bouton structure. *Neuron* **13,** 823–835.

73. Budnik, V., Koh, Y.-H., Guan, B., Hartmann, B., Hough, C., Woods, D., and Gorczyca, M. (1996). Regulation of synapse structure and function by the *Drosophila* tumor suppressor gene *dlg*. *Neuron* **17,** 627–640.

74. Tejedor, F. J., Bokhari, A., Rogero, O., Gorczyca, M., Zhang, J., Kim, E., Sheng, M., and Budnik, V. (1997). Essential role for *dlg* in synaptic clustering of Shaker K^+ channels *in vivo*. *J. Neurosci.* **17,** 152–159.

75. Thomas, U., Kim, E., Kuhlendahl, S., Koh, Y. H., Gundelfinger, E. D., Sheng, M., Garner, C. C., and Budnik, V. (1997). Synaptic clustering of the cell adhesion molecule fasciclin II by discs-large and its role in the regulation of presynaptic structure. *Neuron* **19,** 787–799.

76. Zito, K., Fetter, R. D., Goodman, C. S., and Isacoff, E. Y. (1997). Synaptic clustering of fasciclin II and Shaker: Essential targeting sequences and the role of *dlg*. *Neuron* **19,** 1007–1016.

77. Bailey, C. H., and Kandel, E. R. (1993). Structural changes accompanying memory storage. *Annu. Rev. Physiol.* **55,** 397–426.

78. Jessel, T. M., and Kandel, E. R. (1993). Synaptic transmission: A bidirectional and self-modifiable form of cell-cell communication. *Neuron* **10**(Suppl.), 1–41.

79. Warmke, J., Drysdale, R., and Ganetsky, B. (1991). A distinct potassium channel polypeptide encoded by the *Drosophila eag* locus. *Science* **252,** 1560–1564.

80. Chouinard, S. W., Wilson, G. F., Schlimgen, A. K., and Ganetsky, B. (1995). A potassium channel β subunit related to the aldo-keto reductase superfamily is encoded by the *Drosophila Hyperkinetic* locus. *Proc. Natl. Acad. Sci. USA* **92,** 6763–6767.

81. Titus, S. A., Warmke, J. W., and Ganetsky, B. (1997). The *Drosophila erg* K^+ channel polypeptide is encoded by the *seizure* locus. *J. Neurosci.* **17,** 875–881.

82. Wang, X. J., Reynolds, E. R., Deak, P., and Hall, L. M. (1997). The *seizure* locus encodes the *Drosophila* homolog of the HERG potassium channel. *J. Neurosci.* **17,** 882–890.

83. Loughney, K., Kreber, R., and Ganetsky, B. (1989). Molecular analysis of the *para* locus, a sodium channel gene in *Drosophila*. *Cell* **58,** 1143–1154.

84. Kernan, M. J., Kuroda, M. I., Kreber, R., Baker, B. S., and Ganetsky, B. (1991). nap^{ts}, a mutation affecting sodium channel activity in *Drosophila,* is an allele of *mle,* a regulator of X chromosome transcription. *Cell* **66,** 949–959.

85. Feng, G., Deak, P., Chopra, M., and Hall, L. M. (1995). Cloning and functional analysis of TipE, a novel membrane protein that enhances *Drosophila para* sodium channel function. *Cell* **82,** 1001–1011.

86. Budnik, V., Zhong, Y., and Wu, C.-F. (1990). Morphological plasticity of motor axons in *Drosophila* mutants with altered excitability. *J. Neurosci.* **10,** 3754–3768.

87. Jarecki, J., and Keshishian, H. (1995). Role of neural activity during synaptogenesis in *Drosophila*. *J. Neurosci.* **15,** 8177–8190.

88. Zhong, Y., Budnik, V., and Wu, C.-F. (1992). Synaptic plasticity in *Drosophila* memory and hyperexcitable mutants: Role of cAMP cascade. *J. Neurosci.* **12,** 644–651.

89. Mallart, A., Angaut-Petit, D., Bourret-Poulain, C., and Ferrus, A. (1991). Nerve terminal excitability and neuromuscular transmission in *T(X;Y)V7* and *Shaker* mutants of *Drosophila melanogaster*. *J. Neurogenet.* **7,** 75–84.

90. Angaut-Petit, D., Ferrus, A., and Faille, L. (1993). Plasticity of motor nerve terminals in *Drosophila T(X,Y)V7* mutant: Effect of deregulation of the novel calcium-binding protein frequenin. *Neurosci. Lett.* **153,** 227–231.

91. Goodman, C. S. (1996). Mechanisms and molecules that control growth cone guidance. *Annu. Rev. Neurosci.* **19,** 341–377.
92. Keshishian, H., Chang, T. N., and Jarecki, J. (1994). Precision and plasticity during *Drosophila* neuromuscular development. *FASEB J.* **8,** 731–737.
93. Cremer, H., Lange, R., Christoph, A., Plomann, M., Vopper, G., Roes, J., Brown, R., Baldwin, S., Kraemer, P., Scheff, S., Barthels, D., Rajewsky, K., & Wille, W. (1994). Inactivation of the NCAM gene in mice results in size-reduction of the olfactory bulb and deficits in spatial learning. *Nature* **367,** 455–459.
94. Luthl, A., Laurent, J.-P., Figurov, A., Muller, D., and Schachner, M. (1994). Hippocampal long-term potentiation and neural cell adhesion molecules L1 and NCAM. *Nature* **372,** 777–779.
95. Mayford, M., Barzilai, A., Keller, F., Schacher, S., and Kandel, E. R. (1992). Modulation of an NCAM-related adhesion molecule with long-term synaptic plasticity in Aplysia. *Science* **256,** 638–644.
96. Zinn, K., McAllister, L., and Goodman, C. S. (1988). Sequence analysis and neuronal expression of fasciclin I in grasshopper and *Drosophila. Cell* **53,** 577–587.
97. McAllister, L., Goodman, C. S., and Zinn, K. (1992). Dynamic expression of the cell adhesion molecule fasciclin I during embryonic development in *Drosophila. Development* **115,** 267–276.
98. Elkins, T., Hortsch, M., Bieber, A. J., Snow, P. M., and Goodman, C. S. (1990). *Drosophila* fasciclin I is a novel homophilic adhesion molecule that along with fasciclin III can mediate cell sorting. *J. Cell Biol.* **110,** 1825–1832.
99. Zhong, Y., and Shanley, J. (1995). Altered nerve terminal arborization and synaptic transmission in *Drosophila* mutants of cell adhesion molecule fasciclin I. *J. Neurosci.* **15,** 6679–6687.
100. Harrelson, A. L., and Goodman, C. S. (1988). Growth cone guidance of insects: Fasciclin II is a member of the immunoglobulin superfamily. *Science* **242,** 700–708.
101. Stewart, B. A., Schuster, C. M., Goodman, C. S., and Atwood, H. L. (1996). Homeostasis of synaptic transmission in *Drosophila* with genetically altered nerve terminal morphology. *J. Neurosci.* **16,** 3877–3886.
102. Schuster, C. M., Davis, G. W., Fetter, R. D., and Goodman, C. S. (1996). Genetic dissection of structural and functional components of synaptic plasticity. II. Fasciclin II controls presynaptic structural plasticity. *Neuron* **17,** 655–667.
103. Montminy, M. (1997). Transcriptional regulation by cyclic AMP. *Annu. Rev. Biochem.* **66,** 807–822.
104. Yin, J. C. P., Wallach, J. S., DelVecchio, M., Wilder, E. L., Zhou, H., Quinn, W. G., and Tully, T. (1994). Induction of a dominant negative CREB transgene specifically blocks long-term memory in *Drosophila. Cell* **79,** 49–58.
105. Yin, J. C. P., DelVecchio, M., Zhou, H., and Tully, T. (1995). CREB as a memory modulator: Induced expression of a dCREB2 activator isoform enhances long-term memory in *Drosophila. Cell* **81,** 107–115.
106. Bartsch, D., Ghirardi, M., Skehel, P. A., Karl, K. A., Herder, S. P., Chen, M., Bailey, C. H., and Kandel, E. R. (1995). *Aplysia* CREB2 represses long-term facilitation: Relief of repression converts transient facilitation into long-term functional and structural change. *Cell* **83,** 979–992.
107. Bourtchuladze, R., Frenguelli, B., Blendy, J., Cioffi, D., Schutz, G., and Silva, A. J. (1994). Deficient long-term memory in mice with a targeted mutation of the cAMP-responsive element-binding protein. *Cell* **79,** 59–68.
108. Davis, G. W., Schuster, C. M., and Goodman, C. S. (1996). Genetic dissection of structural and functional components of synaptic plasticity. III. CREB is necessary for presynaptic functional plasticity. *Neuron* **17,** 669–679.

MECHANISMS OF NEUROTRANSMITTER RELEASE

J. Troy Littleton,* Leo Pallanck,† and Barry Ganetzky*

*Department of Genetics, University of Wisconsin, Madison, Wisconsin 53705, and
†Department of Genetics, University of Washington, Seattle, Washington 98195

I. Introduction
II. The SNARE Hypothesis
III. Measuring Synaptic Function in *Drosophila*
IV. Mutational Analysis of the SNARE Complex
V. Synaptotagmin and Ca^{2+} Regulation of SNARE Function
VI. Additional Components of the Release Apparatus
VII. Conclusions
References

I. Introduction

The primary form of intercellular communication within the nervous system is mediated by chemical transmission at synapses. Upon propagation of an action potential into the nerve terminal, there is an influx of Ca^{2+} through voltage-activated Ca^{2+} channels that triggers the fusion of docked synaptic vesicles with the presynaptic membrane. Neurotransmitters are then released into the synaptic cleft and subsequently bind to postsynaptic receptors. It has become evident that the molecular mechanisms of constitutive and regulated secretion are functionally and phylogenetically conserved. The identification of proteins altered by yeast secretory mutations revealed that many of these proteins have homologs present at synapses. These results suggest that neurotransmitter release shares many components with general cellular secretory processes such as vesicular trafficking from the endoplasmic reticulum (ER) to the Golgi apparatus and from the Golgi to the plasma membrane. Biochemical experiments have suggested a general mechanism of vesicle trafficking in which membrane proteins on the transport vesicle interact directly with proteins on the target membrane. These protein–protein interactions presumably ensure correct targeting of the vesicle to the fusion site and provide binding sites for cytosolic proteins that participate in membrane fusion. Synaptic transmission has

0074-7742/99 $30.00

evolved as a highly specialized form of vesicle trafficking capable of rapid Ca^{2+}-triggered exocytotic cycles. It is likely that numerous steps are required for Ca^{2+}-evoked release, including the movement and maintenance of synaptic vesicles at the synapse, the morphological docking of vesicles at the active zone, and additional priming reactions prior to fusion. These steps in vesicle trafficking are likely to be completed prior to Ca^{2+}-triggered fusion, as electrophysiological findings demonstrate that fusion occurs within 200 msec following Ca^{2+} influx.[1] This time constraint suggests a fusion mechanism that functions through Ca^{2+}-induced conformational changes in a preassembled fusion complex.

Genes encoding a number of proteins involved in neurotransmitter release have been disrupted in mice, *Caenorhabditis elegans,* and *Drosophila.* As shown in Table I, the conservation between *Drosophila* synaptic proteins and their mammalian counterparts is extremely high. The absence in *Drosophila* of large protein families that occur in mammals makes the interpretation of mutant phenotypes identified by genetic approaches easier. Moreover, the ability to target defined proteins for mutagenesis and to generate numerous alleles, combined with the availability of electrophysiological techniques to characterize synaptic function at the neuromuscular junction, have combined to make *Drosophila* an outstanding organism for deciphering the *in vivo* mechanisms of synaptic vesicle trafficking.

II. The SNARE Hypothesis

Regulated exocytosis of neurotransmitters requires proteins similar or identical to those that function in the general secretory pathway of all eukaryotic cells (see Ref. 2 for a review). These proteins include the presynaptic membrane proteins SNAP-25 and syntaxin, the synaptic vesicle proteins rab3A and synaptobrevin (VAMP), and the cytosolic proteins NSF, $\alpha/\beta/\gamma$-SNAPs, and rop (n-sec1/munc18). In addition, synaptic counterparts to a group of yeast proteins that assemble into a large secretory complex, termed the exocyst, have also been identified.[3,4] As the function of the exocyst complex remains poorly characterized, it will not be discussed further. Other proteins implicated in neurotransmitter release, such as synaptotagmin, synapsin, synaptophysin, complexins, rabphilin, munc13, and cysteine string protein, have no yeast homologs and may be required exclusively for Ca^{2+}-regulated secretion in higher organisms. Proteins thought to play an important role in synaptic exocytosis are depicted schematically in Fig. 1 (see color insert).

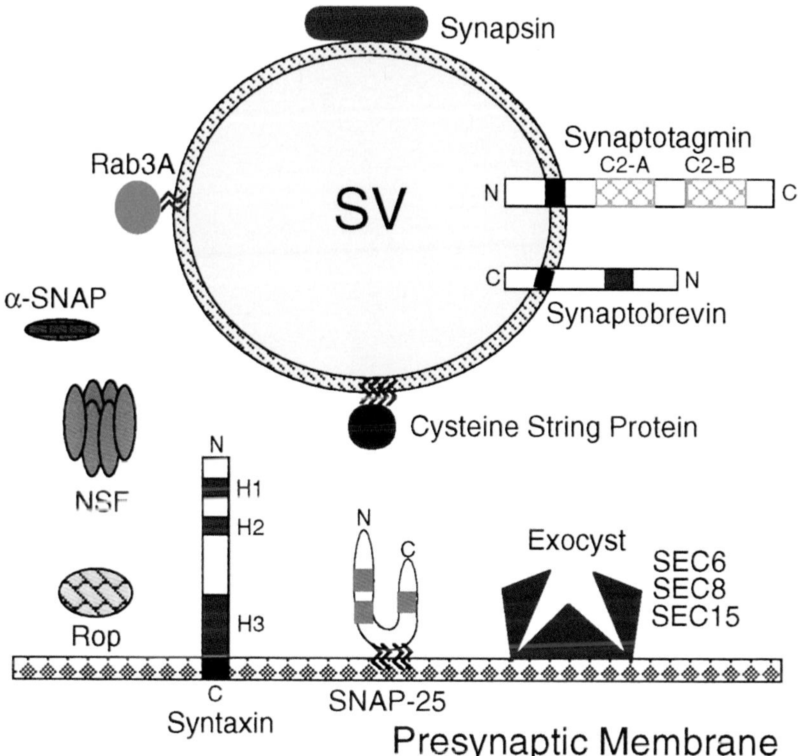

FIG. 1. Schematic diagram of synaptic proteins. Proteins on the synaptic vesicle include synaptotagmin, synaptobrevin, rab3A, synapsin, and cysteine string protein. Cytosolic protein involved in fusion include α-SNAP, NSF, rop, and exocyst complex. Both rop and members of the exocyst can also be found as peripheral membrane proteins. Components of the presynaptic membrane include syntaxin and SNAP-25. A number of motifs have been found within these protein families. Synaptobrevin, syntaxin, and SNAP-25 all have regions within their sequence that form coiled coils (gray boxes depicted within the proteins) that serve to form tight interactions among these three proteins. Synaptotagmin contains two Ca^{2+}-dependent phospholipid-binding motifs known as C2 domains. Both rab3A and NSF have domains involved in nucleotide binding and hydrolysis (GTP for rab3A and ATP for NSF). SNAP-25 and cysteine string proteins have cysteine residues that are palmitoylated and serve to attach the proteins to the presynaptic membrane and synaptic vesicle membrane, respectively.

TABLE I
MEMBRANE TRAFFICKING PROTEINS[a]

Protein							
Yeast	Mammals	*Drosophila*	%ID	Size (amino acid)	Cellular location[b]	Mutant	Suggested function
SEC1	n-Sec1/Munc18	Rop	65	597	C, PM	*rop*	Regulator of syntaxin availability Activating and inhibitory roles
SS01/SS02	Syntaxin	Syntaxin	70	291	PM, SV	*syx*	Fusion protein
SNC1/SNC2	VAMP/synaptobrevin	n-Syb	65	181	SV	Tetanus toxin	Evoked fusion
SEC9	SNAP-25	SNAP-25	61	212	PM	None	Docking/fusion
SEC17	α-SNAP	α-SNAP	62	292	C	*α-SNAP*	NSF attachment to 7S complex
SEC18	NSF	dNSF-1 dNSF-2	62 63	745 752	C C	*comt* *dNSF-2*	Disassembly of 7S complex
YPT1	Rab3A	Rab3A	78	221	C, SV	None	Modulation of SNARE assembly
—	Synaptotagmin	Syt	57	474	SV	*syt*	Fusion clamp, Ca^{2+} sensor
—	Cysteine string protein	CSP	52	249	SV	*csp*	Chaperone, Ca^{2+} channel modulation
—	Synapsin	Synapsin	50[c]	626	SV	None	Vesicle availability

[a] The amino acid number is for the *Drosophila* homolog and identity is between *Drosophila* and mammals.
[b] SV, synaptic vesicle; PM, plasma membrane; C, cytosol.
[c] Identity only in domain C, a conserved core region among the synapsins.

The current model for vesicular membrane trafficking based on biochemical interactions described *in vitro* is termed the SNARE hypothesis.[5] This model predicts that vesicular integral membrane proteins (termed v-SNAREs for vesicle SNAP receptors) provide targeting specificity through interactions with proteins present in the target membrane (t-SNAREs). Pairing of the synaptic vesicle v-SNARE synaptobrevin with the neuronal t-SNAREs syntaxin and SNAP-25 forms a stable 7S complex (based on its sedimentation coefficient) that is resistant to SDS solubilization.[6,7] The 7S complex is proposed to position or dock synaptic vesicles at membrane fusion sites and to serve as a receptor for the cytosolic $\alpha/\beta/\gamma$ SNAPs (soluble NSF attachment proteins). SNAP binding recruits the multimeric ATPase, NSF (*N*-ethylmaleimide-sensitive fusion protein), generating a 20S particle. Hydrolysis of ATP by NSF disrupts this protein complex and is thought to be an important step in vesicle trafficking.

While biochemical studies underlying the SNARE hypothesis have provided a testable model for vesicle trafficking, data from several systems have challenged fundamental aspects of the SNARE model. These data include results suggesting that NSF and α-SNAP play a role prior to docking and fusion[8–10] and that v- and t-SNAREs may function downstream of vesicle docking.[11,12] These studies have raised important unresolved questions. For example, what roles are played by NSF, SNAPs, and SNAREs if they do not mediate bilayer fusion and vesicle targeting/docking, respectively? What proteins mediate vesicle docking and fusion? In addition, how is Ca^{2+} regulation imposed on the constitutive secretory components at the synapse? Many of these questions are now being addressed through genetic manipulations and electrophysiological studies in *Drosophila*. *Drosophila* is an ideal model system for such studies as it allows a combination of *in situ* electrophysiological analysis to be coupled with "classical genetic" (phenotype-to-gene), "reverse genetic" (gene-to-phenotype), transgenic, and ultrastructural approaches. As molecular and functional properties of synaptic transmission appear to be similar in *Drosophila* and mammals, studies of neurotransmitter release mechanisms in *Drosophila* should have general significance.

III. Measuring Synaptic Function in *Drosophila*

The long-standing approach used to measure vesicle release at the synapse has been the application of electrophysiological methods. The postsynaptic membrane typically contains an assortment of ligand-gated receptors whose activity rapidly and reliably detects transmitter release. At some *Drosophila* synapses the release of a single synaptic vesicle, termed a quantum, can be detected as a miniature excitatory synaptic potential (mESP). Synchronized release of larger numbers of vesicles driven by an action potential can also be elicited and constitutes a Ca^{2+}-evoked synaptic potential (ESP). The most common electrophysiological recording technique used in analyzing synaptic mutations in *Drosophila* is intracellular voltage or current clamp recordings at the neuromuscular junction of third instar larvae. This preparation provides easily identifiable isopotential muscle fibers that can be impaled with glass microelectrodes to allow the recording of postsynaptic events in solutions of defined composition.[13] Nerves can be cut from the ventral ganglion and directly stimulated with a suction electrode to evoke action potentials in motor axons and the subsequent fusion of synaptic vesicles at motor terminals. The release of transmitter (L-glutamate) and the activation of postsynaptic glutamate receptors can be directly measured. Since larval muscles do not usually produce action potentials, one can record excitatory junctional potentials (EJPs) that provide an estimate of the number of quanta released during a particular stimulus. For greater precision, a two electrode voltage clamp can be used to measure the excitatory junctional current (EJC) arising directly from the activation of postsynaptic receptors. The third instar larval preparation has been successfully used to analyze mutations in *synaptotagmin* (*syt*), *rop,* and *cysteine string protein* (*csp*).[14–18] However, there are limitations to the usefulness of the preparation. If the function of a gene essential for neurotransmitter release is completely lost, the animal dies at the end of embryogenesis. Therefore one is restricted to analyzing partial loss-of-function or conditional mutations and mutations in nonessential genes that survive to the third instar larval stage. While such analysis can be informative, the phenotypes conferred by partial loss-of-function mutations might also reflect developmental or degenerative defects that could lead to difficulties in interpretation of data.

These drawbacks have been partially addressed by patch clamp recordings from embryonic neuromuscular junctions.[19,20] This technique makes it possible to conduct detailed functional studies at the embryonic stage of development. However, the small size of embryos makes the preparation substantially more difficult to manipulate than the larval preparation. In addition, the muscle fibers are electrically coupled at this stage of development, leading to filtering of the electrical measurements. Because embryonic recordings involve a population of immature cells undergoing active remodeling, their properties may not fully reflect those of more mature synapses. Nonetheless, this technique has proven to be a very useful addition to the electrophysiological repertoire in *Drosophila* as demonstrated by its application to studies of embryonic lethal mutations of *syt, synaptobrevin* (*n-syb*), and *syntaxin* (*syx*).[21–23]

Electrophysiological methods are also available for studies of synaptic function in *Drosophila* adults. These include recordings of the electroretinogram (ERG) from the adult eye and intracellular recordings from the dorsal longitudinal muscle fibers (DLMs) to monitor synaptic activity in the flight motor pathway. ERG recordings involve measurements of light-induced responses of the compound eye via an extracellular electrode inserted beneath the cornea. The ERG contains three components: a sustained depolarization produced by photoreceptors coincident with a light pulse and the on and off transients at the onset and termination of the light stimulus.[24,25] These transients represent synaptic transmission between the photoreceptor cells and the bipolar neurons in the lamina. Loss of on/off transients has been found in several synaptic mutations, including *comatose* (*comt*), *shibire*(*shi*), *syt, rop, csp,* and *syx.* The other major type of electrophysiological analysis in adults is the intracellular recording of postsynaptic responses in DLMs evoked by electrical stimulation of the polysynaptic giant fiber pathway.[26] This type of analysis has been applied to the study of *comt* and *shi* mutations.[27] Both techniques are limited in their application to mutants that survive to adulthood. In addition, the recordings are often more qualitative that quantitative, as the sensitivity of the measurements is usually less than those obtained at embryonic and larval neuromuscular junctions.

A correlation between synaptic function and synaptic structure can be obtained through the use of transmission electron microscopy (TEM). In *Drosophila,* the readily releasable pool of synaptic vesicles can be identified near presynaptic specializations known as T bars, the invertebrate equivalent of active zones.[28] TEM can be used to monitor the quantity and distribution of synaptic vesicles in the nerve terminal in a mutant background, thereby allowing insight into the functional role of the gene under consideration.

In addition to electrophysiological and ultrastructural analysis, optical imaging techniques are beginning to be applied to the larval neuromuscular junction. The uptake and removal of the styryl dye FM1-43 have been used to monitor vesicle recycling.[29,30] In addition, laser-scanning confocal microscopy has been used to visualize the redistribution of compartment-specific markers (docked and reserve pools of synaptic vesicles versus presynaptic membrane proteins) in the temperature-sensitive endocytotic mutant *shi.*[31] Although these methods are still in initial stages of refinement, they should become important tools in following vesicular trafficking at the synapse. Finally, the application of classical biochemical fractionation techniques that allow isolation of subsynaptic compartments are now being applied to *Drosophila*. These techniques make it possible to distinguish synaptic vesicle proteins from cytosolic or presynaptic membrane proteins and to follow their distribution in various synaptic mutants.[23,32] A current limitation to these biochemical approaches is the need for large numbers of viable adults as starting material, but they represent a complementary approach to the other techniques described.

IV. Mutational Analysis of the SNARE Complex

Among those mutations that have been generated in genes encoding components of the SNARE complex (synaptobrevin, syntaxin, SNAP-25, NSF, and α-SNAP), those in the gene encoding the t-SNARE, syntaxin, have been most extensively characterized.[11,23,33,34] Syntaxins are integral membrane proteins that are cleaved by the proteolytic activity of the clostridial neurotoxin, botulinum C, which blocks synaptic transmission.[35] Syntaxins are present on membrane targets for vesicle fusion, including the Golgi, lysosome, plasma membrane, and the presynaptic membrane. Syntaxin interacts with a number of secretory proteins, including SNAP-25, synaptobrevin, synaptotagmin, α-SNAP, rop, sec8, munc13, and presynaptic Ca^{2+} channels, placing it at the center of a number of interacting pathways (for a review see Ref. 36). In yeast, homologs of syntaxin are required for numerous trafficking steps, including ER to Golgi, Golgi to plasma membrane, and plasma membrane to vacuole.[37–39] Two members of the syntaxin family have been cloned in *Drosophila:* sed5,[40] a Golgi-specific syntaxin, and syntaxin 1A,[23] which is present at the plasma membrane of many cell types, including the midgut, epidermis, garland gland, and nervous system. Within the nervous system syntaxin 1A is localized to synapses and axons. A small amount of syntaxin can also be found in purified synaptic vesicle preparations, although the functional importance of this vesicular pool of syntaxin is unknown.[23,41] Complete removal of *syntaxin 1A* (*syx*) causes embryonic lethality with mutant embryos displaying secretory defects in multiple tissues including the gut, the epidermis, and the nervous system.[23,33,34] In addition, maternal syntaxin protein is required for cellularization of the early embryo, a process requiring the extensive addition of membrane.[33,34]

Patch clamp recordings at embryonic neuromuscular junctions of *syx* null mutants show a complete absence of evoked and spontaneous synaptic vesicle fusion, providing strong evidence that syntaxin is required for synaptic transmission.[11,23] To date, mutations in *syx* are the only synaptic mutations that abolish mESPs, suggesting that syntaxin is likely to be a component of the fusion machinery. Defects in *syx* mutants are presynaptic as application of glutamate to the postsynaptic muscle results in a normal depolarization. Synapse formation and synaptic vesicle localization to the synapse are also not affected in these mutants, suggesting that syntaxin is not essential for the transport of vesicles or their maintenance at synapses. Indeed, synaptic vesicles can be found near active zones in close proximity to the presynaptic membrane. The number of vesicles within five vesicle diameters of a presynaptic T bar (termed vesicle clustering by the authors) in *syx* null mutants was the same as in wild type, whereas the number of docked vesicles (defined as vesicles within one vesicle diameter of a T bar) was slightly increased compared to wild type, suggesting a postdocking role for syntaxin in vesicle trafficking.[11] The combined secretory defects in *syx* mutants in neuronal as well as in nonneuronal cells provide strong support for the conclusion that neurotransmitter release represents a conserved cellular secretory pathway that neurons have modified to allow rapid Ca^{2+}-activated fusion and suggests that syntaxin is likely to be an essential component of the fusion machinery itself.

The other proposed membrane t-SNARE is SNAP-25, a palmitoylated peripheral membrane protein that is the target of botulinum toxins A and E.[42] A neurally expressed SNAP-25 isoform with 61% identity to mammalian SNAP-25 has been cloned from *Drosophila,*[43] but no mutations in SNAP-25 have been published, so a comparison of the activities of the two t-SNAREs cannot be made. Genetic manipulations removing both syntaxin and SNAP-25 will be required to determine if redundant roles for the two t-SNAREs account for the persistence of docked synaptic vesicles in *syx* null mutants.

Synaptobrevins are a protein family anchored on synaptic vesicles through a C-terminal transmembrane domain and have been shown to interact with the t-SNAREs syntaxin and SNAP-25. The role of synaptobrevin in neuronal secretion is supported by the finding that several botulinum and tetanus toxins cleave synaptobrevin and block neurotransmission.[44] Two synaptobrevin genes encoding proteins with 70% amino acid identity to one another have been identified in *Drosophila.*[45,46] One gene is expressed ubiquitously (*c-syb*),[47] whereas its counterpart (*n-syb*) is restricted to the nervous system.[45] The function of synaptobrevin has been addressed genetically through the targeted expression of tetanus toxin in transgenic *Drosophila.*[11,22] Tetanus toxin cleaves n-syb, but not c-syb, and completely abolishes evoked neurotransmitter release. Tetanus toxin expression also reduces spontaneous vesicle fusion by 50%. The residual spontaneous vesicle fusion in synaptobrevin mutants indicates that synaptobrevin, unlike syntaxin, is not an essential part of the fusion machinery, but rather facilitates the process. Alternatively, tetanus toxin may not completely cleave n-syb, or additional synaptobrevins, such as the ubiquitous c-syb, may be able to weakly substitute for n-syb in vesicle fusion. As was observed for *syx* null mutations, synaptic vesicles appear morphologically docked at active zones in tetanus toxin expressing embryos, implying a postdocking role for synaptobrevin in synaptic transmission.[11] Characterization of a new temperature-sensitive paralytic allele of *syx* indicates that the mutation disrupts the ability of syntaxin to bind synaptobrevin and form the 7S complex at nonpermissive temperatures. This result correlates with a dramatic increase in both docked and undocked vesicles in the mutant (J. T. Littleton *et al.,* unpublished data), suggesting an important role for v-/t-SNARE interactions in neurotransmitter release after vesicle docking.

Additional proteins central to the SNARE hypothesis include the cytosolic NSF and SNAPs. NSF and SNAPs were initially shown to function at multiple points of the constitutive secretory pathway of mammals and yeast, possibly in vesicle targeting or fusion.[48–53] Further characterization of NSF showed that it binds integral membrane proteins in concert with SNAPs. Binding and release from these membrane receptors are governed by NSF-mediated ATP hydrolysis.[54–57] Using NSF and SNAP as affinity ligands, these SNAP receptors were subsequently identified from neuronal tissue as synaptobrevin, syntaxin, and SNAP-25. These results led directly to the formulation of the SNARE hypothesis.[5]

Previous work on *Drosophila* NSF function has revealed a pair of neurally expressed *Drosophila* NSF genes, encoding proteins with 84% amino acid identity to one another (designated *dNSF-1* and *dNSF-2*).[58–60] Mutations in the *comt* gene, including both lethal and temperature-sensitive paralytic alleles, were shown to disrupt *dNSF-1*.[61] Electrophysiological analysis in *comt* adults revealed a temperature-dependent failure in neuromuscular transmission in the flight motor pathway[27] and a loss of ERG on/off transients (J. T. Littleton *et al.,* unpublished data), indicating an important role for NSF in synaptic transmission. Ultrastructural studies of *comt* mutants have found that synaptic vesicles accumulate at release sites at restrictive temperatures. In addition to the general increase in vesicle density surrounding release sites, there is also a large increase in the number of docked vesicles in *comt* mutants at nonpermissive temperature. Excess accumulation of the 7S complex is also observed in *comt* mutants at nonpermissive temperatures, confirming that NSF is required to disassemble the syntaxin–SNAP–25-synaptobrevin complex at the presynaptic membrane for efficient synaptic transmission to occur (L. Pallanck *et al.,* unpublished data). Thus, while work on vacuole fusion in yeast indicates that NSF is required before vesicle docking,[9] these results suggest that dNSF-1 plays a functional role in neurotransmitter release after vesicle docking. These functional differences may be related to differences inherent in homotypic vacuole fusion versus vectorial transport associated with the regulated release of neurotransmitters.

Although dNSF-2 remains less well characterized than dNSF-1, mutational analysis of *dNSF-2* has shown that this gene is required during the early larval stage of development, indicating nonredundant roles for the two *Drosophila* NSF isoforms (L. Pallanck *et al.,* unpublished data). *dNSF-2* is expressed at similar levels in the embryonic nervous system and during larval and adult development.[59,60] In contrast, dNSF-1 is expressed at high levels in the adult and appears to play its primary role at this stage (L. Pallanck *et al.,* unpublished data). Current work on *Drosophila* NSF function supports the idea that dNSF-2 plays a role in neurotransmitter release during the larval stage of development and that multiple NSF proteins function in adults. It will be important to determine if dNSF-1 and dNSF-2 mediate specific aspects of vesicle trafficking within a single cell or whether the two proteins are used in distinct cell populations.

The SNARE hypothesis predicts that SNAP proteins are required at a similar step in the vesicle cycle as is NSF. This prediction may not be entirely true, as the requirement of α-SNAP in yeast vacuolar fusion can be satisfied by incubating vesicles with α-SNAP and NSF prior to docking.[9] In addition, NSF may have an additional function in fusion independent of SNAPs as NSF remains attached to membranes after disassembly of the 20S complex and has been implicated in a postdocking prefusion step after the requirement of α-SNAP.[8,9,62] To date, two SNAP isoforms have been identified in *Drosophila:* α-SNAP[58] and γ-SNAP (J. T. Littleton *et. al.,* unpublished data). α-SNAP is expressed abundantly in the nervous system[58] and mutations in α-SNAP result in larval lethality (L. Pallanck *et al.,* unpublished data). In addition, heterozygosity for the α-SNAP mutation results in lethality if the flies are also mutant for any of several viable temperature-sensitive *comt* alleles, confirming that α-SNAP and NSF function are intimately tied together (L. Pallanck *et al.,* unpublished data). Null mutations in *rop, syx, γ-SNAP,* and *syt* have failed to show similar dominant interactions with *comt* mutants. Work is underway to define more precisely where SNAPs are required in neurotransmitter release.

V. Synaptotagmin and Ca^{2+} Regulation of SNARE Function

Additional proteins are likely to participate at various stages in the formation and disassembly of the SNARE complex. The functions of these proteins are likely to include controlling the assembly of the 7S complex, providing Ca^{2+}-dependent activation of the fusion reaction, and regulating various aspects of vesicle release and recycling via phosphorylation or second messenger systems. One candidate protein for providing Ca^{2+}-

dependent activation of vesicle fusion is synaptotagmin. Synaptotagmin is an integral synaptic vesicle membrane protein that contains two copies of a Ca^{2+}-dependent phospholipid binding motif (C2 domain) in its cytosolic C terminus.[63] The C2 domain was originally identified in protein kinase C and has emerged as an important Ca^{2+}/phospholipid-binding motif present in a wide range of proteins (currently numbering greater than 50), including several implicated in synaptic vesicle trafficking (rabphilin, doc2, munc13).[64] The presence of two C2 domains in an integral synaptic vesicle membrane protein has made synaptotagmin an attractive candidate for a Ca^{2+} sensor for activating vesicle fusion. Synaptotagmin has been reported to bind multiple proteins at the synapse, including syntaxin, SNAP-25, SV2, presynaptic Ca^{2+} channels, AP-2, neurexin, and β-SNAP (for a review see Ref. 65). Ca^{2+}-regulated interactions attributed to synaptotagmin include its ability to bind acidic phospholipids (EC_{50} = 4–6 μM Ca^{2+}),[66] polyphosphoinositides,[67] and syntaxin (EC_{50} = >200 μM Ca^{2+}),[68] as well as its ability to oligomerize (EC_{50} = 10–100 μM Ca^{2+}).[69]

A single *synaptotagmin* gene has been cloned in *Drosophila,* whereas a large *synaptotagmin* family (10 members) has been identified in mammals.[70,71] Synaptotagmin is expressed abundantly in the nervous system during synaptogenesis and localizes to synaptic contacts after stage 17 of embryogenesis.[72] A collection of 20 mutations in *syt* have now been generated in *Drosophila.*[14,15,73,74] Null mutations in *syt* result in lethality at the embryonic/first instar larval boundary with mutant embryos showing normal development of the nervous system but very little residual peristaltic muscle contractions.[75] Many *syt* alleles also show intragenic complementation, where mutations affecting one domain of the protein in heterozygous combination with mutations affecting another domain result in some restored function, confirming that synaptotagmin forms a multimer with independent functional domains.[15] Recordings from the embryonic neuromuscular junction in null mutants,[21] from the larval neuromuscular junction in hypomorphic mutants,[14,15] or in animals with reduced synaptotagmin levels[16] have all revealed dramatic defects in neurotransmitter release. At low external Ca^{2+} levels, where the wild-type neuromuscular junction functions quite well, evoked release is abolished in *syt* mutants.[14] At higher Ca^{2+} levels, complete loss-of-function mutants exhibit a 95% reduction of the Ca^{2+}-evoked release present in wild type, with 70% of nerve stimulations resulting in failures of transmission.[12] The residual Ca^{2+}-sensitive release in *syt* null mutants suggests that there are likely to be additional Ca^{2+} sensors in neurotransmitter release, although synaptotagmin is clearly required for the majority of Ca^{2+}-evoked release.

Evidence implicating synaptotagmin as a Ca^{2+} sensor has also come from recordings from partial loss-of-function mutations.[15] Some heteroallelic combinations of syt mutations, including those deleting the second C2 domain, change the order of the Ca^{2+} dependence of neurotransmitter release from 3.6 to 1.8. Other heteroallelic combinations of *syt* mutations still display a normal slope of 3.6 over the same range of Ca^{2+}. These results are consistent with synaptotagmin controlling the number of vesicles that fuse with the membrane in response to Ca^{2+} influx. When these recordings are done at very low external Ca^{2+} levels where the probability of release is exceedingly small (less than 0.005 quanta per synapse), the Ca^{2+} cooperativity curve is unchanged in *syt* mutants compared with wild type, again suggesting the existence of a residual Ca^{2+}-sensitive process that can support very limited rates of release at low Ca^{2+} in *syt* mutants.[76] In addition to the decreases in evoked response, the rate of spontaneous vesicle fusions is increased in *syt* mutants,[14,16,21] suggesting that synaptotagmin plays a role in preventing vesicle fusions in the absence of Ca^{2+}. Such a role for synaptotagmin as a negative regulator of neurotransmitter release has also been suggested on the basis of biochemical interactions of synaptotagmin with components of the SNARE complex.[6] These data provide an attractive model whereby the constitutive fusion pathway provided by the SNARE complex is prevented from operating in the absence of Ca^{2+} through synaptotagmin's Ca^{2+}-independent interaction with SNAP-25. Upon Ca^{2+} influx, the microdomains of high Ca^{2+} surrounding the Ca^{2+} channel promote the binding of synaptotagmin to syntaxin and membrane lipids, possibly providing the final conformational changes necessary for the rapid activation of vesicle fusion. In conclusion, genetic dissection of the role of synaptotagmin suggests that the protein may function as a Ca^{2+}-sensitive activator of neurotransmitter release and an inhibitor of spontaneous vesicle fusion. Other roles have also been proposed, including maintaining synaptic vesicles in the docked state[16] and endocytosis.[77,78] Many of these models are based on the known binding of synaptotagmin to other proteins. A model of synaptotagmin serving as a docking protein was championed on the basis of its interaction *in vitro* with the neurexin protein family.[79,80] However, localization of neurexins to glial cells and axons,[28,81] combined with the genetic analysis of *syt* and *neurexin*[28] mutants demonstrating unique roles for both proteins, has provided strong evidence against an *in vivo* role for a neurexin–synaptotagmin interaction. Current data suggest a role for neurexins in axonal–glial interactions instead of synaptic vesicle fusion.[82] Further electrophysiological and ultrastructural studies of the many *syt* partial loss-of-function mutations, together with biochemical analysis of the underlying defects, should provide one of the most detailed functional characterizations of a synaptic protein yet obtained.

VI. Additional Components of the Release Apparatus

Mutations in several genes encoding proteins that likely modulate the assembly, stability, and function of the SNARE complex have been characterized. These include mutations in rop in *Drosophila,* rab3A and synapsins in mice, and cysteine string protein in *Drosophila.* rop (ras opposite) was identified in *Drosophila* by its unique promoter, which is shared with a member of the rab family, ras2, and was found to encode a protein homologous with the yeast secretory protein SEC1.[83] SEC1 homologs have been identified as cytosolic proteins with essential functions in many intracellular trafficking pathways including ER to Golgi, Golgi to plasma membrane, and Golgi to vacuole. Blockage of SEC1 function in yeast results in the accumulation of vesicles derived from the Golgi.[30] Clues to the function of the SEC1 family have come from biochemical experiments demonstrating that these proteins interact with syntaxin and regulate formation of the 7S complex.[84] The rop protein is expressed in a variety of tissues, including the gut, epidermis, and nervous system. Within the nervous system, rop is found in axons and at synapses. Subcellular fractionation suggests that the protein is found in two compartments: the cytosol and attached to the plasma membrane.[32,85] Null mutations in *rop* lead to embryonic lethality and secretory defects in a variety of nonneuronal tissues.[32] Temperature-sensitive *rop* mutations also have defects in synaptic transmission in the visual system with a loss of ERG on/off transients. Electrophysiological analysis of *rop* partial loss-of-function mutations at the larval neuromuscular junction has demonstrated two classes of hypomorphic alleles: those with an increased evoked response and those with a decreased evoked response.[17] Mutations that decrease the evoked response also decrease the frequency of spontaneous fusion events without affecting the amplitude of these mESPs. These findings indicate that rop has both positive and negative roles in vesicle trafficking that can be genetically separated. Synaptic transmission also appears to be sensitive to the absolute levels of rop as removing a single copy of the gene results in a 50% reduction in evoked release, an effect not seen with other synaptic proteins such as syntaxin or synaptotagmin.[17] These results hint that regulation of rop function may be an important target for processes involved in synaptic plasticity. Given that both rop and vertebrate SEC1 homologs bind tightly to syntaxin[84] (J. T. Littleton *et al.,* unpublished data), it is possible that rop is required to generate an active syntaxin t-SNARE to form the 7S complex. Overexpression of rop in *Drosophila* also suggests that the protein plays an inhibitory role in neurotransmitter release.[85] Both spontaneous and evoked responses are dramatically decreased by rop overexpression, although the number of docked vesicles identified by TEM is unaltered.[17] Biochemical data have shown that an nSec1–syntaxin interaction prevents association of syntaxin with SNAP-25 and synaptobrevin.[84] Thus, by overexpressing rop, a negative role for an undissociated rop–syntaxin complex has been revealed. rop overproduction may saturate a normal modulatory mechanism used by synapses to control the rop–syntaxin interaction, reducing the amount of free syntaxin available to participate in fusion. The reduction in neurotransmission caused by rop overexpression can be overcome by simultaneously overexpressing syntaxin, confirming that these two proteins directly interact *in vivo* to regulate secretion.[17]

Interestingly, overexpression of rop in *Drosophila* results in another phenotype that is also observed in mice that lack rab3A.[86] Rab3A is a synaptic vesicle-associated GTP-binding protein that has been shown to cycle between the vesicle and the cytosol during repetitive stimulation. Rab3A is a nonessential gene in the mouse and its absence causes no obvious defects in responses to a single stimulus.[86] However, upon repetitive stimulation, there is a gradual reduction in the evoked response, indicating a defect in replenishing docked synaptic vesicles. rop overexpression causes a similar decrease in evoked responses upon repetitive stimulation.[85] The similarity of the phenotype associated with rab3A loss-of-function mutations and rop gain-of-function mutations is consistent with the possibility that rab3A normally disrupts the syntaxin–rop interaction, thus freeing an active form of syntaxin to participate in vesicle fusion. The occurrence of genetic interactions between rab3A and rop homologs in yeast further supports a model in which these two proteins function in the same pathway.[87] A neuronally expressed *Drosophila* rab3A homolog has been cloned.[45,88] Mutational analysis of rab3A has been undertaken, but no lethal mutations have been isolated yet (T. Schwarz, personal communication), indicating that rab3A is likely to be nonessential in flies as well as in mice. Thus, a nonessential regulatory role for small GTP-binding proteins in synaptic trafficking appears likely.

Another protein that may participate in the assembly or function of the SNARE complex is the cysteine string protein (csp). Cysteine string proteins are synaptic vesicle proteins containing a novel cysteine motif and a domain found in dnaJ proteins that are involved in the assembly of multimeric complexes. Null mutations in *csp* cause the majority of homozygous mutant embryos to remain in the egg case.[89] However, adult escapees can be recovered, indicating that csp is not absolutely required for exocytosis. Escapees are paralyzed when exposed to elevated temperatures and usually die within 24 hr posteclosion. Thus, the absence of csp uncovers a temperature sensitivity for evoked neurotransmitter release that is not present normally. These results are consistent with a model in which csp participates in the stabilization of synaptic protein complexes, such that in its absence, elevated temperatures disrupt the protein interactions required for triggering evoked neurotransmitter release. Larval neuromuscular junction recordings in *csp* null mutants reveal a 50% reduction in the amplitude of EJPs evoked at room temperature, but a normal Ca^{2+} dependence of release. At nonpermissive temperatures there is a complete loss of evoked responses.[18] The frequency of mEJPs is unaltered at either temperature. Interestingly, the block in release at elevated temperatures can be overcome by manipulations that bypass the Ca^{2+} channel, including the application of latrotoxin or Ca^{2+} ionophores.[90] These observations suggest that csp may be required for regulation or localization of Ca^{2+} channels, consistent with the finding that csp can modulate Ca^{2+} channel function when coexpressed in oocytes.[91] This model predicts that calcium channels fail to open properly in the absence of csp, and at elevated temperatures this defect or its effect on local Ca^{2+} concentration is exaggerated and vesicle release is not triggered. Manipulations that bypass the calcium channel would then allow fusion to proceed. In summary, csp appears important in the regulation of exocytosis, although it is still not clear how csp facilitates the process. An analysis of the stability of synaptic complexes, combined with genetic manipulations that affect syntaxin and synaptotagmin, which are also proposed to regulate presynaptic calcium channels, should elucidate whether csp serves as a chaperone to stabilize vesicular trafficking proteins or is more intimately involved in regulating presynaptic calcium channels.

Synapsins, peripheral membrane-associated synaptic vesicle proteins, are also thought to regulate some aspect of neurotransmitter release. These proteins were identified as major phosphoprotein constituents of synaptic vesicles that bind components of the cytoskeleton.[92] These studies have predicted a role for synapsins in controlling the availability of synaptic vesicles for release. Mice homozygous for synapsin I and II deletions are viable, but have an increased incidence of seizures.[93] Although synapsins are not required for exocytotic release in mice, mutants show synaptic depression during repetitive stimulations. These results imply that synapsins function by controlling the availability of fusion-ready vesicles during repetitive stimulations, but are not required for the basic exocytotic machinery. A synapsin homolog, which is expressed abundantly at synapses, has been cloned from *Drosophila*.[94] Mutational analysis of *Drosophila synapsin* is underway and preliminary results indicate that, as in mice, *synapsin* null mutations are not lethal (Godenschwege *et al.*, personal communication). Further studies will be required to determine the role of this protein in regulating vesicle availability at *Drosophila* synapses.

It is clear that many additional proteins are likely to be important in the series of protein–protein interactions responsible for vesicle fusion.

New proteins have been identified in mammalian systems through their interactions with known synaptic proteins, while ongoing analysis of both yeast secretory and *C. elegans* unc mutants are increasing the number of suspected candidate proteins involved in synaptic transmission. With the advent of the *Drosophila* genome project, many of the *Drosophila* counterparts are being identified in expressed sequence tag (EST) sequencing efforts. For example, homologs of munc13, sec8, sec10, Hrs2, uso1, p115, and vap33 have been identified in the *Drosophila* EST database. It is only a matter of time before genetic analyses of these new candidate genes are completed and a greater understanding of their role in secretion is revealed. Interestingly, a number of mammalian synaptic proteins have not yet been found in *Drosophila* or *C. elegans,* including complexins, SV2, synaptoporin, and synaptophysin. Determination of whether these proteins emerged after the divergence of the invertebrate–vertebrate lineage and are specific modulators of vesicle trafficking in higher organisms or if proteins bearing only a distant relationship with their vertebrate counterparts perform similar functions in *Drosophila* and other invertebrates will require more complete sequence analysis of the *Drosophila* genome.

VII. Conclusions

The genetic approach to dissecting neurotransmission, complemented with electrophysiological and biochemical analysis, has yielded important insights into the function of many synaptic proteins. A working model in which biochemical, genetic, and electrophysiological data have been integrated is shown in Figure 2. Protein components such as the exocyst complex and the cytoskeleton may be essential for recruiting vesicles to the active zone. A rop–syntaxin interaction at the presynaptic membrane might prime syntaxin to function as a t-SNARE (Fig. 2A). Displacement of rop, possibly by rab3A or other modulators, would free syntaxin, allowing it to bind SNAP-25 and form a functional t-SNARE to which synaptobrevin subsequently binds, forming the 7S complex (Fig. 2B). The assembly of the 7S complex would allow cytosolic SNAPs and NSF to bind. Hydrolysis by NSF disrupts the 7S complex and is predicted to generate an activated fusion-competent vesicle (Fig. 2C). Current evidence suggests that both assembly and disassembly of the 7S complex at the presynaptic membrane represent priming steps after vesicles have morphologically docked at the active zone. A synaptotagmin–syntaxin/SNAP25 interaction may be essential in maintaining this activated state and in preventing spontaneous fusion in the absence of Ca^{2+}. Vesicle fusion would proceed via an unknown

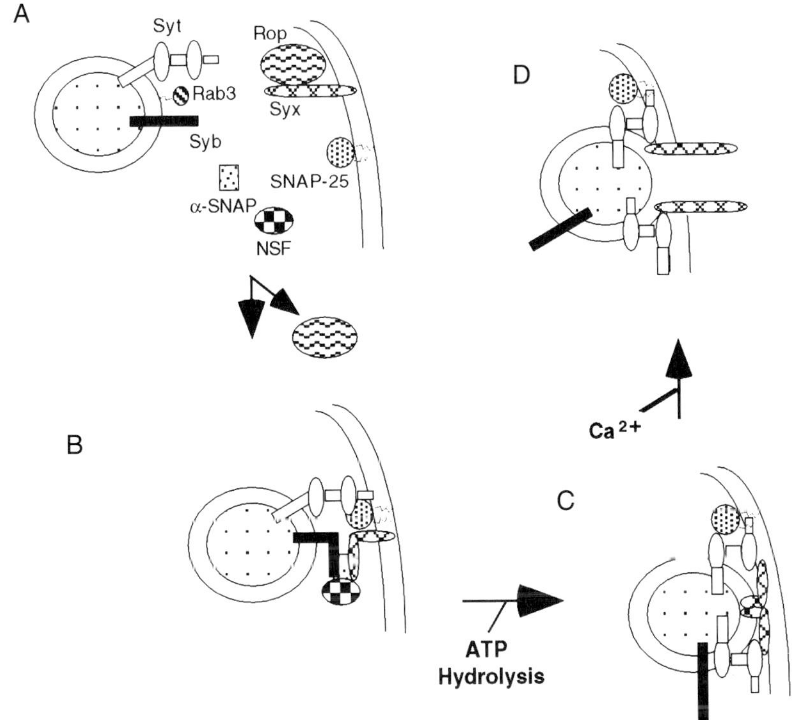

FIG. 2. Model of protein interactions underlying synaptic vesicle fusion. (A) An interaction between rop and syntaxin would be required to generate an activated syntaxin capable of assembling into the fusion complex as shown in B. Overexpression of rop would function at this step to bind free syntaxin, preventing its incorporation into the fusion complex and resulting in decreases in evoked and spontaneous fusion. Absence of rop would lead to reduced levels of activated syntaxin and thus decrease the number of docked/fusion ready vesicles, resulting in decreases in evoked and spontaneous fusion. Mutations in *syx* would also block vesicle fusion at this stage by preventing assembly of the fusion complex, consistent with the loss of both spontaneous and evoked vesicle release in *syx* mutants. (B) Displacement of rop by GTP-binding proteins such as rab3A and additional modulators may free syntaxin to form a multimeric complex with the presynaptic membrane protein SNAP-25 and the synaptic vesicle protein synaptobrevin. This v-SNARE–t-SNARE interaction is likely to be required in a postdocking step. The absence of synaptobrevin would prevent this complex from responding rapidly to a calcium signal and thus abolish evoked response, with some spontaneous fusion events proceeding in its absence. Recognition of the SNARE complex by the cytosolic SNAPs and consequent NSF binding could form an activated fusion complex. Synaptotagmin may serve to maintain this activated complex and prevent vesicle fusion in the absence of Ca^{2+} influx. Mutations in *syt* would result in an increase in spontaneous vesicle fusion due to an absence of its fusion clamp properties. (C) ATP hydrolysis by NSF would disrupt the SNARE complex and form an activated prefusion complex that would only require Ca^{2+} entry to trigger fusion. (D) Following Ca^{2+} influx, synaptotagmin would activate vesicle fusion through an enhanced interaction with syntaxin and the presynaptic membrane. The absence of synaptotagmin's ability to activate the coordinated fusion of docked vesicles would result in decreases in evoked neurotransmission in *syt* mutants. Syt, synaptotagmin; Syx, syntaxin; NSF, *N*-ethylmaleimide sensitive fusion protein; Syb, synaptobrevin; SNAP, soluble NSF attachment proteins.

mechanism that likely requires synaptotagmin-induced conformational changes in syntaxin (Fig. 2D).

Isolation of additional mutations with defects in synaptic transmission and further characterization of the existing mutations should help to define the exact function of many of these proteins. The preexisting synaptic mutations will no doubt allow enhancer/suppressor screens to be carried out to identify novel proteins required for vesicle release. In addition, the generation of double and triple mutants that remove multiple components of the exocytotic machinery should allow the question of functional redundancy to be addressed at the genetic level. Finally, the ability to create transgenic flies containing defined mutations in synaptic proteins that disrupt targeted protein–protein interactions will allow an *in vivo* analysis of the importance of specific biochemical properties of each protein. By integrating genetic, electrophysiological, and biochemical approaches, the processes underlying neurotransmitter release *in vivo* should be unraveled. It is clear from the initial studies in this field that *Drosophila* biologists will play an important role in the genetic dissection of neurotransmitter release.

References

1. Smith, S. J., and Augustine, G. J. (1988). Calcium ions, active zones and synaptic transmitter release. *TINS* **11,** 458–464.
2. Bennett, M. K., and Scheller, R. H. (1993). The molecular machinery for secretion is conserved from yeast to neurons. *Proc. Natl. Acad. Sci. USA* **90,** 2559–2563.
3. TerBush, D. R., Maurice, T., Roth, D., and Novick, P. (1996). The exocyst is a multiprotein complex required for exocytosis in *Saccharomyces cerevisiae. EMBO J.* **15,** 6483–6494.
4. Hsu, S. C., Ting, A. E., Hazuka, C. D., Davanger, S., Kenny, J. W., Kee, Y., and Scheller, R. H. (1996). The mammalian brain rsec6/8 complex. *Neuron* **17,** 1209–1219.
5. Sollner, T., Whiteheart, S. W., Brunner, M., Erdjument-Bromage, H., Geromanos, S., Tempst, P., and Rothman, J. E. (1993). SNAP receptors implicated in vesicle targeting and fusion. *Nature* **362,** 318–324.
6. Sollner, T., Bennett, M. K., Whiteheart, S. W., Scheller, R. H., and Rothman, J. E. (1993). A protein assembly–disassembly pathway in vitro that may correspond to sequential steps of synaptic vesicle docking, activation, and fusion. *Cell* **75,** 409–418.
7. Hayashi, T., McMahon, H., Yamasaki, S., Binz, T., Hata, Y., Sudhof, T. C., and Niemann, H. (1994). Synaptic vesicle membrane fusion complex: Action of clostridial neurotoxins on assembly. *EMBO J.* **13,** 5051–5061.
8. Banerjee, A., Barry, V. A., DasGupta, B. R., and Martin, T. F. J. (1996). *N*-Ethylmaleimide-sensitive factor acts at a prefusion ATP-dependent step in Ca^{2+}-activated exocytosis. *J. Biol. Chem.* **271,** 20223–20226.
9. Mayer, A., Wickner, W., and Haas, A. (1996). Sec18p (NSF)-driven release of Sec17p (α-SNAP) can precede docking and fusion of yeast vacuoles. *Cell* **1996,** 83–94.
10. Nichols, B. J., Ungermann, C., Pelham, H. R. B., Wickner, W. T., and Haas, A. (1997). Homotypic vacuolar fusion mediated by t- and v-SNAREs. *Nature* **387,** 199–202.

11. Broadie, K., Prokop, A., Bellen, H. J., O'Kane, C. J., Schulze, K. L., and Sweeney, S. T. (1995). Syntaxin and synaptobrevin function downstream of vesicle docking in *Drosophila*. *Neuron* **15,** 663–673.
12. Hunt, J. M., Charlton, M. P., Kistner, A., Habermann, E., Augustine, G. J., and Betz, H. (1994). A post-docking role for synaptobrevin in synaptic vesicle fusion. *Neuron* **12,** 1269–1279.
13. Jan, L. Y., and Jan, Y. N. (1976). Properties of the larval neuromuscular junction in *Drosophila melanogaster*. *J. Physiol.* **262,** 189–214.
14. Littleton, J. T., Stern, M., Schulze, K., Perin, M., and Bellen, H. J. (1993). Mutational analysis of *Drosophila synaptotagmin* demonstrates its essential role in Ca^{2+}-activated neurotransmitter release. *Cell* **74,** 1125–1134.
15. Littleton, T. J., Stern, M., Perin, M., and Bellen, H. J. (1994). Calcium dependence of neurotransmitter release and rate of spontaneous vesicle fusions are altered in *Drosophila synaptotagmin* mutants. *Proc. Natl. Acad. Sci. USA* **91,** 10888–10892.
16. DiAntonio, A., and Schwarz, T. L. (1994). The effect on synaptic physiology of *synaptotagmin* mutations in *Drosophila*. *Neuron* **12,** 909–920.
17. Wu, M. N., Littleton, J. T., Bhat, M. A., Prokop, A., and Bellen, H. J. (1998). ROP, the *Drosophila* Sec1 homolog, interacts with syntaxin and regulates neurotransmitter release in a dosage-dependent manner. *EMBO J.* **17,** 127–139.
18. Umbach, J. A., Zinsmaier, K. E., Eberle, K. K., Buchner, E., Benzer, S., and Gundersen, C. B. (1994). Presynaptic dysfunction in *Drosophila* csp mutants. *Neuron* **13,** 899–907.
19. Broadie, K. S., and Bate, M. (1993). Development of the embryonic neuromuscular synapse of *Drosophila melanogaster*. *J. Neurosci.* **13,** 144–166.
20. Broadie, K., and Bate, M. (1993). Synaptogenesis in the *Drosophila embryo:* Innervation directs receptor synthesis and localization. *Nature* **361,** 350–353.
21. Broadie, K., Bellen, H. J., DiAntonio, A., Littleton, J. T., and Schwarz, T. L. (1994). The absence of synaptotagmin disrupts excitation-secretion coupling during synaptic transmission. *Proc. Natl. Acad. Sci. USA* **91,** 10727–10731.
22. Sweeney, S. T., Broadie, K., Keane, J., Niemann, H., and O'Kane, C. J. (1995). Targeted expression of tetanus toxin light chain in *Drosophila* specifically eliminates synaptic transmission and causes behavioural defects. *Neuron* **14,** 341–351.
23. Schulze, K., Broadie, K., Perin, M., and Bellen, H. J. (1995). Genetic and electrophysiological studies of *Drosophila* syntaxin-1A demonstrate its role in nonneuronal secretion and neurotransmission. *Cell* **80,** 311–320.
24. Pak, W. L., Grossfield, J., and White, N. V. (1969). Nonphototatctic mutants in a study of vision in *Drosophila*. *Nature* **222,** 351–354.
25. Hotta, Y., and Benzer, S. (1969). Abnormal electroretinograms in visual mutants of *Drosophila*. *Nature* **222,** 354–356.
26. Tanouye, M. A., Wyman, R. J. (1980). Motor outputs of giant nerve fiber in *Drosophila*. *J. Neurophysiol.* **44,** 405–421.
27. Siddiqi, O., and Benzer, S. (1976). Neurophysiological defects in temperature-sensitive paralytic mutants of *Drosophila melanogaster*. *Proc. Natl. Acad. Sci. USA* **73,** 3253–3257.
28. Baumgartner, S., Littleton, J. T., Broadie, K., Bhat, M. A., Harbecke, R., Lengyel, J. A., Chiquet-Ehrismann, R., Prokop, A., and Bellen, H. J. (1996). A *Drosophila* neurexin is required for septate junction and blood-nerve barrier formation and function. *Cell* **87,** 1059–1068.
29. Ramaswami, M., Krishnan, K. S., and Kelly, R. B. (1994). Intermediates in synaptic vesicle recycling revealed by optical imaging of *Drosophila* neuromuscular junctions. *Neuron* **13,** 363–375.

30. Gonzalez-Gaitan, M., and Jackle, H. (1997). Role of *Drosophila* a-Adaptin in presynaptic vesicle recycling. *Cell* **88,** 767–776.
31. Estes, P. S., Roos, J., van der Bliek, A., Kelly, R. B., Krishnan, K. S., and Ramaswami, M. (1996). Traffic of dynamin within individual *Drosophila* synaptic boutons relative to compartment-specific markers. *J. Neurosci.* **16,** 5443–5456.
32. Harrison, S. D., Broadie, K., van de Goor, J., and Rubin, G. M. (1994). Mutations in the *Drosophila Rop* gene suggest a function in general secretion and synaptic transmission. *Neuron* **13,** 555–566.
33. Schulze, K. L., and Bellen, H. J. (1996). Drosophila syntaxin is required for cell viability and may function in membrane formation and stabilization. *Genetics* **144,** 1713–1724.
34. Burgess, R. W., Deitcher, D. L., and Schwarz, T. L. (1997). The synaptic protein syntaxin 1 is required for cellularization of *Drosophila* embryos. *J. Cell Biol.* **138,** 861–875.
35. Blasi, J., Chapman, E. R., Yamaski, S., Binz, T., Niemann, H., and Jahn, R. (1993). Botulinum neurotoxin C1 blocks neurotransmitter release by means of cleaving HPC-1/ syntaxin. *EMBO J.* **12,** 4821–4828.
36. Linial, M., and Parnas, D. (1996). Deciphering neuronal secretion: Tools of the trade. *Biochim. Biophys. Acta* **1286,** 117–152.
37. Banfield, D. K., Lewis, M. J., and Pelham, H. R. (1995). A SNARE-like protein required for traffic through the Golgi complex. *Nature* **375,** 806–809.
38. Aalto, M. K., Ronne, H., Keranen, S. (1993). Yeast syntaxins Sso1p and Sso2p belong to a family of related membrane proteins that function in vesicular transport. *EMBO J.* **12,** 4095–4104.
39. Piper, R. C., Whitters, E. A., and Stevens, T. H. (1994). Yeast Vps45p is a Sec1p-like protein required for the consumption of vacuole-targeted, post-Golgi transport vesicles. *Eur. J. Biol.* **65,** 305–318.
40. Banfield, D. K., Lewis, M. J., Rabouille, C., Warren, G., and Pelham, H. R., (1994). Localization of Sed5, a putative vesicle targeting molecule, to the cis-Golgi network involves both its transmembrane and cytoplasmic domains. *J. Cell Biol.* **127,** 357–371.
41. Walch-Solimena, C., Blasi, J., Edelmann, L., Chapman, E. R., von Mollard, G. F., and Jahn, R. (1995). The t-SNAREs syntaxin 1 and SNAP-25 are present on organelles that participate in synaptic vesicle recycling. *J. Cell Biol.* **128,** 637–645.
42. Blasi, J., Chapman, E. R., Link, E., Binz, T., Yamasaki, S., De Camilli, P., Sudhof, T. C., Niemann, H., and Jahn, R. (1993). Botulinum neurotoxin A selectively cleaves the synaptic protein SNAP-25. *Nature* **365,** 160–163.
43. Risinger, C., Blomqvist, A. G., Lundell, I., Lambertsson, A., Nassel, D., Pieribone, V., Brodin, L., and Larhammar, D. (1993). Evolutionary conservation of synaptosome-associated protein 25 kDa (SNAP-25) shown by *Drosophila* and *Torpedo* cDNA clones. *J. Biol. Chem.* **268,** 24408–24414.
44. Schiavo, G., Benfenati, F., Poulain, B., Rossetto, O., Polverino de Laureto, P., DasGupta, B. R., and Montecucco, C. (1992). Tetanus and botulinum-B neurotoxins block neurotransmitter release by proteotytic cleavage of synaptobrevin. *Nature* **359,** 832–835.
45. DiAntonio, A., Burgess, R. W., Chin, A. C., Deitcher, D. L., Scheller, R. H., Schwarz, T. L. (1993). Identification and characterization of *Drosophila* genes for synaptic vesicle proteins. *J. Neurosci.* **13,** 4924–4935.
46. Sudhof, T. C., Baumert, M., Perin, M. S., and Jahn, R. (1989). A synaptic vesicle membrane protein is conserved from mammals to *Drosophila. Neuron* **2,** 1475–1481.
47. Chin, A. C., Burgess, J. W., Wong, B. R., Schwartz, T. L., and Scheller, R. H. (1993). Differential expression of transcripts from *syb,* a *Drosophila melanogaster* gene encoding a VAMP (synaptobrevin) that is abundant in non-neuronal cells. *Gene* **131,** 175–181.

48. Novick, P., Field, C., and Schekman, R. (1980). Identification of 23 complementation groups required for post-translational events in the yeast secretory pathway. *Cell* **21,** 205–215.
49. Beckers, C. J., Block, M. R., Glick, B. S., Rothman, J. E., and Balch, W. E. (1989). Vesicular transport between the endoplasmic reticulum and the Golgi stack requires NEM-sensitive fusion protein. *Nature* **339,** 397–398.
50. Diaz, R., Mayorga, L. S., Weidman, P. J., Rothman, J. E., and Stahl, P. D. (1989). Vesicle fusion following receptor-mediated endocytosis requires a protein active in Golgi transport. *Nature* **339,** 398–400.
51. Orci, L., Malhotra, V., Amherdt, M., Serafini, T., and Rothman, J. E. (1989). Dissection of a single round of vesicular transport: Sequential intermediates for intercisternal movement in the Golgi stack. *Cell* **56,** 357–368.
52. Graham, T. R., and Emr, S. D. (1991). Compartmental organization of Golgi-specific protein modification and vacuolar protein sorting events defined in a yeast sec 18 (NSF) mutant. *J. Cell Biol.* **114,** 207–218.
53. Wilson, D. W., Wilcox, C. A., Flynn, G. C., Chen, E., Kuang, W.-J., Henzel W. J., Block, M. R., Ullrich, A., and Rothman, J. E. (1989). A fusion protein required for vesicle-mediated transport in both mammalian cells and yeast. *Nature* **339,** 355–359.
54. Weidman, P. J., Melancon, P., Block, M. R., and Rothman, J. E. (1989). Binding of an N-ethylmaleimide-sensitive fusion protein to Golgi membranes requires both a soluble proteins(s) and an integral membrane receptor. *J. Cell Biol.* **108,** 1589–1596.
55. Wilson, D. W., Whiteheart, S. W., Wiedman, M., Brunner, M., and Rothman, J. E. (1992). A multisubunit particle implicated membrane fusion. *J. Cell Biol.* **117,** 531–538.
56. Griff, I. C., Schekman, R., Tothman, J. E., and Kaiser, C. A. (1992). The yeast Sec 17 gene product is functionally equivalent to mammalian α-SNAP protein. *J. Biol. Chem.* **267,** 12106–12115.
57. Whiteheart, S. W., Griff, I. C., Brunner, M., Clary, D. O., Mayer, T., Buhrow, S. A., and Rothman, J. E. (1993). SNAP family of NSF attachment proteins includes a brain-specific isoform. *Nature* **362,** 353–355.
58. Ordway, R. W., Pallanck, L., and Ganetzky, B. (1994). Neurally expressed *Drosophila* genes encoding homologs of the NSF and SNAP secretory proteins. *Proc. Natl. Acad. Sci. USA* **91,** 5715–5719.
59. Pallanck, L., Ordway, R. W., Ramaswami, M., Chi, W. Y., Krishnan, K. S., and Ganetzky, B. (1995). Distinct roles for *N*-ethylmaleimide-sensitive fusion protein (NSF) suggested by the identification of a second *Drosophila* NSF homolog. *J. Biol. Chem.* **270,** 18742–18744.
60. Boulianne, G. L., and Trimble, W. S. (1995). Identification of a second homolog of N-ethylmaleimide-sensitive fusion protein that is expressed in the nervous system and secretory tissues of *Drosophila. Proc. Natl. Acad. Sci. USA* **92,** 7095–7099.
61. Pallanck, L., Ordway, R. W., and Ganetzky, B. (1995). A *Drosophila* NSF mutant. *Nature* **376,** 25.
62. Barlowe, C. (1997). Coupled ER to golgi transport reconstituted with purified cytosolic proteins. *J. Cell Biol.* **139,** 1097–1108.
63. Perin, M. S., Fried, V. A., Mignery, G. A., Jahn, R., and Südhof, T. C. (1990). Phospholipid binding by a synaptic vesicle protein homologous to the regulatory region of protein kinase C. *Nature* **345,** 260–263.
64. Brose, N., Hofmann, E., Hata, Y., and Sudhof, T. C. (1995). Mammalian homologues of *Caenorhabditis elegans* unc-13 gene define novel family of C2 domain proteins. *J. Biol. Chem.* **270,** 25273–25280.
65. Littleton, J. T., and Bellen, H. J. (1995). Synaptotagmin controls and modulates synaptic-vesicle fusion in a Ca^{2+} dependent manner. *TINS* **18,** 177–183.

66. Davletov, B. A., and Sudhof, T. C. (1993). A single C2 domain from synaptotagmin I is sufficient for high affinity Ca^{2+} phospholipid binding. *J. Biol. Chem.* **268,** 26386–26390.
67. Schiavo, G., Gu, Q. M., Prestwich, G. D., Sollner, T. H., and Rothman, J. E. (1996). Calcium-dependent switching of the specificity of phosphoinositide binding to synaptotagmin. *Proc. Natl. Acad. Sci. USA* **93,** 13327–13332.
68. Chapman, E. R., Hanson, P. I., An, S., and Jahn, R. (1995). Ca^{+2} regulates the interaction between synaptotagmin and syntaxin 1. *J. Biol. Chem.* **270,** 23667–23671.
69. Chapman, E. R., An, S., Edwardson, J. M., and Jahn, R. (1996). A novel function for the second C2 domain of synaptotagmin: Ca++-triggered dimerization. *J. Biol. Chem.* **271,** 5844–5849.
70. Perin, M. S., Archer, B. T., Ozcelik, T., Francke, U., Jahn, R., and Südhof, T. C. (1991). Structural and functional conservation of synaptotagmin (p65) in *Drosophila* and humans. *J. Biol. Chem.* **266,** 615–622.
71. Li, C., Davletov, B. A., and Sudhof, T. C. (1995). Distinct Ca^{2+} and Sr^{2+} binding properties of synaptotagmins. *J. Biol. Chem.* **270,** 24898–24902.
72. Littleton, J. T., Bellen, H. J., and Perin, M. S. (1993). Expression of synaptotagmin in *Drosophila* reveals transport and localization of synaptic vesicles to the synapse. *Development* **118,** 1077–1088.
73. DiAntonio, A., Parfitt, K. D., and Schwartz, T. L. (1993). Synaptic transmission persists in *synaptotagmin* mutants of *Drosophila. Cell* **73,** 1281–1290.
74. Littleton, T. J., and Bellen, H. J. (1994). Genetic and phenotypic analysis of thirteen essential genes in cytological interval 22F1-2; 23B1-2 reveals novel genes required for neural development in *Drosophila. Genetics* **138,** 111–123.
75. Littleton, J. T., Upton, L., and Kania, A. (1995). Axonal outgrowth, synapse formation and synaptic vesicle localization in *synaptotagmin* mutations. *J. Neurobiol.* **65,** 32–40.
76. Parfitt, K., Reist, N., Li, J., Burgess, R., Deitcher, D., DiAntonio, A., and Schwarz, T. L. (1995). *Drosophila* genetics and the functions of synaptic proteins. *Cold Spring Harb. Symp. Quant. Biol.* **60,** 371–377.
77. Jorgensen, E. M., Hartweig, E., Schuske, K., Nonet, M. L., Jin, Y., and Horvitz, H. R. (1995). Defective recycling of synaptic vesicles in synaptotagmin mutants of *Caenorhabditis elegans. Nature* **378,** 196–199.
78. Zhang, J. Z., Davletov, B. A., Sudhof, T. C., and Anderson, R. G. (1994). Synaptotagmin I is a high affinity receptor for clathrin AP-2: Implications for membrane recycling. *Cell* **78,** 751–760.
79. Petrenko, A. G., Perin, M. S., Bazbek, A., Davletov, B. A., Ushkaryov, Y. A., Geppert, M., and Sudhof, T. C. (1991). Binding of synaptotagmin to the α-latrotoxin receptor implicates both in synaptic vesicle exocytosis. *Nature* **353,** 65–68.
80. Hata, Y., Davletov, B., Petrenko, A. G., Jahn, R., and Sudhof, T. C. (1993). Interaction of synaptotagmin with the cytoplasmic domains of neurexins. *Neuron* **10,** 307–315.
81. Russell, A. B., and Carlson, S. S. (1997). Neurexin is expressed on nerves, but not at nerve terminals, in the electric organ. *J. Neurosci.* **15,** 4734–4743.
82. Littleton, J. T., Bhat, M. A., and Bellen, H. J. (1997). Deciphering the function of neurexins at cellular junctions. *J. Cell Biol.* **137,** 793–796.
83. Salzberg, A., Cohen, N., Halachmi, N., Kimchie, Z., and Lev, Z. (1993). The *Drosophila Ras* 2 and *Rop* gene pair: A dual homology with a yeast Ras-like gene and a suppressor of its loss-of-function phenotype. *Development* **117,** 1309–1319.
84. Pevsner, J., Hsu, S.-C., Braun, J. E. A., Calakos, N., Ting, A. E., Bennett, M. K., and Scheller, R. H. (1994). Specificity and regulation of a synaptic vesicle docking complex. *Neuron* **13,** 353–361.

85. Schulze, K. L., Littleton, J. T., Salzberg, A., Halachmi, N., Stern, M., Lev, Z., and Bellen, H. J. (1994). *rop,* a *Drosophila* homolog of the vertebrate n-Sec1/Munc-18 and yeast Sec1 proteins, is a negative regulator of neurotransmitter release in vivo. *Neuron* **13,** 1099–1108.
86. Geppert, M., Bolshakov, V. Y., Siegelbaum, S. A., Takei, K., De Camilli, P., Hammer, R. E., and Sudhof, T. C. (1994). The role of rab3a in neurotransmitter release. *Nature* **369,** 493–497.
87. Dascher, C., Ossig, R., Gallwitz, D., and Schmitt, H. D. (1991). Identification and structure of four yeast gene (*SLY*) that are able to suppress the functional loss of *YPT1,* a member of the *RAS* superfamily. *Mol. Cell. Biol.* **11,** 872–885.
88. Johnston, P. A., Archer, B. T., III, Robinson, K., Mignery, G. A., Jahn, R., and Südhof, T. C. (1991). Rab3A attachment to the synaptic vesicle membrane mediated by a conserved polyisoprenylated carboxyl-terminal sequence. *Neuron* **7,** 101–109.
89. Zinsmaier, K. E., Eberle, K. K., Buchner, E., Walter, N., and Benzer, S. (1994). Paralysis and early death in cysteine string protein mutants of *Drosophila. Science* **263,** 977–980.
90. Umbach, J. A., and Gundersen, C. B. (1997). Evidence that cysteine string proteins regulate an early step in the Ca++-dependent secretion of neurotransmitter at *Drosophila* neuromuscular junctions. *J. Neurosci.* **17,** 7203–7209.
91. Mastrogiacomo, A., Parsons, S. M., Zampighi, G. A., Jenden, D. J., Umbach, J. A., and Gunderson, C. B. (1994). Cysteine string proteins: A potential link between synaptic vesicles and presynaptic calcium channels. *Science* **263,** 981–982.
92. Bahler, M., and Greengard, P. (1987). Synapsin I bundles F-actin in a phosphorylation-dependent manner. *Nature* **326,** 704–707.
93. Rosahl, T. W., Spillane, D., Missler, M., Herz, J., Selig, D. K., Wolff, J. R., Hammer, R. E., Malenka, R. C., and Sudhof, T. C. (1995). Essential functions of synapsins I and II in synaptic vesicle regulation. *Nature* **375,** 488–493.
94. Klagges, B. R. E., Heimbeck, G., Godenschwege, T. A., Hofbauer, A., Pflugfelder, G. O., Reifegerste, R., Reisch, D., Schaupp, M., Buchner, S., and Buchner, E. (1996). Invertebrate synapsins: A single gene codes for several isoforms in *Drosophila. J. Neurosci.* **16,** 3154–3165.

VESICLE RECYCLING AT THE *DROSOPHILA* NEUROMUSCULAR JUNCTION

Daniel T. Stimson and Mani Ramaswami

Department of Molecular and Cellular Biology and Arizona Research Laboratories Division of Neurobiology, University of Arizona, Tucson, Arizona 85721

I. Introduction
II. Origins of Synaptic Vesicle Recycling Studies in *Drosophila*
III. Molecular and Phenotypic Analysis of *shi* Mutants
   A. The *shi* Product Dynamin Is Required for Vesicle Fission during Synaptic Vesicle Endocytosis
   B. Pleiotropic Effects of *shibire*
IV. Cell Biology of Synaptic Vesicle Recycling in *Drosophila*
   A. Intermediate Compartments during Recovery of Vesicle-Depleted *shi* Nerve Terminals
   B. Ca^{2+} Dependence of Synaptic Vesicle Recycling
   C. Alternative Pathways and Active Zones for Synaptic Vesicle Recycling
V. Molecules Involved in Synaptic Vesicle Recycling
VI. Assays for Synaptic Vesicle Recycling in *Drosophila*
   A. Synaptic Depression during High Frequency Stimulation
   B. A Dye Uptake Assay to Measure the Rate of Membrane Internalization
   C. Measurements of Dead Time
   D. Capacitance Measurements
VII. Direct and Indirect Effects
VIII. Conclusions
   References

I. Introduction

After exocytosis, membrane proteins of synaptic vesicles are recycled to form new synaptic vesicles. As the total complement of synaptic vesicles at a motor nerve terminal can be exhausted within 2–3 min, this local recycling of vesicle proteins is an important and necessary cellular function. It is now established that synaptic vesicle membrane proteins are retrieved from presynaptic membrane by endocytosis, and new synaptic vesicles are reformed at the nerve terminal from retrieved constituents. During endocytosis, specific membrane proteins are sorted into regions of plasma membrane that form invaginated pits; the contents of these pits are internalized as endocytic vesicles after membrane fission at the "neck" of the invagi-

0074-7742/99 $30.00

nation. New synaptic vesicles are formed from recycled membrane proteins, and neurotransmitter molecules synthesized in the cytoplasm are pumped into the newly formed synaptic vesicles by proton-driven transporters present on the vesicle membrane. Regulation of synaptic vesicle recycling could control transmitter release by directly affecting the releasable pool of synaptic vesicles or by indirectly affecting the composition of the presynaptic plasma membrane.

Molecules potentially involved in synaptic vesicle recycling are being identified by biochemical and genetic approaches. However, direct involvement of these molecules in reutilization of synaptic vesicle membrane has been demonstrated for only a very few. In addition, spatial features of synaptic vesicle endocytosis and subsequent events in recycling also remain mysterious. It is not known how much the membrane composition of recycling endocytic vesicles differs from that of synaptic vesicles: whether they retain all of the peripheral proteins required for synaptic vesicle function and whether they contain normally-plasma membrane proteins. Also unknown is whether endocytic vesicles must, after internalization, necessarily fuse with an intermediate membrane compartment, the early endosome, before the formation of new synaptic vesicles. It remains possible that vesicles retrieved from the axolemma may immediately uncoat and/or mature into recycled synaptic vesicles.[1,2] Mechanisms and pathways for synaptic vesicle recycling at nerve terminals may also be involved in the endocytosis of presynaptic proteins such as cell adhesion molecules that mediate inhibitory constraints on synapse expansion or receptors for postsynaptically released growth factors such as neurotrophins.

This review first discusses contributions made by *Drosophila* genetics to the current understanding of synaptic vesicle recycling. It then outlines, in the context of tools available at the *Drosophila* larval neuromuscular synapse, what should be the most promising avenues to pursue in *Drosophila* to enhance the understanding of the mechanisms involved in the endocytic recycling of synaptic vesicles.

II. Origins of Synaptic Vesicle Recycling Studies in *Drosophila*

Classic studies at the frog neuromuscular synapse established the existence of an active pathway for synaptic vesicle recycling. Physiological and morphological studies at the frog neuromuscular junction demonstrated that the number of synaptic vesicles at a frog motor terminal is far fewer than the number of quanta it can subsequently release. The number of available vesicles does not change when the synapse is stimulated at a moderate rate, but is depleted by high frequency tetanic stimulation.[3] In 1973, Heuser and Reese demonstrated that synaptic vesicle depletion at the frog neuromuscular junction is accompanied by an expansion of the presynaptic plasmalemma. Subsequent recovery of the normal complement of synaptic vesicles on cessation of stimulation was accompanied by a corresponding contraction of the presynaptic plasma membrane to its original surface area. If horseradish peroxidase (HRP) was present in the medium during the recovery process, it could be detected subsequently in a large fraction of synaptic vesicles that appeared following recovery.[4] These studies first laid out the current model for synaptic vesicle recycling: after exocytosis, synaptic vesicles are retrieved via endocytosis, along with an incidental bit of extracellular fluid, and reutilized to form new exocytosis-competent synaptic vesicles. Also in 1973, Suzuki and colleagues reported the isolation of a collection of reversible temperature-sensitive paralytic mutants in *Drosophila*. Six of these were temperature-sensitive alleles of a gene named *shibire* (after the Japanese word meaning "paralyzed," as well as for the impolite double entendre of the abbreviation *shi*[ts]). Analysis and use of *shi*[ts] mutants that are paralyzed due to a conditional block in synaptic vesicle endocytosis[5,6] currently comprise the major contributions of *Drosophila* to the field of synaptic vesicle recycling. Lessons from analyses of *shi* may also serve to illustrate the unique virtues and limitations of studies of vesicle recycling at the *Drosophila* neuromuscular junction.

III. Molecular and Phenotypic Analysis of *shi* Mutants

Molecular cloning of the *shi* gene demonstrated that it encoded a *Drosophila* homolog of dynamin, a protein identified as a microtubule-activated GTPase that induced microtubule-bundling and could stimulate the relative sliding of microtubule filaments *in vitro*.[7–10] Based on these *in vitro* studies, dynamin was believed to be a third class of a microtubule-based molecular motor. This microtubule–motor activity of dynamin does *not* reflect its *in vivo* function, which appears to occur exclusively in membrane internalization.[11,12] Our current knowledge of the *in vivo* function of dynamin derives from the phenotypic analysis of *shi* mutants.

A. The *shi* Product Dynamin Is Required for Vesicle Fission during Synaptic Vesicle Endocytosis

Early physiological analyses of *shi* mutants provided hints of *shi* function. Synaptically derived on and off transients in electroretinograms[13,14] as well as evoked synaptic responses in adult flight muscles were found to be abolished in *Drosophila shi*^{ts1} mutants at 29°C.[15] Failure of adult flight muscle synapses requires prior neural activity.[16] In 1979, Poodry and Edgar provided a cell biological interpretation of these electrophysiological studies on *shi*^{ts1} mutants.[5] Using electron microscopy, they demonstrated that synapses on adult tibial muscle of *shi*^{ts1} mutants were depleted of synaptic vesicles when the flies were paralyzed by a 5-min exposure to 29°C. This depletion could be prevented by prior treatment with tetrodotoxin. They suggested in their landmark study that *shi*^{ts} mutants may show conditional defects in synaptic vesicle recycling.[5] The specific stage at which recycling is blocked in *shi*^{ts1} mutants was demonstrated by Kosaka and Ikeda in 1983, when they reported that synaptic vesicle depletion in these mutants is accompanied by the accumulation of structures on the plasma membrane that they termed "collared pits."[6] These structures (usually about 60 nm in diameter, but frequently much larger) appear to be nascent endocytic vesicles arrested at a stage prior to vesicle fission.[6] They have distinct necks connected to plasma membrane that are surrounded by an electron-dense collar containing presumed components of a membrane fission machinery. Similar collared pits have also been observed at the larval neuromuscular synapses of *shi*^{ts1} mutants.[17] Molecular characterization of several *shi*^{ts} alleles indicates that their conditional defects derive from mutations in the GTPase domain of *Drosophila* dynamin.[7,18] Thus, a block in some essential function of the GTPase domain results in the arrest of endocytosis at this collared pit stage, when the vesicle lumen remains connected to the extracellular space.[6,19,20] Studies in mammalian synaptosomes where similar collared pit structures may be arrested by treatment with a nonhydrolyzable GTP analog indicate that the *shi* product dynamin is associated with the electron-dense collars of forming endocytic vesicles.[12,21] As the first protein identified as part of a potential membrane fission machinery, dynamin has served as a linchpin for biochemical studies to identify new components of the endocytic machinery.[2]

B. Pleiotropic Effects of *shibire*

Detailed phenotypic studies on *shi* mutants reveal features that must be anticipated for other genes with roles in vesicle recycling. First, *shi* mutants show highly pleiotropic phenotypes. When heat pulsed at different stages of development, a whole range of phenotypes is observed in almost every tissue that has been examined.[22–24] At elevated temperatures, membrane internalization has been found to be inhibited in *shi*^{ts} mutants in essentially every context where this has been directly examined.[6,20,25–29] Null mutants in *shi* result in embryonic lethality, although gross disorganization is not seen in the carcasses, probably due to a large maternal contribution of dynamin to the developing embryo. For analyzing mutants with pleiotropic effects such as *shi,* classical loss-of-function and null mutant analyses may be only marginally informative due to the combination of potential early developmental defects, pleiotropy, and the presence of maternal product. Tight and reversible conditional alleles such as shi^{ts1} cannot be expected for the vast majority of genes. Before discussing potential ways in which other genes involved in recycling may be analyzed in *Drosophila,* the following section considers insights that the study of shi^{ts} mutants has allowed into the cell biology of synaptic vesicle recycling.

IV. Cell Biology of Synaptic Vesicle Recycling in *Drosophila*

For analysis of cellular pathways involved in vesicle recycling, a unique advantage of *Drosophila* is the availability of shi^{ts} mutants. Following the sustained stimulation of neuromuscular synapses at nonpermissive temperatures, synaptic vesicle membrane proteins destined for endocytosis are arrested on plasma membrane, while endocytic compartments downstream of internalization are cleared. Restoring the mutants to permissive temperatures results in a synchronized wave of endocytic traffic.[17,20,30] Observations at different time points of this wave of endocytosis allow analysis of spatial locations and the temporal sequence of membrane compartments involved in endocytic recycling of synaptic vesicles. While several features of vesicle recovery in *shi* mutants, such as a requirement of dynamin function, are obviously essential for synaptic vesicle recycling, two caveats are important to observe in these studies of *shi* synapse recovery. First, synaptic vesicle recycling under physiological rates of stimulation may occur via an alternative recycling pathway.[31] Thus, observations of *shi* synapse recovery may not always be relevant to synaptic vesicle recycling events that occur during low rates of nerve stimulation. The second caveat is the possible existence of two populations of synaptic vesicles at neuromuscular synapses on larval body wall. Studies suggest that a population of vesicles that are located at the periphery of presynaptic varicosities participate in exocytic and endocytic events under moderate levels of stimulation. A second population of vesicles (about 50% of the total population) comprises a more centrally located reserve pool that is mobilized for exocytosis on depletion of the active pool.[32,33] Recovery of *shi* synapses after the complete depletion of vesicles may not distinguish between these two pools, whose rates and requirements for recycling may differ in significant ways. Despite these caveats, vesicles do recycle efficiently following synaptic vesicle depletion in *shi* mutants and this process may be uniquely analyzed at *Drosophila* synapses by a variety of methodologies.

A. Intermediate Compartments during Recovery of Vesicle-Depleted *shi* Nerve Terminals

Koenig and Ikeda depleted coxal neuromuscular junctions as well as retinula cell synapses of shi^{ts1} mutants by stimulation at elevated temperatures and observed membrane compartments that formed and then disappeared during synaptic vesicle recovery.[30,34] Collared pits assembled on plasma membrane were apparent following vesicle depletion at 29°C. Upon recovery at 19°C, a progression of membrane compartments was observed. Within 3 min, large cisternae pinch off from the plasma membrane and are visible in the presynaptic terminal. In addition, coated vesicles as well as several tubular structures are also present. This medley of vesicles, tubules, and cisternae resolves within about 20 min into recycled synaptic vesicles that are presumed to arise by fragmentation or budding from the tubular or cisternal structures. This pathway for synaptic vesicle recycling has several features in common with similar analyses of frog motor terminals stimulated with high K^+ saline in the presence of the K^+ channel blocker 4-amino-pyridine for prolonged presynaptic depolarization. Here too, synaptic vesicle recovery on cessation of stimulation proceeds via large endosomal structures.[4,35,36]

While the studies of Koenig and Ikeda describe the qualitative progression of synaptic vesicle recycling, from collared pits through vesicles and cisternae to recycled vesicles, quantitative and biochemical relationships between these different compartments remain mysterious. First, the number of collared pits on plasma membrane is far fewer than expected for the number of synaptic vesicles that have fused. This anomaly may be explained by a limited availability of some component of endocytosis required for collared pit formation. Thus, at coxal synapses, where the machinery for collared pit formation may maximally accommodate perhaps 10% of synaptic vesicles, the excess synaptic vesicle membrane may remain on the axonal plasmalemma. After internalization of these preformed collared pits, vesicle budding machinery may be recycled to form new collared pits until eventually all synaptic vesicle membrane proteins have been internalized. A second feature revealed by EM studies on *shi* terminals depleted of synaptic vesicles is the apparent lack of clathrin coats on most collared pit structures: this suggests that clathrin coats may not be essential for endocytic vesicle formation during synaptic vesicle recycling. However, in light of several studies that clearly link clathrin with synaptic vesicle endocytosis, alternative possibilities, such as enhanced sensitivity of presynaptic clathrin cages to aldehyde fixatives, should not be completely ruled out.[31] A third intriguing observation made by Koenig and Ikeda is the apparent internalization of a large fraction of membrane as large cisternae, rather than as small endocytic vesicles. This observation suggests a dynamin-dependent pathway of bulk membrane internalization that does not require the assembly of small endocytic vesicles. It is possible that the large cisternal structures that accumulate in recovering *shi* synapses are biochemically identical to plasma membrane. Consistent with this interpretation, vesicles that bud from these cisternal structures may also be seen to have collars similar to those arrested on plasma membrane.[30,31] In the framework of this model, proteins involved in endocytic vesicle assembly, rather than in membrane fission, may not be required for membrane internalization during the recovery of *shi*^{ts1} mutant synapses, but could be required for subsequent events in vesicle recycling. An alternative interpretation for the observation that cisternae, rather than endocytic vesicles, arise early in recovery of *shi* synapses has been discussed.[37] This proposal that dynamin may be required for membrane invagination rather than membrane fission at synapses, while a valid possibility, is not easily reconciled with studies of dynamin function in other cellular contexts.[2,12] A final conundrum raised by studies of Koenig and Ikeda is the apparent lack of expansion of the presynaptic plasma membrane under conditions where synaptic vesicles (with total surface area exceeding total plasma membrane area) have been completely depleted. This observation is inconsistent with current models of membrane trafficking. Although considered unlikely by the authors, the most simple explanation is a flow of bulk membrane further back into the axon than examined in this study.[30] It would be of great interest to couple the pioneering EM studies of Koenig and Ikeda with the use of gold-conjugated antibodies against synaptic membrane markers to gain further insights into the traffic of synaptic vesicle membrane during vesicle recycling.

B. Ca^{2+} Dependence of Synaptic Vesicle Recycling

The availability of *shi*^{ts1} mutants to temporally separate synaptic vesicle exocytosis from endocytosis allows one to isolate and study molecular and ionic requirements for stages in vesicle recycling subsequent to the *shi*^{ts1} block. This feature of vesicle recycling in *Drosophila* has been very sparsely utilized to date. Ramaswami, Krishnan, and Kelly demonstrated that neither extracellular Ca^{2+} nor elevated intracellular Ca^{2+} was essential for recovery of *shi*^{ts1} mutant synapses.[20] In their study, the authors used a fluorescent styryl dye, FM1-43,[38] to optically monitor endocytosis at third-instar larval motor terminals. Synaptic vesicles, depleted by stimulation of *shi*^{ts1} preparations at 34°C, were recycled even if synapses were incubated for 5 min in Ca^{2+}-free saline prior to shifting to a permissive (22°C) temperature.[20] Continued endocytosis, indicated by FM1-43 uptake, clearly demonstrated that continued large increases in intracellular Ca^{2+} were not required for processes subsequent to the *shi*^{ts1} block in synaptic vesicle recycling.

Kuromi, Yoshihara, and Kidokoro showed that measurements of FM1-43 uptake during *shi* synapse recovery may be combined with specific pharmacological agents to investigate requirements for synaptic vesicle recycling events downstream of *shibire*.[39] A clear interpretation of their data describing how endocytosis during *shi* synapse recovery is affected by inhibiting calcineurin, a Ca^{2+}-dependent phosphatase (see Table I), must await careful quantitation of endocytic rates during the recovery of vesicle-depleted *shi*^{ts1} synapses (these methods are discussed later in this review). However, Kuromi and colleagues' use of *shi* mutants combined with inhibitors of molecules that could function downstream of *shi* highlight a potential approach to studying synaptic vesicle recycling in *Drosophila*. In addition to pharmacological agents to inhibit downstream molecules, second-site genetic mutations could be studied in combination with *shi* to investigate their effects on downstream events in vesicle recycling. Finally, these studies describe the use of fluorescence measurements to observe endocytosis at living synapses, a technique that should be of considerable importance for future research on vesicle recycling in *Drosophila*.[20,39]

C. Alternative Pathways and Active Zones for Synaptic Vesicle Recycling

The existence of alternative pathways for synaptic vesicle recycling has been suggested and debated since the mid-1970s.[36] One school of thought pioneered by Heuser and Reese holds that a synaptic vesicle collapses completely into plasma membrane during exocytosis, a process clearly demonstrated in freeze fracture analysis at the frog neuromuscular junction.[4] Following fusion, vesicle membrane proteins diffuse to distinct sites away from the active zone where they are internalized. Endocytic vesicles fuse to form an intermediate membrane compartment, the early endosome, before the reformation of new synaptic vesicles. The strongest argument in favor of this pathway for synaptic vesicle recycling is that it has been unequivocally demonstrated. In general, if a very large amount of synaptic vesicle membrane is deposited on the plasmalemma, synaptic vesicle membrane transits through intermediate endosomes prior to the formation of recycled synaptic vesicles.[4,30] However, under low rates of exocytosis, the alternative view that vesicles may operate by a ''kiss and run'' mechanism where they fuse transiently with plasma membrane has not been ruled out. Capacitance measurements of endocytic rates in goldfish retinal bipolar synapses have shown two pathways distinguished by their rate constants: at low Ca^{2+} concentrations, a rapid retrieval mechanism with a time constant of 2 sec operates, while at elevated Ca^{2+} concentrations, a slower component with a time constant of 30 sec predominates.[40] Finally, studies at the frog neuromuscular junction have shown that under some conditions, vesicle collapse into plasma membrane, monitored by a loss of FM1-43 from dye-loaded synaptic vesicles, does not occur during stimulus-evoked neurotransmitter release.[41] To reconcile all of these studies, it appears essential to accept that at least two, and possibly more, independent pathways for vesicle recycling may operate at presynaptic terminals (reviewed in Ref. 2). In the simplest form, a rapid pathway may operate at the site of vesicle fusion, but a slower pathway may operate at sites distinct from vesicle fusion sites.

Indirect evidence for two such pathways exists at *Drosophila* motor terminals.[17,34] Koenig and Ikeda provided good evidence for two pathways of synaptic vesicle recycling in *Drosophila:* a relatively rapid pathway at the active zone for vesicle fusion and a slower pathway away from the active zone.[34] If *shi*^{ts1} retinula synapses depleted by stimulation at 29°C were shifted to 26°C, a temperature where dynamin activity is only partially restored, long tubular invaginations were seen emanating from the active zone within 1 min. These invaginations were seen in almost 100% of active zones that were analyzed. At this time point, relatively few endocytic pits or invaginations were visible away from the active zone. However, at later time points, accumulations of collared pits and branched tubular invaginations of plasma membrane were visible at sites away from the active zone. These temporally and spatially separated endocytic pathways suggest distinct pathways for synaptic vesicle recycling.[34] More detailed studies provide evidence consistent with a model that the active zone pathway contributes to an active zone-associated pool of synaptic vesicles, while the slower pathway may contribute to a reserve pool of vesicles, distant from sites of fusion. If recovery is allowed to occur for 1 min at 19°C instead of 26°C, then recycled synaptic vesicles, instead of tubular invaginations, are seen associated with active zones. Remarkably, no synaptic vesicles are evident in other regions of the nerve terminal at this time point. Specialized regions for synaptic vesicle endocytosis are also suggested by immunofluorescence studies at the larval neuromuscular junction.[17] Large presynaptic terminals at the larval neuromuscular synapse allow a significant amount of subsynaptic detail to be resolved by confocal microscopy. In *shi*^{ts1} terminals depleted of synaptic vesicles, synaptic vesicle membrane proteins may be seen relatively evenly dispersed on presynaptic plasma membrane. In contrast, dynamin, which should mark sites of endocytosis, is found sharply concentrated in spots on the plasma membrane that may represent active zones for endocytosis.[17] However, the location of these dynamin hot spots relative to active sites for vesicle fusion must await immuno-EM studies or good markers for T bars, which mark active zones of exocytosis.

V. Molecules Involved in Synaptic Vesicle Recycling

Several reviews have outlined current models for synaptic vesicle endocytosis and recycling, and readers are directed to these for more detailed information, little of which derives directly from studies in *Drosophila*.[1,2] Synaptic vesicle internalization probably involves clathrin and clathrin-associated proteins that are also required for endocytosis in nonsynaptic contexts. The current model for vesicle recycling suggests that after exocytosis, the cytosolic domain of synaptotagmin serves as a high-affinity receptor for the clathrin-associated AP2 adaptor complex. Low-affinity interactions among synaptic vesicle membrane proteins,[42] as well as between AP2 and other synaptic vesicle membrane proteins, may target them into clathrin-coated endocytic pits. In addition to AP2, synapse-specific clathrin-binding proteins such as AP180 may function to increase the speed and fidelity of synaptic vesicle endocytosis.[43] Vesicle fission may be catalyzed directly by constriction of an oligomeric ring of dynamin around the vesicle neck. However, several other proteins such as the SH3 domain containing protein amphiphysin and the AP2-binding protein Eps15, as well as the Eps15- and amphiphysin-binding protein synaptojanin, may participate in neck formation or vesicle fission. In particular, amphiphysin has been shown to be required for vesicle fission: injection into lamprey reticulospinal synapses of peptides that interfere with amphiphysin–dynamin interactions results in a block in endocytosis, as well as the arrest of endocytic clathrin-coated pits that lack dynamin collars.[44] Thus, amphiphysin may recruit dynamin to vesicle necks. The amphiphysin-binding protein synaptojanin is an inositol-5-phosphatase with a C-terminal domain homologous to the yeast Sac1 protein, identified as a suppressor of an actin mutant.[45] In addition, amphiphysin is similar to yeast Rvs167, also implicated in actin function.[46] The involvement of synaptojanin and amphiphysin in synaptic vesicle recycling, as well as several other lines of evidence, suggests that phosphoinositides, phosphoinositide modifying proteins, and the actin cytoskeleton play important roles in regulating vesicle budding.[2,47]

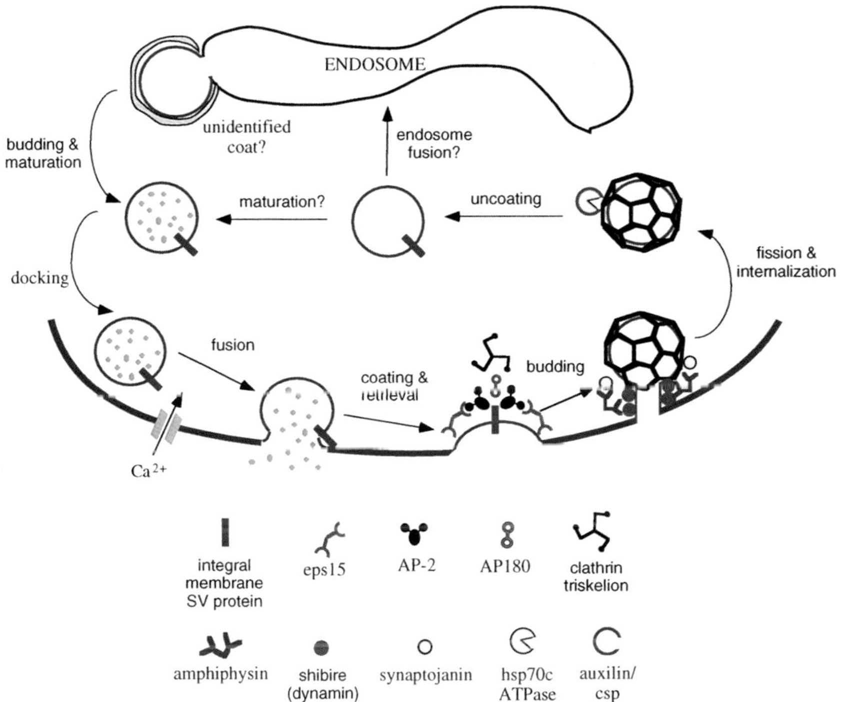

FIG. 1. Schematic of the synaptic vesicle cycle, highlighting proteins with well-established roles in synaptic vesicle recycling. Two models of postinternalization events are shown: internalized vesicles may uncoat and immediately reenter the synaptic vesicle pool or they may uncoat and fuse with endosomes, where their membrane components could be sorted and packaged into new synaptic vesicles. While clathrin coats mediate budding from the plasma membrane, budding from endosomes occurs via an unknown mechanism. Targeting and assembly of clathrin on the plasma membrane depend on adaptors (AP2 and AP180) and may also involve Eps 15. Adaptors may be targeted to the plasma membrane by synaptic vesicle proteins: AP2 binds to synaptotagmin (see Table I), while AP180 may bind synaptobrevin. Coated vesicles appear to be pinched off by oligomeric rings of dynamin, which may be recruited to vesicle necks by amphiphysin. Amphiphysin also binds synaptojanin, which may allow the modulation of endocytosis by phosphoinositides and actin. Vesicles are uncoated by hsp70c ATPase, which acts in concert with dnaJ domain proteins (auxilin and/or csp). Where possible, physical interactions among these proteins have been indicated diagramatically. For a more detailed description of binding interactions, refer to Table I and references cited in the text.

Subsequent to membrane internalization, the fate of endocytic vesicles is unclear. Clathrin coats may be removed by an uncoating ATPase, hsc70, in concert with auxilin or cysteine string protein, two synaptic molecules that interact with hsc70. These uncoated vesicles may pass through an intermediate endosomal compartment, a process that could require proteins involved in membrane fusion. Exit from endosomes may involve novel vesicle coat proteins, as well as other molecules known to be involved in endocytic trafficking events in the *Drosophila* compound eye, such as Hook and Deep Orange.[48]

It is likely that a large number of proteins involved in synaptic vesicle recycling remain to be identified. These molecules are likely to exist in dynamic complexes similar to those demonstrated for synaptic vesicle exocytosis. Identifying sequential events in protein assembly and disassembly that occur during vesicle recycling is an important goal for the future. Figure 1 (see color insert) shows an outline of the current working model for synaptic vesicle recycling, and Table I contains a list of molecules with potential roles in vesicle recycling, an indication of the status of their analysis in *Drosophila,* and a brief summary with references of their postulated functions during synaptic vesicle recycling.

VI. Assays for Synaptic Vesicle Recycling in *Drosophila*

A major facility in *Drosophila* is the availability of a well-defined neuromuscular synapse accessible to detailed functional and morphological characterization. This allows one, by analysis of mutant synapses, to test the role of specific molecules in synaptic functions. For the study of molecular mechanisms of synaptic vesicle recycling, there are several challenges to this general approach, some of which have been discussed earlier in this review. Synaptic vesicle recycling is invisible to conventional electrophysiological techniques that record postsynaptic responses to transmitter release. However, because synaptic vesicle recycling depends on prior exocytosis, measurements of endocytosis must generally be interpreted in the context of exocytic activity. As discussed earlier, endocytosis is completely blocked in *shi*^{ts} mutants at an elevated temperature, allowing temporal separation of exocytosis and endocytosis. A selective defect in endocytosis may be demonstrated by the complete loss of dye (FM1-43) uptake during synapse stimulation at elevated, but not permissive, temperatures. The majority of mutations that affect vesicle recycling are unlikely to yield tight, rapidly reversible, temperature-sensitive phenotypes like the conditional *shi* alleles. More likely, mutants in synaptic vesicle recycling will be embryonic lethal

TABLE I
Molecules Potentially Involved in Synaptic Vesicle Recycling

Gene product	Identified in fly?	Mutant phenotype	Proposed function
Dynamin I	One gene, six alternatively spliced isoforms[7,8]	Reversible temperature-sensitive block in synaptic vesicle endocytosis results in active terminals depleted of vesicles,[5] leading to synaptic depression[49] and failure.[15] Terminals exhibit collared pits on presynaptic PM[6] and do not take up FM1-43[20]	May form constricting rings around neck of budding vesicle to promote fission from PM[2,12,21]
Clathrin			
Heavy chain	50	Four alleles, three lethal, one semilethal. Male sterility in some contexts[50]	Trimeric triskelions form coats on vesicle and may provide mechanical force for budding;[51,52] may regulate coat assembly and disassembly
Light chain	C. Bazinet, unpublished	—	
Clathrin-associated proteins			
hsp70c ATPase	K. Zinsmaier, unpublished	—	Strips clathrin from coated vesicles.[53,54] Binds to cysteine string protein[55,56]
Auxilin	—	—	Recruits hsp70c to coated vesicles via chaperone-like dnaJ domain[57]
Adaptor proteins			
(AP2) α-adaptin	58	Reduced FM1-43 uptake by active terminals, synaptic vesicle depletion, and PM deformation[58]	Facilitates formation of coated vesicles from plasma membrane by linking clathrin to target proteins in plasma membrane[60] and by promoting clathrin cage assembly.[51] Fly β-adaptin may be a hybrid of β1 and β2 adaptin
β2-adaptin	59		
μ2-adaptin (AP50)	C. O'Kane, unpublished	—	
σ2-adaptin (AP17)	M. Ramaswami, unpublished	—	
AP180	B. Zhang and B. Ganetzky, unpublished	—	A synapse-specific[61,62] clathrin,[63] α-adaptin,[64] and phosphoinositide-binding protein.[65] Assembles clathrin into homogeneously sized cages.[63] AP180 affects synaptic transmission *in vivo*[66]

Stoned B	Produced from ORF2 of *stoned*[67]	—	Probable vesicle coat protein based on homology to AP50.[67] Contains EH domain-binding motifs present on Eps15-binding proteins[68]
Eps15	—	—	Binds to α-adaptin and may recruit several endocytic proteins to PM, including clathrin, AP2,[69,70] and synaptojanin[68,71]
Amphiphysin	C. O'Kane unpublished	—	Binds AP2, may recruit dynamin to coated pits via SH3 domain, homologous to yeast Rvs167 implicated in actin function[2,44]
Dap160	J. Roos and R. B. Kelly, unpublished	—	Dynamin-binding protein purified from *Drosophila* brain. Contains an EH and multiple SH3 domains, may serve functions analogous to EH proteins and amphiphysin. Is present at larval motor nerve terminals
Synaptojanin I	H. Bellen, unpublished	—	An inositol-5-phosphatase that interacts with amphiphysin[45] and Eps15.[68,71] May regulate local concentration of PPIs to promote membrane fission.[47,72] Contains a domain homologous to yeast SAC1 implicated in actin functions
Endophilin	—	—	Binds to dynamin and synaptojanin via the SH3 domain,[73,74] but is excluded from amphiphysin–dynamin–synaptojanin complexes[75]
Regulators			
PKC	76	Induction of pseudosubstrate results in abnormal learning acquisition[80]	Phosphorylates dynamin I in resting nerve terminals[81]
Calcineurin	77–79	No mutants, but calcineurin inhibitors affect FM1-43 uptake into larval motor terminals[39]	Dephosphorylates dynamin I and reduces its GTPase activity; may act as Ca^{2+}-activated switch to mobilize dynamin–GTP to PM following exocytosis[2,12,82]

(*continues*)

TABLE I (*Continued*)

Gene product	Identified in fly?	Mutant phenotype	Proposed function
Proteins that regulate SV exocytosis			
Synaptotagmin	83	In shi^{ts1}–*syt* double mutants, recovery from paralysis is slow and the rate of synaptic depression is enhanced over shi^{ts1} [84]	May act as high affinity receptor for AP2 to promote SV retrieval from PM[85,86]
csp	87	shi^{ts1} rescues lethality of csp^{X1} (K. S. Krishnan, unpublished)	Binds hsc70 involved in uncoating clathrin-coated vesicles.[55] May play an additional role in sorting recycled SV proteins via chaperone-like dnaJ domain. However, see Ref. 88
Stoned A	Produced from ORF1 of *stoned*[67,89]	stn^{ts2} is synthetically lethal with shi^{ts1}; *stoned* mutants show redistribution of synaptic vesicle proteins within motor terminals.[89]	May promote SV membrane retrieval by direct interaction with synaptotagmin (A. M. Phillips and L. E. Kelly, unpublished)
Proteins implicated in endosomal traffic			
ARFs	91,92	—	May regulate the production of vesicles from recycling endosomes.[93,94] Also may regulate targeting of adaptors to PM[2,95] (ARF activity has not yet been investigated in neurons)
rabs	96,97	Dominant-negative rab1 transgenics exist[97]	rab5 may regulate kinetics of vesicle–endosome, endosome–endosome fusion[98] in synaptic terminals[99]
Hook	100	R7 fails to concentrate the sevenless ligand boss into MVBs; MVBs fail to mature[48,101]	May regulate endosome formation and/or endosomal trafficking of SV proteins[48,100,101]
Deep orange	*Drosophila* homolog of yeast VPS18[102]	Very pleiotropic phenotypes, probably affects pigment granule formation in the compound eye	Physically interacts with Hook and is involved in the endocytic pathway to lysosomes[48]

when null and behaviorally sluggish when the gene function is only partially impaired. Such a situation is illustrated by *ada* mutants in *Drosophila* α-adaptin.[58] Null *ada*[3] mutants show a severe reduction in the number of synaptic vesicles and increased plasma membrane infoldings in their embryonic central synapses. Partial loss-of-function *ada*[1] mutants show reduced uptake of the endocytic tracer FM1-43 in response to the high K^+ stimulation of third instar larval neuromuscular preparations. However, it is unclear from these dye uptake data alone whether fewer vesicles have fused during stimulation or if endocytosis has been specifically altered in the mutants. Thus, while the phenotypes of *ada* mutants do not directly demonstrate defects in vesicle recycling, they provide *in vivo* evidence consistent with extensive biochemical data that suggest a role for α-adaptin in vesicle recycling.[58]

For mutations in unknown or novel genes, more stringent criteria are required to assess their effects on synaptic vesicle recycling. This section considers some assays that, when taken together, should allow one to directly assess endocytic efficiency at *Drosophila* synapses. These approaches are generally discussed in the context of type I synapses on muscles 6 and 7 of the larval body wall. However, similar approaches may also eventually be useful for analysis of the diverse collection of synapses accessible in the larval neuromuscular preparation.[103,104] While some of the assays discussed have been tested in other species,[105] none of them have yet been optimally utilized at *Drosophila* neuromuscular junctions. Because mutants have their own particular idiosyncrasies, the following methods may not always be optimal; in addition, there will undoubtedly be other, possibly more appropriate, ways not considered here to assess endocytosis in unusual mutants.[88] Finally, unlike transmitter release, synaptic vesicle recycling has yet to be visualized at the embryonic synapse. For this reason, to analyze mutations in genes essential for synaptic vesicle recycling, weak or conditional alleles are currently required so that mutant synapses may be studied in the third instar stage of larval development.

A. Synaptic Depression during High Frequency Stimulation

Although synaptic depression can occur by other mechanisms, depression observed in *shi*[ts] synapses at elevated temperatures, under conditions where wild-type synapses are not affected, almost certainly reflects synaptic vesicle depletion.[49] Depression can be observed even under conditions where vesicle recycling is not completely blocked. Larval neuromuscular synapses in *shi*[ts1] mutants show stable neuromuscular transmission at 28°C for up to 10 min when stimulated at 0.5 Hz, but show depression at higher frequencies. However, at 30°C with 0.5 Hz stimulation, EJPs measured at *shi*^{ts1} synapses decline from a value corresponding to about 200 quanta to a stable level corresponding to about 10 quanta in 15 min.[84] Synaptic vesicle recycling is not completely blocked at 30°C in these mutants, and the steady-state EJP at this temperature probably corresponds to a steady-state number of synaptic vesicles. This situation is reached when the rate of vesicle depletion by exocytosis is equally matched by the rate of synaptic vesicle recycling. It is possible that *shi*^{ts1} synapses will be sensitized to small alterations in the efficiency of synaptic vesicle recycling. In this scenario, weak mutant alleles of genes thought to be involved in endocytosis may alter the rate of synaptic depression in *shi*^{ts1} synapses without altering the efficiency of neurotransmitter release measured by EJC recordings. Thus, by double mutant analysis, altered synaptic vesicle recycling by a specific mutation may be revealed by its effects on the temperature, stimulation frequency, and extent of depression of *shi*^{ts1} synapses. The time course of EJC recovery in *shi*^{ts1} synapses restored to permissive temperature after synaptic vesicle depletion[106] reflects a variety of events in synaptic vesicle recycling that occur downstream of the *shi*^{ts1} block. Double mutant analyses to measure the effect of specific mutations on the rate of EJC recovery in *shi*^{ts1} synapses may provide interpretable information on the role of the mutated gene in downstream events in synaptic vesicle recycling. However, if for reasons discussed earlier vesicle budding components must be utilized multiple times during *shi* recovery, then mutations that affect membrane internalization events "upstream" of the *shi*^{ts1} block may also alter the rate of recovery of *shi*^{ts1} synapses. Thus, the effect of mutations on the onset or recovery of synaptic depression in *shi*^{ts1} mutants may be used to investigate their consequence on synaptic vesicle recycling. However, there are two limitations of this analysis: (1) depression may occur in the mutant by mechanisms other than synaptic vesicle depletion and (2) mutants with defects in exocytosis as well as endocytosis may be complicated to analyze by electrophysiological techniques alone.

B. A Dye Uptake Assay to Measure the Rate of Membrane Internalization

A quantitative optical assay for synaptic vesicle cycling devised by Bill Betz in 1992 allowed, for the first time, direct observation and quantification of synaptic vesicle recycling at living nerve terminals. The assay takes advantage of FM1-43, a dye whose unusual physical properties allow it to be internalized into recycling vesicles and subsequently released upon exocytosis.[38] If synapses are stimulated in the presence of FM1-43 and washed free of surface-bound dye, recycled synaptic vesicles loaded with FM1-43 can be imaged within synaptic terminals. Quantitative fluorescence measurements of FM1-43 within nerve terminals provide a quantitative estimate for the fraction of dye-filled synaptic vesicles.[38] A protocol to determine the time required for the synaptic vesicle membrane to be internalized after exocytosis has been used to study endocytosis at the frog neuromuscular junction, as well as in synapses formed among cultured hippocampal cells.[105,107] If FM1-43 is applied to synapses at varied times after a burst of exocytic activity has ceased, the total FM1-43 uptake into recycling synaptic vesicles is observed to decline as the time interval between stimulation and dye application is increased. This occurs because FM1-43 uptake into nerve terminals occurs only while synaptic vesicle membrane continues to be retrieved from the presynaptic plasma membrane. The duration between stimulation and dye application that produces half-maximal dye uptake provides a direct measure for the time taken for the internalization of the synaptic vesicle membrane. This time is roughly 30 sec in cultured hippocampal synapses and about 60 sec at the frog neuromuscular synapse.[105]

At the *Drosophila* larval motor terminal, a burst of exocytosis may be induced by a 30-Hz stimulation for 1 min (buzz) in the presence of a postsynaptic blocker. If FM1-43 is added to the preparation just after the buzz, significant dye uptake into recycling vesicles can be seen. However, if dye is added 5 min after the buzz, no synaptic labeling is visible. Thus, in wild-type synapses, the synaptic vesicle membrane is almost completely retrieved within 5 min after the buzz (D. T. Stimson and M. Ramaswami, unpublished). Quantitative measurements of dye uptake at intermediate time points should allow direct measurements of membrane internalization rates in wild type and mutant *Drosophila* larval nerve terminals. An important point to note is that the *half time* for membrane internalization in synapses of a given mutant is independent of the extent of synaptic vesicle exocytosis during the buzz.[108] The extent of vesicle fusion will affect the overall brightness of terminals, but should not alter the time dependence of endocytosis. Thus, this assay may offer significant advantages for the analysis of mutants in genes such as *synaptotagmin* and *cysteine string protein,* which may participate in both exocytic and endocytic arms of the synaptic vesicle cycle.[109–111]

C. Measurements of Dead Time

A variation of the dye uptake technique described earlier has been used by Bill Betz to estimate the time required for an entire synaptic vesicle cycle at the frog neuromuscular junction. Frog terminals may be loaded with

FM1-43 by stimulation in the presence of the dye. They may be subsequently unloaded during synaptic vesicle exocytosis induced by nerve stimulation. Betz and coworkers[112] showed that in vertebrate terminals imaged during unloading, the initial fractional dye loss plotted as a function of time exactly parallels the total amount of transmitter released because transmitter and dye are released from the same pool of synaptic vesicles. However, within about a minute ("dead time"), fractional dye loss deviates significantly from total transmitter released because the pool of loaded fluorescent vesicles is diluted by newly recycled vesicles that do not contain FM1-43. The time taken for noticeable divergence between the curves for fractional dye loss and total transmitter released reflects the time it takes for newly recycled vesicles to join the releasable pool. This provides an estimate for the rate of recycling within stimulated terminals.[38,112] Some technical problems need to be overcome before this assay can be adapted for use in *Drosophila*. First, muscle contractions during unloading make fluorescence imaging impossible and so unloading must be observed in the presence of a blocker of muscle contraction. While curare allows one to achieve a complete block in vertebrate muscle contraction without abolishing EJPs, such a situation is difficult to achieve with existing *Drosophila* glutamate receptor blockers. A second problem is that type I motor terminals lie embedded in subsynaptic reticulum to which free dye binds quite tightly, and extensive washing is required to view purely presynaptic fluorescence. However, if these technical details are overcome, it should be possible to measure the time required for an entire synaptic vesicle cycle in *Drosophila*.

D. Capacitance Measurements

Capacitance measurement is a particularly exciting technique for measuring rates of exocytosis and endocytosis. Because capacitance relates directly to membrane surface area, increases in capacitance correspond to membrane expansion resulting from exocytosis whereas decreases correspond to endocytic events. Using these measurements, several novel insights have been made into mechanisms of membrane retrieval. Capacitance measurements are very rapid, and patch electrodes offer access to the cellular cytoplasm, allowing an unusual range of manipulations.[40] If presynaptic terminals in *Drosophila* could be patch-clamped, a unique set of experiments would be feasible to investigate mechanisms of synaptic vesicle recycling. Patch clamp recordings from presynaptic varicosities have been made from type III peptidergic terminals on the larval body wall muscle.[113] While this first study focused exclusively on characterizing presynaptic ion channels, it holds promise for the analysis of exocytic and endocytic events at type III terminals. Patch clamp recordings from type I fast synaptic terminals would be particularly exciting. However, these are extremely difficult in wild-type larvae because of the surrounding subsynaptic reticulum. Nevertheless, it is conceivable that mutants with reduced subsynaptic reticulum[114] and increased bouton size[113] may allow access to these terminals.

VII. Direct and Indirect Effects

Functional analysis of *Drosophila* mutants is a powerful way to study the function of specific gene products *in vivo.* Thus, by complementing studies of biochemical interactions observed *in vitro,* measurement of vesicle recycling in *Drosophila* mutant synapses has the potential to rival contributions made by studies of mutants in neurotransmitter release. However, just as *in vitro* studies cannot stand alone, analysis of mutant phenotypes has its limitations. For novel gene products such as those of *stoned,* for which little biochemical information is available, or for genes such as *clathrin heavy chain,* which may be required for several intracellular processes, including vesicle budding at the Golgi complex, interpretations of mutant phenotypes can be difficult. Detailed biochemical studies demonstrating, for instance, binding of a *stoned* gene product to synaptotagmin may support the argument that *synaptotagmin*-like phenotypes of *stoned* mutants reflect common functions of the two gene products. Thus, biochemical techniques to purify and study synaptic protein complexes are especially important to develop in *Drosophila* in order to complement the phenotypic analysis of mutations in novel genes (L. Pallanck and B. Ganetzky, unpublished; J. Roos and R. B. Kelly, unpublished).

Defects observed at nerve terminals may derive not only from direct inhibition of the synaptic functions of these gene products, but also indirectly from altered synapse composition and morphology. For this reason, techniques to achieve conditional inhibition of presynaptic molecules after normal development of synapses would be especially useful. The utility of conditional inhibition is demonstrated by analysis of *shi*^{ts} mutants in *Drosophila,* as well as of the effects of amphiphysin inhibition by peptide injections into the lamprey synapse.[44] In both cases, endocytosis of synaptic vesicle membrane is blocked by the acute (immediate) inhibition of dynamin or amphiphysin. This not only allows clear demonstration of the *in vivo* function for a gene product, but also the trapping and observation of specific intermediate stages in endocytosis. In *shi*^{ts} mutants, a collared pit intermediate is arrested at elevated temperatures, whereas when amphiphysin–dynamin interactions are inhibited, pits with no collars may be observed to accumulate. Arresting intermediate stages of vesicle recycling provides the opportunity to study the disposition of several different presynaptic molecules at each intermediate stage. Unfortunately, fast onset, conditional mutant alleles will never be obtained for the vast majority of genes, and the small size of presynaptic terminals makes the injection of inhibitors very difficult. Thus alternative approaches would be useful. We suggest that for the analysis of synaptic vesicle recycling two little-used approaches, which have their own handicaps, may prove particularly useful. First, the induction of dominant-negative forms (generally interacting domains) of specific proteins under heat shock control may provide a means to inhibit particular presynaptic proteins after synapses have developed normally. If these (generally truncated) forms of the proteins have synapse targeting signals,[115] then it should take no more than 10 min for the proteins to be transported 1 mm, the distance between motor neuron soma and nerve terminal. A second method to achieve conditional inhibition of a specific protein is suggested by new methods to transport dominant-negative peptides across cell membranes. A 16 amino acid peptide derived from the third helix of Antennapedia, termed penetratin, has the ability to cross mammalian[116,117] and *Drosophila* (D. Sandstrom and M. Ramaswami, unpublished) plasma membrane. If small interfering peptides could be designed with the specificity of those used for lamprey amphiphysin or for Ca^{2+}–calmodulin-dependent protein kinase inhibition,[118] then perhaps acute and conditional phenotypes may be achieved by simple bath application of penetratin fusion peptides.

VIII. Conclusions

This review has attempted to give an account of specific contributions and features of *Drosophila* relevant to the analysis of synaptic vesicle recycling. Because the study of this phenomenon is still in its relative infancy, we have tried, perhaps to slight excess, to speculate on the future of this field in *Drosophila*. Independent of our speculations and suggestions, there is little doubt that with existing tools and gaining momentum, continued major contributions may be expected from *Drosophila* to this burgeoning field.

Acknowledgments

Reviewing an explosively growing area of research when faced with deadlines inevitably results in some errors of omission and inclusion. To those whose work or opinions we may have overlooked, we offer our apologies. We are grateful to Chris Bazinet, Hugo Bellen, Pietro De Camilli, Barry Ganetzky, Cahir O'Kane, Len Kelly, Regis Kelly, K. S. Krishnan, Leo Pallanck, A. Marie Phillips, Jack Roos, Dave Sandstrom, Tom Schwarz, Bing Zhang, and Konrad Zinsmaier for permission to cite their unpublished results. We also thank Patricia Estes, Leona Mukai, Dave Sandstrom, Robin Staples, and Jane Robinson for their comments on the manuscript. Work in the authors' lab is supported by grants from the NINDS, NSF, HFSPO, and the McKnight and Sloan Foundations. Dan Stimson acknowledges support from Developmental Neuroscience training grants at the University of Arizona, funded by the Flinn Foundation and the NIH (NS34889).

References

1. De Camilli, P., and Takei, K. (1996). Molecular mechanisms in synaptic vesicle endocytosis and recycling. *Neuron* **16,** 481–486.
2. Cremona, O., and De Camilli, P. (1997). Synaptic vesicle endocytosis. *Curr. Opin. Neurobiol* **7,** 323–330.
3. Ceccarelli, B., Hurlbut, W. P., and Mauro, A. (1972). Depletion of vesicles from frog neuromuscular junctions by tetanic stimulation. *J. Cell Biol.* **54,** 30.
4. Heuser, J., and Reese, T. (1973). Evidence for recycling of synaptic vesicle membrane during transmitter release at the frog neuromuscular junction. *J. Cell Biol.* **57,** 315–344.
5. Poodry, C. A., and Edgar, L. (1979). Reversible alterations in the neuromuscular junctions of *Drosophila* melanogaster bearing a temperature-sensitive mutation, *shibire. J. Cell Biol.* **81,** 520–527.
6. Kosaka, T., and Ikeda, K. (1983). Possible temperature-dependent blockage of synaptic vesicle recycling induced by a single gene mutation in *Drosophila. J. Neurobiol.* **14**(3), 207–225.
7. van der Bliek, A. M., and Meyerowtiz, E. M. (1991). Dynamin-like protein encoded by the *Drosophila shibire* gene associated with vesicular traffic. *Nature* **351,** 411–414.
8. Chen, Y. S., Obar, R. A., Schroeder, C. C., Austin, T. W., Poodry, C. A., Wadsworth, S. C., and Vallee, R. B. (1991). Multiple forms of dynamin are encoded by the shibire, a *Drosophila* gene involved in endocytosis. *Nature* **351,** 583–586.
9. Shpetner, H., and Vallee, R. (1989). Identification of dynamin, a novel mechanochemical enzyme that mediates interactions between microtubules. *Cell* **59,** 421–432.
10. Obar, R., Collins, C., Hammarback, J., Shpetner, H., and Vallee, R. (1990). Molecular cloning of the microtubule-associated mechanochemical enzyme dynamin reveals homology with a new family of GTP-binding proteins. *Nature* **347,** 256–261.
11. Vallee, R. B. (1992). Dynamin: Motor protein or regulatory GTPase? *J. Muscle Res. Cell Motil.* **13,** 493–496.
12. Warnock, D. E., and Schmid, S. L. (1996). Dynamin GTPase, a force-generating molecular switch. *BioEssays* **18**(11), 885–893.
13. Kelly, L. E., and Suzuki, D. T. (1974). The effects of increased temperature on electroretinograms of temperature-sensitive paralysis mutants of *Drosophila melanogaster. Proc. Natl. Acad. Sci. USA* **71,** 4906–4909.
14. Suzuki, D. (1974). Behaviour in *Drosophila melanogaster:* A geneticist's view. *Can. J. Genet. Cytol.* **16,** 713–735.
15. Ikeda, K., Ozawa, S., and Hagiwara, S. (1976). Synaptic transmission reversibly conditioned by a single-gene mutation in *Drosophila melanogaster. Nature* **259,** 489–491.

16. Salkoff, L., and Kelly, L. (1978). Temperature-induced seizure and frequency-dependent neuromuscular block in a ts mutant of *Drosophila. Nature* **273,** 156–158.
17. Estes, P., Roos, J., van der Bliek, A., Kelly, R. B., Krishnan, K. S., and Ramaswami, M. (1996). Traffic of dynamin within single *Drosophila* synaptic boutons relative to compartment-specific presynaptic markers. *J. Neurosci.* **16,** 5443–5456.
18. Grant, D., Unadkat, S., Katzen, A., Krishnan, K. S., and Ramaswami, M. (1998). Probable mechanisms underlying interallelic complementation and temperature sensitivity of mutations at the *shibire* locus of *Drosophila melanogaster. Genetics* **149,** 1019–1030.
19. van der Bliek, A. M., Redelmeier, T. E., Damke, H., Tisdale, E. J., Meyerowitz, E. M., and Schmid, S. L. (1993). Mutations in human dynamin block an intermediate stage in coated vesicle formation. *J. Cell Biol.* **122,** 553–563.
20. Ramaswami, M., Krishnan, K. S., and Kelly, R. B. (1994). Intermediates in synaptic vesicle recycling revealed by optical imaging of *Drosophila* neuromuscular junctions. *Neuron* **13,** 363–375.
21. Takei, K., McPherson, P. S., Schmid, S., and DeCamilli, P. (1995). Tubular membrane invaginations coated by dynamin rings are induced by GTPγS in nerve terminals. *Nature* **374,** 186–190.
22. Poodry, C. A., Hall, L., and Suzuki, D. (1973). Developmental properties of *shibire*^{ts1}: A pleiotropic mutation affecting larval and adult locomotion and development. *Dev. Biol.* **32,** 373–386.
23. Poodry, C. A. (1990). *shibire:* A neurogenic mutant of *Drosophila. Dev. Biol.* **138,** 464–472.
24. Ramaswami, M., Rao, S., van der Bliek, A., Kelly, R. B., and Krishnan, K. S. (1993). Genetic studies on dynamin function in *Drosophila. J. Neurogenet.* **9,** 73–87.
25. Tsuruhara, T., Koenig, J., and Ikeda, K. (1990). Synchronized endocytosis studied in the oocyte of a temperature-sensitive mutant of *Drosophila melanogaster. Cell Tissue Res.* **259,** 199–207.
26. Koenig, J. H., and Ikeda, K. (1990). Transformational process of the endosomal compartment in nephrocytes of *Drosophila melanogaster. Cell Tissue Res.* **262,** 233–244.
27. Masurm, S. K., Kim, Y. T., and Wu, C. F. (1990). Reversible inhibition of endocytosis in cultured neurons from the *Drosophila* temperature-sensitive mutant shibire ts1. *J. Neurogenet.* **6,** 191–206.
28. Kramer, H., Cagan, R. L., and Zipursky, S. L. (1991). Interaction of bride of sevenless membrane-bound ligand and the sevenless tyrosine-kinase receptor. *Nature* **352,** 207–212.
29. Tabata, T., and Kornberg, T. B. (1994). Hedgehog is a signaling protein with a key role in patterning *Drosophila* imaginal discs. *Cell* **76,** 89–102.
30. Koenig, J. H., and Ikeda, K. (1989). Disappearance and reformation of synaptic vesicle membrane upon transmitter release observed under reversible blockage of membrane retrieval. *J. Neurosci.* **9**(11), 3844–3860.
31. Takei, K., Mundgihl, O., Daniell, L., and DeCamilli, P. (1996). The synaptic vesicle cycle: A single vesicle budding step involving clathrin and dynamin. *J. Cell Biol.* **133,** 1237–1250.
32. Kuromi, H., and Kidokoro, Y. (1998). Two distinct pools of synaptic vesicles in single presynaptic boutons revealed in a temperature-sensitive *Drosophila* mutant, shibire. *Neuron* **20,** 917–925.
33. Pieribone, V. A., Shupliakov, O., Brodin, L., Hilfiker-Rothenfluh, S., Czernik, A. J., and Greengard, P. (1995). Distinct pools of synaptic vesicles in neurotransmitter release. *Nature* **375,** 493–497.
34. Koenigm, J. H., and Ikeda, K. (1996). Synaptic vesicles have two distinct recycling pathways. *J. Cell Biol.* **135,** 797–808.
35. Heuser, J. H., and Reese, T. (1981). Structural changes after transmitter release at the frog neuromuscular junction. *J. Cell Biol.* **88,** 564–580.

36. Heuser, J. (1989). The role of coated vesicles in recycling of synaptic vesicle membrane. *Cell Biol. Int. Rep.* **13,** 1063–1076.
37. Roos, J., and Kelly, R. B. (1997). Is dynamin really a "pinchase." *Trends Cell Biol.* **7**(7), 257–259.
38. Betz, W. J., Mao, F. M., and Bewick, G. S. (1992). Activity-dependent fluorescent staining and destaining of living vertebrate motor nerve terminals. *J. Neurosci.* **12,** 363–375.
39. Kuromi, H., Yoshihara, M., and Kidokoro, Y. (1997). An inhibitory role of calcineurin in endocytosis of synaptic vesicles at nerve terminals of *Drosophila* larvae. *Neurosci. Res.* **27,** 101–113.
40. Matthews, G. (1996). Synaptic exocytosis and endocytosis: Capacitance measurements. *Curr. Opin. Neurobiol.* **6,** 358–364.
41. Henkel, A., and Betz, W. J. (1995). Staurosporine blocks evoked release of FM1-43 but not acetylcholine from frog motor nerve terminals. *J. Neurosci.* **15,** 8246–8258.
42. Bennett, M. K., Calakos, N., Kreiner, T., and Scheller, R. H. (1992). Synaptic vesicle membrane proteins interact to form a multimeric complex. *J. Cell Biol.* **116,** 761–75.
43. Morris, S. A., and Schmid, S. L. (1995). The ferrari of endocytosis. *Curr. Biol.* **5,** 113–115.
44. Shupliakov, O., Low, P., Grabs, D., Gad, H., Chen, H., David, C., Takei, K., and De Camilli, P. (1997). Synaptic vesicle endocytosis impaired by disruption of dynamin-SH3 domain interactions. *Science* **276,** 259–263.
45. McPherson, P. S., Garcia, E. P., Slepnev, V. I., David, C., Zhang, X., Grabs, D., Sossin, W. S., Bauerfeind, R., Nemoto, Y., and De Camilli, P. (1996). A presynaptic inositol-5-phosphatase. *Nature* **379**(6563), 353–357.
46. Munn, A. L., Stevenson, B. J., Geli, M. I., and Reizman, H. (1995) end5, end6, and end7: Mutations that cause actin delocalization and block the internalization step of endocytosis in *Saccharomyces cerevisiae. Mol. Biol. Cell.* **6,** 1721–1742.
47. De Camilli, P., Emr, S. D., McPherson, P. S., and Novick, P. (1996). Phosphoinositides as regulators in membrane traffic. *Science* **271,** 1533–1539.
48. Lloyd, V., Ramaswami, M., and Krämer, H. (1998). Not just pretty eyes: *Drosophila* eye color mutations and lysosomal delivery. *Trends Cell Biol.* **8,** 257–259.
49. Koenig, J., Kosaka, T., and Ikeda, K. (1989). The relationship between the number of synaptic vesicles and the amount of neurotransmitter released. *J. Neurosci.* **9,** 1937–1942.
50. Bazinet, C., Katzen, A. L., Morgan, M., Mahowald, A. P., and Lemmon, S. K. (1993). The *Drosophila* clathrin heavy chain gene: Clathrin function is essential in a multicellular organism. *Genetics* **134,** 1119–1134.
51. Robinson, M. S. (1994). The role of clathrin, adaptors and dynamin in endocytosis. *Curr. Opin. Cell Biol.* **6,** 538–544.
52. Robinson, M. S. (1997). Coats and vesicle budding. *Trends in Cell Biol.* **7,** 99–102.
53. Schlossman, D. M., Schmid, S. L., Braell, W. A., and Rothman, J. E. (1984). An enzyme that removes clathrin coats: Purification of an uncoating ATPase. *J. Cell Biol.* **99,** 723–733.
54. Ungewickell, E., Ungewickell, H., and Holstein, S. E. (1997). Functional interaction of the auxilin J domain with the nucleotide- and substrate-binding modules of Hsc70. *J. Biol. Chem.* **272,** 19594–19600.
55. Braun, J., Wilbanks, S. M., and Scheller, R. H. (1996). The cysteine string secretory vesicle protein activates Hsc70 ATPase. *J. Biol. Chem.* **271,** 25989–25993.
56. Chamberlain, L. H., and Burgoyne, R. D. (1997). Activation of the ATPase activity of heat-shock proteins Hsc70/Hsp70 by cysteine-string protein. *Biochem. J.* **322,** 853–858.
57. Ungewickell, E., Ungewickell, H., Holstein, S. E. H., Lindner, R., Prasad, K., Barouch, W., Martin, B., Greene, L. E., and Eisenberg, E. (1995). Role of auxilin in uncoating clathrin-coated vesicles. *Nature* **378,** 632–635.

58. Gonzales-Gaitan, M., and Jackle, H. (1997). Role of *Drosophila* α-adaptin in presynaptic vesicle recycling. *Cell* **88,** 767–776.
59. Camidge, D. R., and Pearsc, B. M. (1994). Cloning of *Drosophila* β-adaptin and its localization and expression in mammalian cells. *J. Cell Sci.* **107,** 709–718.
60. Robinson, M. S. (1992). Adaptins. *Trends Cell Biol.* **2,** 293–297.
61. Sousa, R., Tannery, N. H., Zhou, S., and Lafer, M. (1992). Characterization of a novel synapse-specific protein. I. Developmental expression and cellular localization of the F1-20 protein and mRNA. *J. Neurosci.***12**(6), 2130–2143.
62. Zhou, S., Tannery, N. H., Yang, J., Puszkin, S., and Lafer, E. M. (1993). The synapse-specific phosphoprotein F1-20 is identical to the clathrin assembly protein AP-3. *J. Biol. Chem.* **268**(17), 12655–12662.
63. Ye, W., and Lafer, E. M. (1995). Clathrin binding and assembly activities of expressed domains of the synapse-specific clathrin assembly protein AP-3. *J. Biol. Chem.* **270**(18), 10933–10939.
64. Wang, L. H., Sudhof, T. C., and Anderson, R. G. (1995). The appendage domain of alpha-adaptin is a high affinity binding site for dynamin. *J. Biol. Chem.* **270,** 10079–10083.
65. Hao, W., Tan, Z., Prasad, K., Reddy, K. K., Chen, J., Prestwich, G. D., Falck, J. R., Shears, S. B., and Lafer, E. M. (1997). Regulation of AP-3 function by inositides. *J. Biol. Chem.* **272**(10), 6393–6398.
66. Lafer, E. M., Womack, M., Zhao, X., Prasad, K., and Augustine, G. (1997). Microinjection of the clathrin assembly domain of AP-3 enhances synaptic transmission at the squid giant synapse. *Soc. Neurosci.* **23**(1), 361. [Abstract]
67. Andrews, J., Smith, M., Merakovsky, J., Coulson, M., Hannan, F., and Kelly, L. E. (1996). The *stoned* locus of *Drosophila melanogaster* produces a dicistronic transcript and encodes two distinct polypeptides. *Genetics* **143,** 1699–1711.
68. Salcini, A. E., Confalonieri, S., Doria, M., Santolini, E., Tassi, E., Minenkova, O., Cesareni, G., Pelicci, P. G., and Di Fiore, P. P. (1997). Binding specificity and in vivo targets of the EH domain, a novel protein-protein interaction module. *Genes Dev.* **11,** 2239–2249.
69. Tebar, F., Sorkina, T., Sorkin, A., Ericsson, M., and Kirchhausen, T. (1996). Eps15 is a component of clathrin-coated pits and vesicles and is located at the rim of coated pits. *J. Biol. Chem.* **271**(46), 28727–28730.
70. van Delft, S., Schumacher, C., Hage, W., Verkleij, A. J., and van Bergen en Henegouwen, P. M. (1997). Association and colocalization of Eps15 with adaptor protein-2 and clathrin [published erratum appears in *J. Cell Biol.* **137**(1), 259]. *J. Cell Biol.* **136**(4), 811–21.
71. Haffner, C., Takei, K., Chen, H., Ringstad, N., Hudson, A., Butler, M., Salcini, A. E., Di Fiore, P. P., and De Camilli, P. (1998). Synaptojanin I: Localization on coated endocytic intermediates in nerve terminals and interaction of its 170 kDa isoform with Eps15. Submitted for publication.
72. Martin, T. F. J. (1997). Phosphoinositides as spatial regulators of membrane traffic. *Curr. Opin. Neurobiol.* **7,** 331–338.
73. Ringstad, N., Nemoto, Y., and De, C. P. (1997). The SH3p4/Sh3p8/SH3p13 protein family: Binding partners for synaptojanin and dynamin via a Grb2-like Src homology 3 domain. *Proc. Natl. Acad. Sci. USA* **94**(16), 8569–8574.
74. de Heuvel, E., Bell, A. W., Ramjaun, A. R., Wong, K., Sossin, W. S., and McPherson, P. S. (1997). Identification of the major synaptojanin-binding proteins in brain. *J. Biol. Chem.* **272**(13), 8710–8716.
75. Micheva, K. D., Kay, B. K., and McPherson, P. S. (1997). Synaptojanin forms two separate complexes in the nerve terminal: Interactions with endophilin and amphiphysin. *J. Biol. Chem.* **272**(43), 27239–27245.

76. Schaeffer, E., Smith, D., Mardon, G., Quinn, W., and Zuker, C. (1989). Isolation and characterization of two new *Drosophila* protein kinase C genes, including one specifically expressed in photoreceptor cells. *Cell* **57,** 403–412.
77. Guerini, D., Montell, C., and Klee, C. B. (1992). Molecular cloning and characterization of the genes encoding the two subunits of *Drosophila melanogaster* calcineurin. *J. Biol. Chem.* **267**(31), 22542–22549.
78. Warren, W. D., Phillips, A. M., and Howells, A. J. (1996). *Drosophila melanogaster* contains both X-linked and autosomal homologues of the gene encoding calcineurin B. *Gene* **177**(1–2), 149–153.
79. Brown, L., Chen, M. X., and Cohen, P. T. W. (1994). Identification of cDNA encoding a *Drosophila* calcium/calmodulin regulated protein phosphatase, which has its most abundant expression in the early embryo. *FEBS Lett.* **339,** 124–128.
80. Kane, N. S., Robichon, A., Dickinson, J. A., and Greenspan, R. J. (1997). Learning without performance in PKC-deficient *Drosophila. Neuron* **18,** 307–314.
81. McLure, S. J., and Robinson, P. J. (1996). Dynamin, endocytosis and intracellular signalling. *Mol. Membr. Biol.* **13,** 189–215.
82. Liu, J.-P., Sim, A. T. R., and Robinson, P. J. (1994). Calcineurin inhibition of dynamin I GTPase activity coupled to nerve terminal depolarization. *Science* **265,** 970–973.
83. Perin, M. S., Johnston, P. A, Ozcelik, T., Jahn, R., Franke, U., and Sudhof, T. C. (1991). Structural and functional conservation of synaptotagmin (p65) in *Drosophila* and humans. *J. Biol. Chem.* **266,** 615–622.
84. Li, J., Reist, N., DiAntonio, A., and Schwarz, T. (1995). Interaction of *Drosophila synaptotagmin* (*syt*) and *shibire* mutations. *Soc. Neurosci.* **21**(1), 326. [Abstract]
85. Zhang, J. Z., Davletov, B. A., Sudhof, T. C., and Anderson, R. G. W. (1994). Synaptotagmin I is a high affinity receptor for clathrin AP-2: Implications for membrane recycling. *Cell* **78,** 751–760.
86. Sudhof, T. C., and Rizo, J. (1996). Synaptotagmins: C_2-domain proteins that regulate membrane traffic. *Neuron* **17,** 379–388.
87. Zinsmaier, K. E., Hofbauer, A., Heimbeck, G., Pflugfelder, G. O., Buchner, S., and Buchner, E. (1990). Cysteine-string protein is expressed in retina and brain of *Drosophila. J. Neurogenet.* **7,** 15–29.
88. Ranjan, R., Bronk, P., and Zinsmaier, K. (1998). Cysteine string protein is required for calcium-secretion coupling of evoked neurotransmission in *Drosophila* but not for vesicle recycling. *J. Neurosci.* **18,** 956–964.
89. Stimson, D., Estes, P. S., Smith, M., Kelly, L. E., and Ramaswami, M. (1998) A product of the *Drosophila stoned* genetic locus regulates neurotransmitter release. *J. Neurosci.,* in press.
90. Petrovich, T. Z., Merakovsky, J., and Kelly, L. E. (1993). A genetic analysis of the *stoned* locus and its interaction with *dunce, shibire,* and *Suppressor of stoned* variants of *Drosophila melanogaster. Genetics* **133,** 955–965.
91. Murtagh, J. J., Jr., Lee, F.-J.S., Deak, P., Hall, L. M., Monaco, L., Lee, C.-M., Stevens, L. A., Moss, J., and Vaughan, M. (1993). Molecular characterization of a conserved, guanine nucleotide-dependent ADP-ribosylation factor in *Drosophila melanogaster. Biochemistry* **32,** 6011–6018.
92. Lee, F.-J.S., Stevens, L. A., Hall, L. M., Murtagh, J. J., Jr., Kao, Y. L., Moss, J., and Vaughan, M. (1994). Characterization of class II and class III ADP-ribosylation factor genes and proteins in *Drosophila melanogaster. J. Biol. Chem.* **269**(34), 21555–21560.
93. D'Souza-Schorey, C., Guangpu, L., Colombo, M. I., and Stahl, P. D. (1995). A regulatory role of ARF6 in receptor-mediated endocytosis. *Science* **267,** 1175–1178.

94. Faundez, V., Horng, J.-T., and Kelly, R. B. (1997). ADP ribosylation factor 1 is required for synaptic vesicle budding in PC12 cells. *J. Cell Biol.* **138**(3), 505–515.
95. Seaman, M. N. J., Ball, C. L., and Robinson, M. S. (1993). Targeting and mistargeting of plasma membrane adaptors in vitro. *J. Cell Biol.* **123**(5), 1093–1105.
96. Sasamura, T., Kobayashi, T., Kojima, S., Qadota, H., Ohya, Y., Masai, I., and Hotta, Y. (1997). Molecular cloning and characterization of *Drosophila* genes encoding small GTPases of the *rab* and *rho* families. *Mol. Gen. Genet.* **254,** 486–494.
97. Satoh, A. K., Tokunaga, F., and Ozaki, K. (1997). Rab proteins of *Drosophila melanogaster:* Novel members of the Rab-protein family. *FEBS Lett.* **404,** 65–69.
98. Rybin, V., Ullrich, O., Rubino, M., Alexandrov, K., Simon, I., Seabra, C., Goody, R., and Zerial, M. (1996). GTPase activity of Rab5 acts as a timer for endocytic membrane fusion. *Nature* **383**(6597), 266–269.
99. de Hoop, M. J., Huber, L. A., Stenmark, H., Williamson, E., Zerial, M., Parton, R. G., and Dotti, C. G. (1994). The involvement of the small GTP-binding protein Rab5a in neuronal endocytosis. *Neuron* **13,** 11–22.
100. Kramer, H., and Phistry, M. (1996). Mutations in the *Drosophila hook* gene inhibit endocytosis of the boss transmembrane ligand into multivesicular bodies. *J. Cell Biol.* **133,** 1205–1215.
101. Phistry, M., Sunio, A., and Kramer, H. (1998). Submitted for publication.
102. Shestopal, S. A., Makunin, I. V., Belyaeva, E. S., Ashburner, M., and Zhimulev, I. F. (1997). Molecular characterization of the *deep orange* (*dor*) gene of *Drosophila melanogaster. Mol. Gen. Genet.* **253,** 642–648.
103. Keshishian, H., Chiba, A., Chang, T. N., Halfon, M. S., Harkins, E. W., Jareck, J., Wang, L., Anderson, M., Cash, S., Halpern, M. E., and Johansen, J. (1993). Cellular mechanisms governing synaptic development in *Drosophila melanogaster. J. Neurobiol.* **24,** 757–787.
104. Gorczyca, M., Augart, C., and Budnik, V. (1993). Insulin-like receptor and insulin-like peptide are localized at neuromuscular junctions in *Drosophila. J. Neurosci.* **13,** 3692–3704.
105. Betz, W. J., and Wu, L.-G. (1995). Kinetics of synaptic-vesicle recycling. *Curr. Biol.* **5**(10), 1098–1101.
106. Umbach, J., and Gunderson, C. (1997). Evidence that cysteine string proteins regulate an early step in the calcium-dependent secretion of neurotransmitter at *Drosophila* neuromuscular junctions. *J. Neurosci.* **17**(19), 7203–7209.
107. Ryan, T. A., and Smith, S. J. (1995). Vesicle pool mobilization during action potential firing at hippocampal synapses. *Neuron* **14,** 983–989.
108. Ryan, T. A., Li, L., Chin, L. S., Greengard, P., and Smith, S. J. (1996). Synaptic vesicle recycling in synapsin I knock-out mice. *J. Cell Biol.* **134,** 1219–1227.
109. DiAntonio, A., Parfitt, K., and Schwarz, T. L. (1993). Synaptic transmission persists in synaptotagmin mutants of *Drosophila. Cell* **73,** 1281–1290.
110. Littleton, T. J., Stern, M., Schulze, K., Perin, M., and Bellen, H. J. (1993). Mutational analysis of *Drosophila* synaptotagmin demonstrates its essential role in Ca^{2+}-activated neurotransmitter release. *Cell* **74,** 1125–1134.
111. Zinsmaier, K., Eberle, K., Buchner, E., Walter, N., and Benzer, S. (1994). Paralysis and early death in cysteine string protein mutants of *Drosophila. Science* **263,** 977–980.
112. Betz, W. J., and Bewick, G. S. (1993). Optical monitoring of transmitter release and synaptic vesicle recycling at the frog neuromuscular junction. *J. Physiol.* **460,** 287–309.
113. Martinez-Padron, M., and Ferrus, A. (1997). Presynaptic recordings from *Drosophila:* Correlation of macroscopic and single-channel K^+ currents. *J. Neurosci.* **17,** 3412–3424.
114. Lahey, T., Gorczyca, M., Jia, X.-X., and Budnik, V. (1994). The *Drosophila* tumor suppressor gene *dlg* is required for normal synaptic bouton structure. *Neuron* **13,** 823–835.

115. VanBerkum, M., and Goodman, C. (1995). Targeted disruption of Ca(2+)-calmodulin signaling in *Drosophila* growth cones leads to stalls in axon extension and errors in axon guidance. *Neuron* **14,** 43–56.
116. Troy, C. M., Derossi, D., Prochiantz, A., Greene, L. A., and Shelanski, M. L. (1996). Downregulation of Cu/Zn superoxide dismutase leads to cell death via the nitric oxide-peroxynitrite pathway. *J. Neurosci.* **16**(1), 253–261.
117. Hall, H., Williams, E. J., Moore, S. E., Walsh, F. S., Prochiantz, A., and Doherty, P. (1996). Inhibition of FGF-stimulated phosphatidylinositol hydrolysis and neurite outgrowth by a cell-membrane permeable phosphopeptide. *Curr. Biol.* **6**(5), 580–587.
118. Wang, J., Renger, J. J., Griffith, L. C., Greenspan, R. J., and Wu, C.-F. (1994). Concomitant alterations of physiological and developmental plasticity in *Drosophila* CaM kinase II-inhibited synapses. *Neuron* **13,** 1373–1384.

IONIC CURRENTS IN LARVAL MUSCLES OF *DROSOPHILA*

Satpal Singh* and Chun-Fang Wu†

*Department of Biochemical Pharmacology, SUNY at Buffalo, Buffalo, New York 14260, and †Department of Biology, University of Iowa, Iowa City, Iowa 52242

I. Introduction
  A. The Scope of the Review
  B. The Larval Body Wall Muscle Preparation
II. Potassium Currents
  A. The Voltage-Activated K^+ Currents: Transient I_A and Delayed I_K
  B. The Ca^{2+}-Activated K^+ Currents: Fast I_{CF} and Slow I_{CS}
  C. The Corresponding Microscopic Single Channel K^+ Currents
  D. The Gene-Dosage Dependence of I_A
  E. The Roles of K^+ Current in Muscle Membrane Excitability
  F. Developmental Regulation and Second Messenger Modulation of K^+ Channels
  G. The Functional Roles of Modulatory and Auxiliary Subunits of K^+ Channels
III. Calcium Currents
  A. Modulation of Ca^{2+} Currents
IV. Conclusions
  References

I. Introduction

A. The Scope of the Review

Larval body wall muscles of *Drosophila* provide a very useful system for studying K^+ and Ca^{2+} currents. Availability of a number of mutations that affect specific ionic currents have helped greatly in these studies. Ionic currents, especially voltage gated, in the larval body wall muscles are a central theme of this review. Where available, we will discuss the subunit composition of the channels, the genes encoding the subunits, the kinetic properties and voltage dependency of the currents, pharmacological blockers and activators, and the signal transduction pathways regulating or modulating the channel properties. Emphasis will be mainly on physiological aspects of the currents. For details of DNA sequences and related molecular information on the channel structure the reader is referred to the original articles cited in the text and to several reviews.[1–3]

0074-7742/99 $30.00

Muscle currents in adults will be briefly summarized for comparison with currents in the larval muscles. Neuronal currents have been reviewed elsewhere[4] and will be presented only for comparative purposes. The role of different ionic currents in the physiology of muscle fibers will be discussed in experiments involving current clamping and membrane potential measurements.

B. The Larval Body Wall Muscle Preparation

Studies on the function and regulation of ion channels can be aided greatly by the availability of many well-characterized ion channel mutants and cloned and sequenced genes. A sensitive and convenient *in situ* experimental system is required for these studies. For a physiological analysis of ion channels in *Drosophila,* larval body wall muscles provide an excellent preparation. The sensitivity level of detection is superb when physiological phenotypes are examined. Many molecular probes, such as antibody staining or *in situ* hybridization, have provided independent evidence for channel expression but do not always detect low-level but physiologically significant channel expression.[5–7]

Larval body wall muscles are organized in a regular pattern in each segment and are easily identifiable. Passive electrical properties have been studied in these muscles[8,9] and their relatively large size allows an easy impalement by two microelectrodes for voltage clamping.[10,11] It is easy to approach isopotential conditions for the accurate control of membrane potential and measurement of current.[12,13] Physiological and pharmacological properties have been used to clearly resolve various K^+ and Ca^{2+} currents.[11,12,14] The evolutionary fact that invertebrate muscles do not express Na^+ channels[15–17] makes voltage clamping easier in the absence of a fast Na^+ current, and the interaction between inward Ca^{2+} currents and outward K^+ currents is simpler for rigorous biophysical studies. All this helps in analyzing the effect of various mutations on specific ionic currents and muscle membrane excitability.

Despite their impressive success in the elucidation of molecular mechanisms and functional domains of ion channels, ectopic expression or heterologous expression does not necessarily produce the phenotypes observed *in vivo.* The functional role of the channels, i.e., control of membrane potentials, is difficult to study in such systems. In addition, the stoichiometry of subunit assembly of native channels, the functional consequences of channel modulation, and the gene-dosage dependence of channel expression are also difficult to examine in these systems.

Extensive knowledge of properties of the channels in the larval muscles and accessibility to both voltage clamping and current clamping make the system ideal for determination of the functional roles of individual currents. It is also possible to address interesting developmental problems in the system. The embryonic muscle fibers continue to grow through different moltings and eventually reach their maximum size in the third instar larval stage. The adult muscle fibers are generated *de novo* after the histolysis of larval muscles.[18] K^+ currents have been well characterized in the adult flight muscle fibers of *Drosophila,*[1,19,20] although the voltage clamping conditions and pharmacological treatment are less easily controlled in these muscles. The supercontracting larval muscles as compared to the isometric adult indirect flight muscles have different functional requirements and structural constraints.[21,22] A developmental analysis of the currents can lead to insights into how the channels differ in their properties to satisfy the functional challenges. Mutant and multiple-mutant flies are readily constructed for a genetic dissection and reconstruction of functional circuits among interacting ionic currents and structural interdependence among channel subunits. The behavioral consequences of channel mutations can be first studied in this preparation to provide a physiological link. In this manner, a convergence of physiology, pharmacology, and genetics provides the benefit of an integrated approach.

II. Potassium Currents

Whole cell voltage clamp experiments have demonstrated that four distinct K^+ currents are present in larval body wall muscle fibers of *Drosophila.*[11,12,23] The same currents also appear to be present in adult indirect flight muscles and show similar sensitivity to specific mutations, pharmacological blockers, and physiological manipulations.[19,20] A depolarizing voltage step evokes two voltage-activated outward K^+ currents, termed I_A and I_K. The transient current I_A is activated rapidly and is subsequently inactivated during sustained depolarization. In the case of I_K, the activation is slower and relatively little inactivation occurs, conforming to the properties of the classical delayed rectifier.[24] Two Ca^{2+}-activated K^+ currents are evoked by Ca^{2+} influx and their kinetics overlap with those of the two voltage-activated currents. The rapidly activating component, I_{CF}, rises immediately after Ca^{2+} channel opening and declines subsequently. The slowly activating component, I_{CS}, sustains during prolonged depolarization. Several genes encoding the channel subunits underlying these currents have been identified in *Drosophila.* Cloning and sequence analysis of these genes have provided invaluable information about the molecular mechanisms of K^+ channels. *Drosophila* mutants of the genes encoding K^+ channel subunits not only have identified functional motifs in different types of K^+ channel subunits, but also have been the major entry points for the identification of homologous channel subunits in other species, including humans.

A. The Voltage-Activated K^+ Currents: Transient I_A and Delayed I_K

Genetic dissection using specific mutations has been successfully applied to individual K^+ currents in *Drosophila* muscle. The same K^+ current components can also be separated based on their distinct physiological and pharmacological properties.[11,12,23] In Ca^{2+}-free saline, the outward K^+ current consists of only voltage-activated I_A and I_K. Under voltage clamp conditions, I_A is activated by lower voltages of depolarization. It appears as an initial peak of the outward current because of its inactivation properties, i.e., the channel closes during sustained depolarization due to the inactivation mechanism following initial opening (Fig. 1A). At more positive voltages, I_K is activated. The activation is less rapid than that of I_A. I_K channels exhibit little inactivation during sustained depolarization, giving rise to the delayed plateau (Fig. 1A).

The two currents can be further separated by a conditioning voltage prepulse that inactivates I_A, the subsequent test pulse activating only the delayed I_K.[11,13] Similarly, I_K can be observed in isolation from I_A by using a genetic approach to eliminate I_A in *Shaker* (*Sh*) mutants (Fig. 1B).

Interestingly, many *Drosophila* K^+ channel mutants have been isolated based on a simple ether-induced leg-shaking phenotype.[20,25] K^+ channels have been the best studied ion channel types in *Drosophila* because of the availability of such mutants and the ease of studying the currents mediated by these channels in well-established muscle preparations by voltage clamping.[10,11,20,26–28] Subsequent cloning and sequence analysis confirm that these identified genes code for distinct ion channel subunits.[29–34]

Among them, the best studied is the *Sh* gene[31,35,36] and the I_A current mediated by *Sh* channels.[11,37] The *Sh* gene was first identified by classical genetics and is the first K^+ channel gene analyzed by molecular techniques, allowing for the identification of homologous genes in other species.[3] It provides the first molecular structure for a K^+ channel α subunit: an integral membrane protein with six transmembrane domains (S1–S6), plus several distinct molecular motifs, including a pore-forming region (H5 or P) for ion transport and a ball-and-chain domain near the N terminus for channel

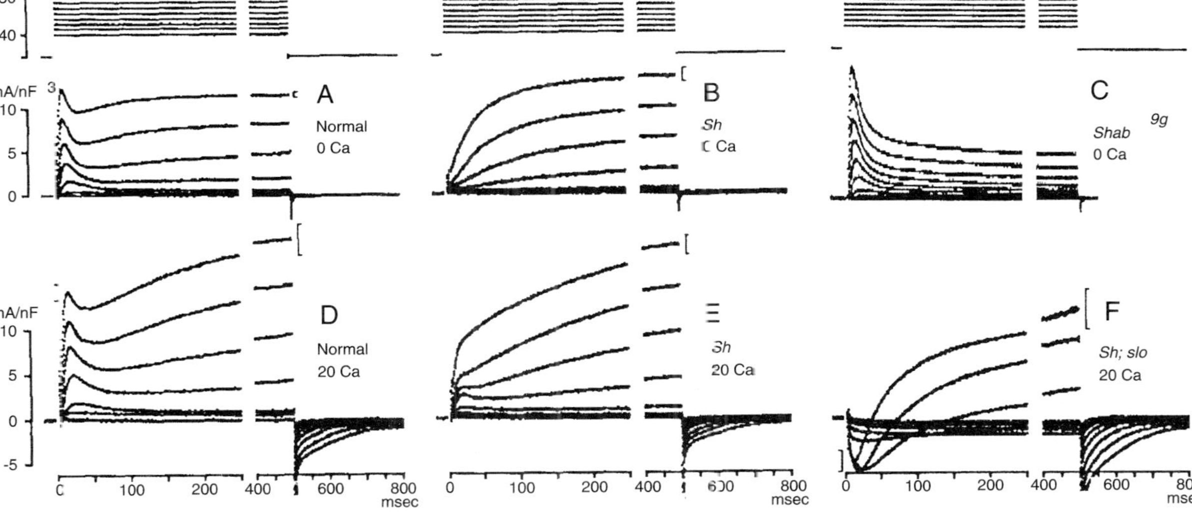

FIG. 1. Whole cell membrane currents from body wall muscle fibers of wild-type and mutant larvae in saline containing 0 or 20 m*M* Ca^{2+}. The early fast and the delayed slow current components are shown at two different time scales with the intermediate parts of traces omitted. (A) Wild-type fibers in Ca^{2+}-free saline display only voltage-activated K^+ currents, transient I_A and the delayed I_K. (B) *Sh*^{KS133} fibers lack I_A and exhibit only I_K in Ca^{2+}-free saline. (C) The *Shab*^{9g} mutation affects a component of I_K without altering I_A. (D) The presence of 20 m*M* Ca^{2+} elicits two additional Ca^{2+}-activated outward K^+ currents, I_{CF} and I_{CS}. Note that activation of I_{CS}, but not that of I_K, gives rise to a strong inward tail current. (E) Elimination of I_A in *Sh*^{KS133} fibers reduces the early peak of the total current in saline containing 20 m*M* Ca^{2+}. (F) Elimination of both I_A and I_{CF} in *Sh*^{KS133}; *slo*^{1} double mutant fibers reveals the inward Ca^{2+} current in saline containing 20 m*M* Ca^{2+}. A, B, D, E, and F were reproduced from Singh and Wu,[12] with permission. C shows unpublished data from Hegde *et al.*

inactivation.[38,39] The *Sh* gene is known to be a complex transcription unit, producing several splicing variants, each capable of forming functional channels when heterologously expressed in *Xenopus* oocytes.[36,40–42]

A gene encoding channel subunits mediating I_K has been identified more recently. Mutations of the gene *Shab* (Chopra *et al.*, unpublished; Hegde *et al.*, unpublished) have been shown to reduce delayed I_K without affecting I_A (Fig. 1C). The *Shab* gene was identified by homology to *Sh* and its sequence implicates a channel subunit structurally homologous to *Sh* subunits[43] (Hegde *et al.*, unpublished). Heterologous expression of *Shab* in *Xenopus* oocytes results in a noninactivating current similar in biophysical properties to the *in vivo* delayed rectifier, I_K.[44] These data are consistent with the reduction of the delayed rectifier K^+ current in embryonic neurons and myotubes by a deletion that spans the *Shab* gene.[45]

Early physiological studies have shown that I_A and I_K are distinct in their pharmacological profiles. In particular, low concentrations of 4-aminopyridine (4-AP) eliminate I_A, whereas low concentrations of quinidine strongly suppress I_K.[27] In addition, dendrotoxin reduces I_A[46] and quinine, tacrine, and cinchonine reduce I_K.[47,48] The physiological and pharmacological distinctions of the two channels are summarized in Table I.

B. The Ca^{2+}-Activated K^+ Currents: Fast I_{CF} and Slow I_{CS}

Like skeletal muscles in other invertebrate species,[15,17,22,49] body wall muscle fibers in *Drosophila* larvae express only Ca^{2+}, but not Na^+, channels. Unlike vertebrate muscle fibers, Ca^{2+} influx is required for insect muscle contraction.[50] The Ca^{2+} influx mediated by the inward Ca^{2+} current also initiates Ca^{2+}-activated K^+ currents. In saline containing 20 m*M* Ca^{2+}, the total membrane current evoked by membrane depolarization contains additional components as compared to the total current obtained in Ca^{2+}-free saline (compare Figs. 1A and 1D). At 20 m*M* external Ca^{2+}, a pronounced delayed component appears during sustained depolarization, which is followed by a strong inward tail current at the offset of the voltage pulse (Fig. 1D). The inward tail current at −80 mV is a hallmark of the slow Ca^{2+}-activated K^+ current, I_{CS},[27] indicating that its reversal potential is different from that of I_K. There is little tail current of I_K at a holding potential of −80 mV because its reversal potential is close to −70 mV and the channels close very fast.[11] The initial outward peak current appears greater in saline containing 20 m*M* Ca^{2+} compared to Ca^{2+}-free saline (compare Figs. 1A and 1D). In *Sh* mutants, a substantial initial peak still remains. Even though

TABLE I

PROPERTIES OF POTASSIUM AND CALCIUM CURRENTS IN LARVAL MUSCLES OF *Drosophila*

Current (activated by)	Whole cell physiology	Single channel physiology	Genes (mutations)/subunit	Blocked by	Modulated by
I_A (voltage)	Fast activation Fast inactivation Low activation threshold (1)	Low conductance (10 pS) (4)	*Shaker*/α (5) *Hyperkinetic*/β (6) *ether-a-go-go*/α (7)	4-AP (11) TEA (11) Dendrotoxin (12)	cAMP (17) Calcium (18) W7 (19)
I_K (voltage)	Slow activation Very slow inactivation Higher activation threshold (1)	Larger conductance (30 pS) (4)	*Shab*/α (8) *ether-a-go-go*/α (7)	Quinidine (13) Quinine (14) Tacrine (15) Cinchonine (14) TEA (11)	cAMP (17) W7 (19) Caffeine (19) Ras-Raf (20) cGMP (19) Dev. Temp. (21)
I_{CF} (Ca^{2+})	Fast activation Fast decay (2)	Large conductance (55 pS) Flickering Activation at 20–50 n*M* Ca^{2+} (4)	*slowpoke*/α (9) ether-a-go-go/α (7)	Charybdotoxin (16) TEA (11)	W7 Caffeine (19)
I_{CS} (Ca^{2+})	Slow activation Slow decay (2)		*ether-a-go-go*/α (7)		cAMP W7 Caffeine (19)
DHP-sensitive (voltage)	High activation threshold Slow inactivation (3)		*l(2)35Fa*/α, (10)	DHPs Diltiazem (3)	cAMP–PKA (22) PLC–DAG–PKC (23)
Amiloride-sensitive (voltage)	High activation threshold Slow inactivation (3)		—	Amiloride (3)	

Numbers given after the properties point to references in the bibliography as follows: (1): 11, 52; (2): 23, 27; (3): 91; (4): 52; (5): 10, 25, 29, 30, 31, 35, 36; (6): 25, 32, 81; (7): 14, 25, 28, 33, 80; (8): 43, Chopra *et al.*, unpublished; Hegde *et al.*, unpublished; (9): 34, 64; (10): 99; (11): 11; (12): 46; (13): 12, 27, 48; (14): 48; (15): 47; (16): 64; (17): 70; (18): 118; (19): 28; (20): 76; (21): 71; (22): Bhattacharya and Singh, unpublished; (23): 73.

I_A is eliminated by some *Sh* mutations, the fast Ca^{2+}-activated current, I_{CF}, remains intact (compare Figs. 1B and 1E).

The leg-shaking behavior was used in the identification of a different gene, *slowpoke* (*slo*), which encodes an α subunit of I_{CF}.[12,20,34] Mutations of the *slo* gene eliminate I_{CF}, and in *Sh; slo* double mutants both I_A and I_{CF} are absent, revealing an inward Ca^{2+} current at the beginning of the voltage pulses in saline containing 20 m*M* Ca^{2+} (Fig. 1F). By the end of the pulse, this Ca^{2+} current is masked by the slowly activating I_{CS}. Molecular cloning of the *slo* gene[34] identified a K^+ channel α subunit that is highly homologous to *Sh* subunits in transmembrane domains and the pore-forming region. Nevertheless, the *slo* polypeptide is distinct in its longer C terminus containing a Ca^{2+}-binding consensus sequence. Expression of the *slo* polypeptide in *Xenopus* oocytes leads to Ca^{2+}-dependent outward currents with properties similar to those observed in large conductance, Ca^{2+}-activated BK channels in other species.[51] The molecular counterpart for I_{CS} channels has not yet been identified.

C. The Corresponding Microscopic Single Channel K^+ Currents

Patch clamp recordings from enzyme-treated larval muscle membranes or artificially induced muscle membrane vesicles reveal a variety of single channel activities.[52,53] Channel openings activated at different membrane potentials can be observed and categorized according to kinetic properties. Rapidly inactivating currents from low conductance single channel activities can be evoked by long depolarizing pulses (Fig. 2A,1), and the time course of their ensemble average over repetitive stimuli resembles the inactivating kinetics of the whole cell current I_A (Fig. 2A,2). A different voltage-sensitive channel type displays longer channel openings during sustained depolarization (Fig. 2B,1), and the time course of their ensemble average correlates well with the whole cell-delayed rectifier, I_K (Fig. 2B,2).

Another frequently encountered single channel activity appears to be associated with channels sensitive to cytosolic Ca^{2+}. These channels are activated at physiological Ca^{2+} concentrations, between 10^{-8} and 10^{-7} *M* (Fig. 2C). A striking feature of this channel type is the interruption of the channel opening by high frequency brief closures, giving rise to the appearance of flickering. The activity of this channel has never been observed in membrane patches of *slo* mutant muscle,[52] suggesting that it represents the microscopic current corresponding to the macroscopic I_{CF}, which is eliminated in *slo* muscles.

In addition to voltage-activated and Ca^{2+}-activated single channel activities, currents flowing through a large conductance channel can be activated

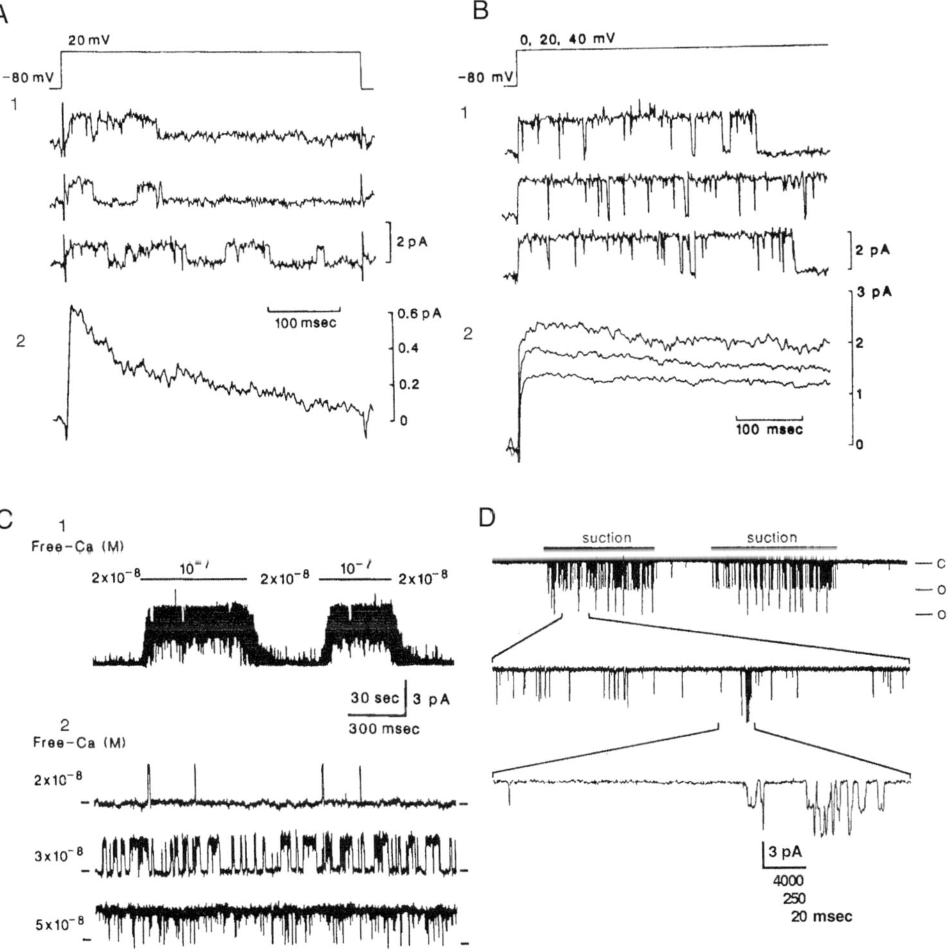

FIG. 2. Single channel activities in larval muscle membrane. (A) Patch clamp recordings of single channel currents corresponding to the macroscopic current, I_A. (1) Current traces in response to voltage pulses, showing brief channel openings and reduced opening probability toward the end of the pulse. (2) Ensemble current average from a large number of traces shown in 1. Note the decay of current over time reflecting channel inactivation. (B) Recordings of single channel currents corresponding to the macroscopic current, I_K. (1) Current traces in response to voltage pulses, showing prolonged channel opening during the pulse. (2) Ensemble average from a large number of single channel current traces. Note the relatively insignificant inactivation over time. (C) Recordings of single channel currents in inside-out cell-free patches corresponding to the macroscopic current I_{CF}. (1) A patch containing four active channels, displaying high levels of activity at 10^{-7} *M* Ca^{+2} but little activity at a concentration of 2×10^{-8} *M* Ca^{2+}. (2) Single channel activity in a different patch at three Ca^{2+} levels, demonstrating the steep Ca^{2+} dependence between 2 and 5×10^{-8} *M*. Note the flickering appearance of the single channel current. (D) Recording of stretch-activated single channel currents. Channel activity is increased on application of negative pressure to an inside-out patch displayed at three different time scales. The recording was obtained with a patch pipette containing high K^+ at 0 mV, hence the inward appearance of the currents. A, B, and C were reproduced from Komatsu *et al.*[52] and D from Gorczyca and Wu,[53] with permission.

by negative pressure applied to the patch pipette (Fig. 2D). These channels are frequently encountered during single channel recordings and represent stretch-activated K^+ channels in *Drosophila.*[53] However, their macroscopic counterpart has not been studied at the whole cell level and their functional significance is unknown.

Detailed analyses of voltage- and stretch-activated single channel activities have also been carried out in embryonic myocytes and myotubes.[54] In this study, single channels with properties similar to those shown in Fig. 2 are well characterized and categorized into A1, KD, and KST. The A1 channels are eliminated in some *Sh* mutants,[55] confirming their correspondence to the macroscopic I_A (cf. Fig. 2A). KD channels are most likely to correlate with the delayed rectifier component that is encoded by *Shab* (cf. Figs. 1C and 2B).

D. The Gene-Dosage Dependence of I_A

Two-electrode voltage clamp experiments on heteroallelic combinations of *Sh* mutants and strains carrying deficiencies or duplications of the *Sh* gene demonstrate the exquisite precision of gene-dosage dependence of the expression of I_A channels.[13] The *Sh* gene resides on the X chromosome and a duplication of the wild-type copy of the *Sh* gene doubles the amplitude of the current in male flies (Fig. 3A). This is exactly what is expected from the dosage compensation mechanism for genes on the X chromosome in male flies. In contrast, an additional wild-type copy of *Sh* increases the I_A amplitude by 50% in female flies, as predicted by the dosage compensation mechanism.[13]

Combining different *Sh* alleles in female heterozygotes results in heteromultimeric channels composed of two different forms of *Sh* subunits. Because of the multiple *Sh* subunits involved in a functional channel, strong dominant-negative effects can be seen in certain *Sh* mutant heterozygotes. For example, a mixture of Sh^{KS133} with the wild-type counterpart produces only about 1/8, instead of 1/2, of the wild-type I_A amplitude. Another mutant allele, Sh^M, which eliminates I_A in homozygotes, as does Sh^{KS133}, produces 1/2 of I_A when combined with the wild-type allele in heterozygotes, indicating that the Sh^M mutation is amorphic (true null)[56] in contrast to the antimorphic Sh^{KS133}.[13]

By the same token, duplication of a wild-type copy of *Sh* in male flies with the Sh^M background produces the same amount of I_A as in wild type, whereas the same duplication in male Sh^{KS133} produces only approximately 1/8 of the I_A in males carrying two wild-type copies of *Sh* (Fig. 3B). The same pattern of gene-dosage dependence of I_A has been confirmed in

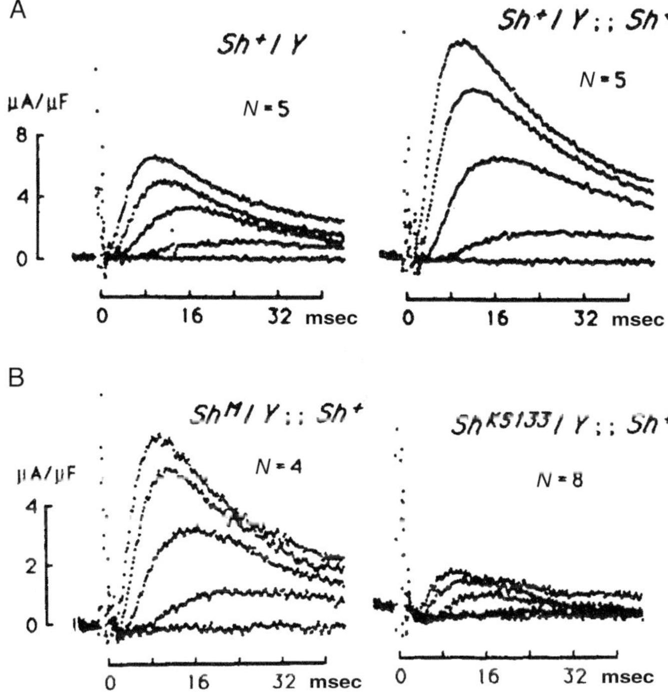

FIG. 3. Gene-dosage dependence of I_A in larval muscle fibers. (A) I_A in *Sh*$^+$ male larvae possessing a duplication of the wide-type (*Sh*$^+$) copy of the *Sh* gene (right) is nearly twice that of normal I_A in wild-type male larvae (left). (B) I_A in *Sh*^{M} male larvae possessing a duplication of *Sh*$^+$ is nearly normal (left), demonstrating the amorphic nature of the *Sh*^{M} mutation. I_A in *Sh*^{KS133} male larvae possessing a duplication of *Sh*$^+$ is nearly 25% of normal (right) demonstrating the dominant negative effect of the *Sh*^{KS133} mutation. Currents elicited by depolarizing steps to −40, −20, 0, 20, and 40 mV from a holding potential, V_h of −80 mV. Reproduced from Haugland and Wu,[13] with permission.

oocyte expression experiments using cRNAs of different *Sh* alleles.[57] Oocyte expression experiments using mixtures of varying proportions of two identified *Sh* polypeptides indicate that four *Sh* subunits most likely are involved in forming a functional channel.[58,59] However, from the gene-dosage dependence *in vivo,* it has been suggested that three *Sh* subunits plus one additional α subunit encoded by a different gene form the native I_A channels.[13,28,60]

Similar gene-dosage dependence studies have been carried out for the *slo* subunit in the formation of I_{CF} channels in heterozygotes. In this case, *slo*/+ heterozygotes produce nearly the same amount of I_{CF} as in wild-type larvae, indicating that other limiting factors, rather than *slo* subunits, determine the density of functional I_{CF} channels.[52]

E. The Roles of K^+ Current in Muscle Membrane Excitability

As mentioned earlier, the larval body wall muscles, as well as adult flight muscles, rely on Ca^{2+} influx for contraction. The excitability of the muscle membrane is crucial to the regulation of Ca^{2+} channel opening for Ca^{2+} influx. The different K^+ channels with distinct kinetic properties and activation mechanisms could, in principle, engender a rich variety of excitability patterns. Adult flight muscles are able to generate all-or-none action potentials and display different spike firing activities.[61–64] In contrast, larval body wall muscles seldom exhibit all-or-none action potentials, probably due to a different proportion of Ca^{2+} to K^+ channel densities. Only after increasing Ca^{2+} influx by raising the external Ca^{2+} concentration (Fig. 4A) or decreasing K^+ efflux by treatment with channel blockers such as TEA can prolonged Ca^{2+} action potentials be evoked by the synaptic potential at the neuromuscular junction or by applying depolarizing currents.[27,65] Differences between the two sets of muscles may reflect the different functional requirements: a slow contraction speed and the capability of supercontraction in larval, but not adult, skeletal muscles.

In normal saline[8] or the hemolymph-like saline HL3,[66] wild-type larval muscles normally display graded excitatory junctional potentials in response to nerve stimulation without action potentials. Perhaps the Ca^{2+} influx during the relatively long-lasting EJPs would better support the slow contraction in these muscles. The Ca^{2+} spikes found in adult flight muscles last a few milliseconds and can be reset for high frequency repetitive firing, presumably more suitable for the isometric, high speed contractions underlying flight. Larval muscles show graded potentials in response to current injection with indication of only a subthreshold regenerative Ca^{2+} spike (Fig. 4A). This is due to the strong repolarizing action of the transient I_A and the fast I_{CF} initiated at the beginning of the current injection. Elimination of I_A in *Sh* mutants allows the expression of Ca^{2+}-dependent regenerative potentials to a greater extent (Fig. 4B). With elimination of I_{CF} in *slo* muscle fibers, an even more pronounced Ca^{2+} spike could be evoked when current injection reaches the threshold level (Fig. 4C). More striking regenerative potentials could be observed when both I_A and I_{CF} are eliminated in *Sh slo* double mutants, similar to the phenomenon found in wild-type muscles treated with the channel blocker TEA.[27,65]

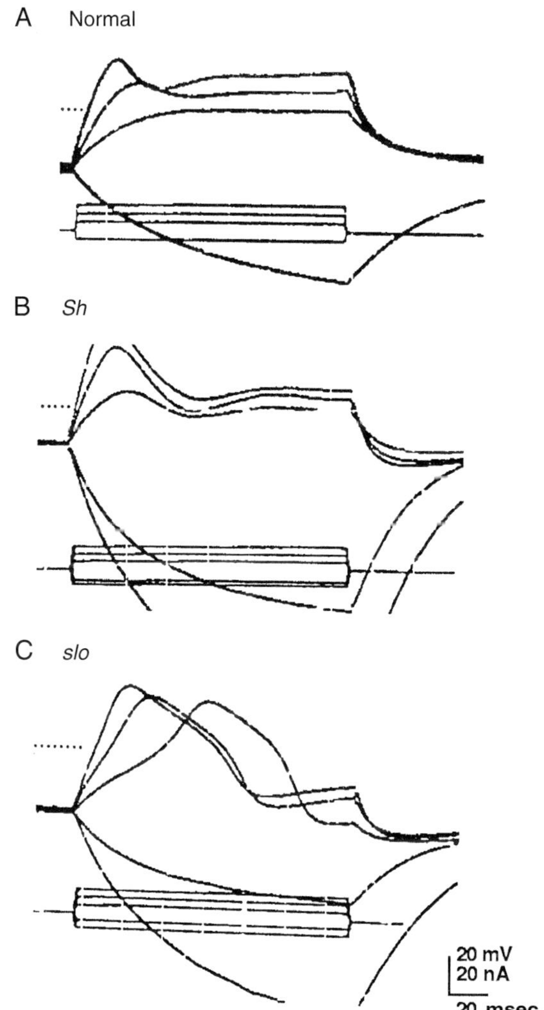

FIG. 4. Membrane potentials in response to current injection in saline containing 20 m*M* Ca^{2+}. Horizontal dotted lines represent the zero potential. Only graded responses to current injection are seen in wild-type (A) and *Sh* (B) muscle fibers. Note that elimination of I_A by *Sh* mutations enhances the depolarization in the early phase but exerts little effect on the later phase during the depolarizing current pulse. Elimination of I_{CF} in *slo* fibers gives rise to all-or-none Ca^{2+} action potentials with a clear threshold voltage (C). Reproduced from Singh and Wu,[27] with permission from The Company of Biologists, LTD.

F. Developmental Regulation and Second Messenger Modulation of K^+ Channels

It is well known that the physiological condition of excitable cells is modulated in order to respond to environmental challenges and to endow the nervous system plasticity for activity-dependent modification underlying conditioned behavior. The major targets of physiological regulation or second messenger-mediated modulation in the nervous system include ion channels.[67,68] The modulation of K^+ channels has been studied extensively in different species.[69] In *Drosophila,* many observed alterations in physiology or behavior induced by mutations of identified genes may be attributed to the long-term effects of the mutations on different biochemical pathways or cellular structures during development. There are ample examples in which the mutational effects are not completely mimicked by acute applications of pharmacological agents that act on the same biochemical pathways.[70]

The developmental conditions can alter the membrane excitability of muscle fibers through the regulation of K^+ currents in *Drosophila.*[71] This can be demonstrated directly by using *slo* larval muscles in which all-or-none Ca^{2+} spikes can be normally elicited. *slo* larvae raised at a higher temperature of 28°C show reduced membrane excitability attributable to reduced I_K. Ca^{2+} spikes initiated by current injection in *slo* larval muscle are reduced in amplitude and duration (Fig. 5A). Enhancement of delayed rectification, due to I_K, can be demonstrated in larvae of different genotypes (Fig. 5B). It will be important to determine whether the expression of I_K channels is enhanced or the conductance of individual channels is increased by second messenger-triggered modulation. This phenomenon can

---

Fig. 5. Developmental regulation and cAMP cascade-mediated K^+ channel modulation. (A) Effects of developmental temperature on muscle action potentials recorded from *slo* larvae in saline containing 20 m*M* Ca^{2+}. Note the full-blown action potentials in larvae raised at 18°C but not at 18°C. (B) Enhancement of I_K in wild-type larvae raised at 28°C as compared to the current amplitude seen in larvae raised at 18°C. (C) Enhancement of I_A by *dnc* mutations. Both dnc^1 and dnc^2 mutations increase the amplitude of I_A without altering its kinetic properties. (D) Enhancement of I_K by *dnc* mutations. The amplitude, but not kinetics of I_K, is altered by dnc^1 and dnc^2 mutations. (E) Effects of W7 on the total current in saline containing 20 m*M* Ca^{2+}. dnc^{M14} and rut^1 mutations do not appear to exert a striking effect on I_{CF} or I_{CS} but clearly alter their responses to treatment with W7. *dnc* mutations differentially enhance the response of I_{CF} and *rut* mutations preferentially increase the response of I_{CS}. Current traces in response to depolarizing steps from a V_h of −80 mV to voltages of −40 to 30 mV in increments of 10 mV (B), to voltages indicated (C, D), or to 10 mV (E). A and B were reproduced from Chopra and Singh[71] and C, D, and E from Zhong and Wu,[70] with permission.

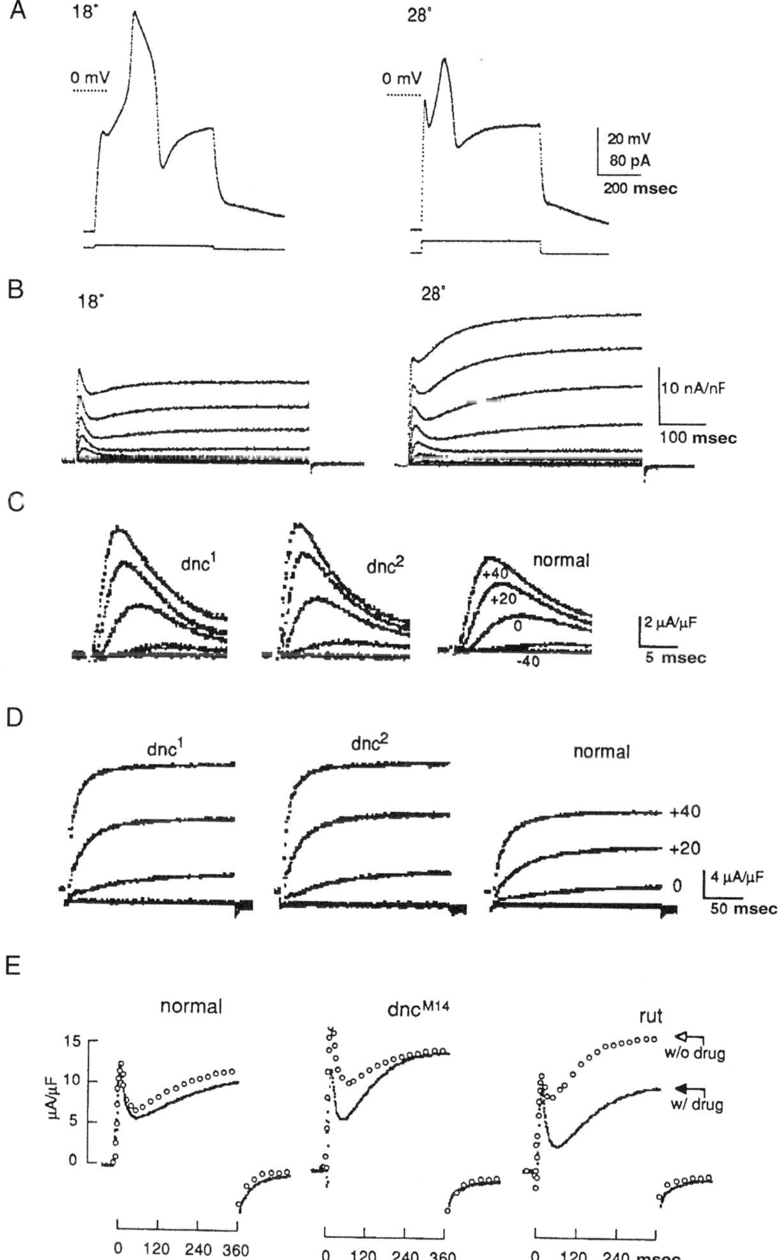
A
18°
28°
0 mV
0 mV
20 mV
80 pA
200 msec
B
18°
28°
10 nA/nF
100 msec
C
dnc1
dnc2
normal
+40
+20
0
-40
2 μA/μF
5 msec
D
dnc1
dnc2
normal
+40
+20
0
4 μA/μF
50 msec
E
normal
dncM14
rut
w/o drug
w/ drug
μA/μF
15
10
5
0
0 120 240 360
0 120 240 360
0 120 240 msec

be studied using mutants of second messenger systems (see, e.g., Refs. 70, 72, and 73) and can be examined at the single channel level.

K^+ channel modulation by the cyclic AMP (cAMP) cascade has been studied in *Drosophila* utilizing the mutations affecting the enzymes regulating cAMP metabolism. Mutations of the *dunce* (*dnc*) gene, which encodes the cAMP degradation enzyme, phosphodiesterase, lead to higher levels of cAMP, whereas mutations of the *rutabaga* (*rut*) gene, which encodes the cAMP synthesis enzyme adenylyl cyclase, result in lowered levels of cAMP.[74,75] Both I_A and I_K are enhanced in larvae of different *dnc* alleles (Figs. 5C and 5D). Interestingly, these effects are not mimicked by acute treatment with permeant cAMP analogs in wild-type muscles. The Ca^{2+}-activated K^+ currents, I_{CS} and I_{CF} in *dnc* and *rut* examined in saline containing high Ca^{2+}, do not show striking defects, but their responses to drugs affecting second messenger pathways are greatly altered.[70] For example, the response to caffeine, which increases intracellular-free Ca^{2+}, is enhanced in both *dnc* and *rut*. The drug W7, which preferentially affects the Ca^{2+}/calmodulin-dependent kinase (CaMKII), differentially enhances I_{CF} in *dnc,* but I_{CS} in *rut* (Fig. 5E). The striking effects of these drugs on larval muscle K^+ currents argue for the possibilities of modulation mechanisms involving CaMKII, cGMP-dependent protein kinase (PKG), and protein kinase C. Mutations affecting these second messenger pathways are available in *Drosophila* and their effect on K^+ currents can be examined in detail.

More recently, modulation of K^+ channels initiated by peptides has been demonstrated in *Drosophila* larval muscle.[76] A striking increase in the outward K^+ currents was found to be induced by high frequency stimulation of the segmental nerve.[77] A similar phenomenon was observed to be induced by a PACAP-like neuropeptide, which is mediated by the coactivation of the cAMP and the Ras signal transduction pathways.[76] Interestingly, a Ras-specific GTP-activated protein, analogous to the human neurofibromatosis type 1 (NF1), is involved in this pathway of strong enhancement of K^+ currents.[78]

G. The Functional Roles of Modulatory and Auxiliary Subunits of K^+ Channels

In addition to *Sh* and *slo,* which encode the α subunits of two different classes of K^+ channels, additional genes that code for K^+ channel subunits, including *ether-a-go-go* (*eag*) and *Hyperkinetic* (*Hk*), have been identified[25] and molecularly characterized[32,33] in *Drosophila.* The *eag* polypeptide is homologous to but distinct from the *Sh* subunit. It consists of the hallmark of six transmembrane domains and a pore-forming region found in different splicing variants of the *Sh* and *slo* subunits. However, *eag* is unique in its C terminus, which contains many putative sites for protein kinase phosphorylation and a cyclic nucleotide-binding site.[33] This structural feature suggests that the *eag* subunit may be important in channel modulation. Indeed, mutations in the *eag* gene alter or eliminate the modulation of different K^+ currents in larval muscles.[28]

Unlike *Sh* or *slo* mutations, *eag* mutations affect all four K^+ currents with strong temperature dependence and allele specificity.[14] Significantly, deletion of the *eag* gene only reduces these currents but does not eliminate any of them.[14] As mentioned earlier, cGMP greatly enhances I_K in wild type but the effect is eliminated in *eag* mutants (Fig. 6A). Similarly, W7, a CaMKII blocker, strongly suppresses both I_{CF} and I_{CS} in wild type, whereas the effects are again lacking in *eag* alleles (Fig. 6B). Different *eag* alleles also show altered response to the action of W7 on I_A (Fig. 6C) to different extents in an allele-dependent manner.

Although the *eag* subunit has been shown to form homomultimeric channels in the oocyte expression system,[79] *in vivo* results, described earlier, indicate that it may participate in channel assembly as a shared subunit common to different K^+ channels.[28,60] This notion is further supported by the observation that combinations of different *eag* and *Sh* alleles in double mutants demonstrate allele-dependent interactions, indicated by new phenotypes not present in either single mutant.[28] It is likely that *eag* coassembles with other channel α subunits to form heteromultimeric channels in which the *eag* subunit confers the effects of second messenger-mediated modulation. Indeed, the C terminus of the *eag* polypeptide has been shown to be an excellent substrate for phosphorylation by CaMKII.[72] Furthermore, coexpression of the *eag* and *Sh* subunits in *Xenopus* oocytes produces novel K^+ currents that cannot be explained by summation of the currents mediated by separate homomultimeric *eag* and *Sh* channels.[80] These novel K^+ currents are likely to be mediated by heteromultimeric channels.

The *Hk* gene codes for an auxiliary β subunit, which is not an integral membrane protein but may coassemble with other α subunits to modify channel expression or function.[32] Examination of larval K^+ channels demonstrates that only I_A, but not other K^+ currents (I_K, I_{CF}, or I_{CS}), is affected by *Hk* mutations.[81] Significantly, elimination of the *Hk* gene alters properties of I_A without deleting the current. The modulatory role of *Hk* is evident because the most pronounced effects on I_A are found at lower membrane potentials near the threshold levels of channel activation. This is important for the maintenance of the quiescent state of the excitable cell because the initiation of a regenerative event could be adjusted by the action of *Hk*. Supporting evidence for this notion is obtained from the mutant effect on neuronal activity[82] and the underlying K^+ currents.[83]

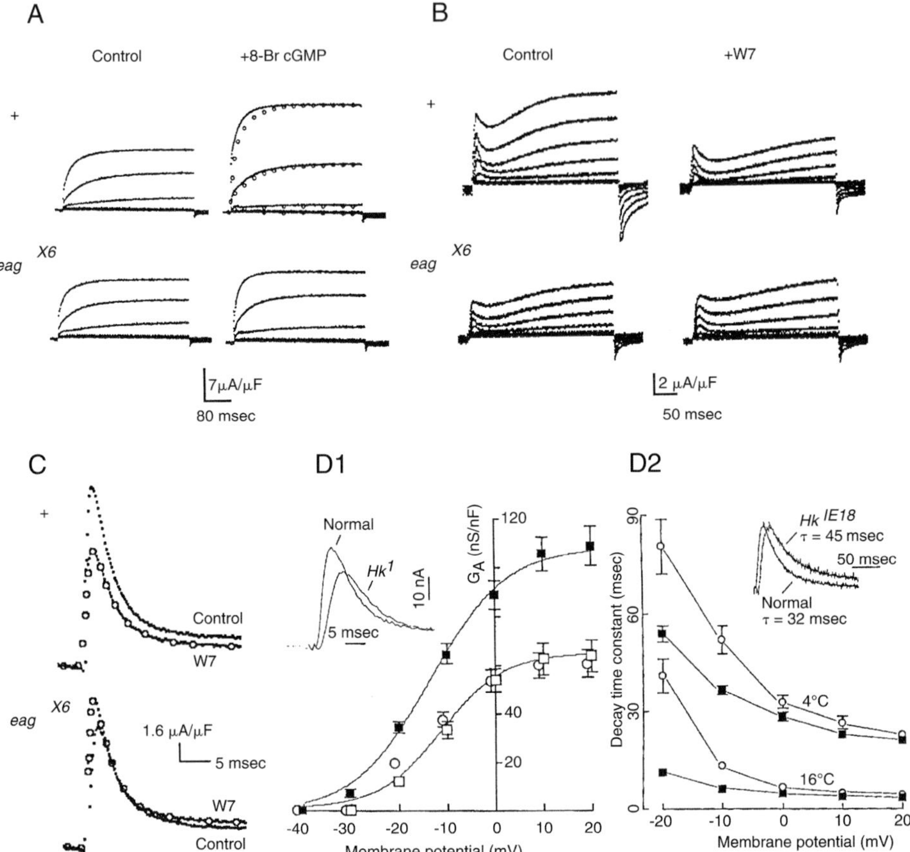

FIG. 6. Alterations in channel modulation in muscle fibers of *eag* and *Hk* larvae. (A) The membrane permeant analog of cGMP greatly enhances I_K in wild-type muscle but the effect is absent in larvae carrying eag^{x6}, a null mutation. Depolarizing steps to −40, −20, 0, 20, and 40 mV from a V_h of −80 mV. (B) W7, which preferentially affects CaMKII, strongly suppresses both I_{CF} and I_{CS} in wild-type fibers in saline containing 0.9 m*M* Ca^{2+}. The eag^{x6} mutation decreases both I_{CF} and I_{CS} and the modulatory effect of W7 becomes minimal. Voltage steps to −40 to 20 mV in 10-mV increments. (C) W7 suppresses I_A in wild-type fibers but slightly increases I_A in eag^{x6} fibers. Currents elicited by depolarization to 20 mV. (D) *Hk* mutations modify the properties of I_A in larval muscle fibers. (1) The muscle membrane conductance G is reduced in both Hk^1 (□) and Hk^{IE18} (○) as compared to wild type (■). (2) Alterations in inactivation kinetics of I_A by *Hk* mutations depend on voltage and temperature. The decay time constant of I_A is increased in Hk^{IE18} fibers (○), which is proportionally more extreme at lower membrane voltages and higher temperatures, as compared to wild type fibers (■). A, B, and C were reproduced from Zhong and Wu[28] and D from Wang and Wu,[81] with permission.

The *Hk* polypeptide is homologous to aldose reductase, suggesting that it may be involved in the regulation of the oxidation–reduction state of the channel. Interestingly, evidence suggests that chemicals modifying the oxidation–reduction state of *Sh* channels expressed in *Xenopus* oocytes, especially at a methionine residue in the N terminus, greatly affect the inactivation mechanism of the channels.[84] The *Hk* subunit is potentially capable of interacting with other types of K^+ channel subunits besides *Sh*. Coexpression of *Hk* with *eag* in *Xenopus* oocytes has been shown to generate K^+ currents different in properties compared to those mediated by homomultimeric *eag* channels, even though expression of *Hk* alone does not produce functional channels.[85]

III. Calcium Currents

Like K^+ channels, there are many types of Ca^{2+} channels that express in neurons and muscles. Ca^{2+} channels play an important role in many cell functions. They have a particularly significant role to play in insect muscles where influx of Ca^{2+} is required for muscle contraction.[22]

Existence of Ca^{2+} channels in larval muscles was initially inferred by analyzing action potentials[86] and by examining Sr^{2+} current.[11] Recordings from embryonic neurons[87–89] and brain membrane preparations[90] showed that pharmacology of Ca^{2+} channels in *Drosophila* may be different from pharmacology of vertebrate Ca^{2+} channels. However, recordings of Ca^{2+} channel currents in the larval muscles[91] and experiments on the heart rate[92] in *Drosophila* show a considerable overlap in the pharmacology of the *Drosophila* and the vertebrate Ca^{2+} channels. Voltage-dependent Ca^{2+} channels of the body wall muscles of larvae have been resolved into two subtypes.[91] One subtype shares pharmacological properties with the vertebrate L-type Ca^{2+} channels and one with the vertebrate T-type Ca^{2+} channels. Although more comprehensive biophysical analysis of the two larval currents requires further investigation, some interesting information is available on them.

The most studied among the vertebrate Ca^{2+} channels are the L-type channels. These channels activate at a high threshold and inactivate slowly. The L-type current in the larval muscles of *Drosophila*[91] also activates at similarly high voltages and inactivates slowly. As the cell is depolarized more and more, the current increases as more channels open. When the depolarization approaches the reversal potential (30–40 mV), the current starts decreasing (Figs. 7A and 7B). The maximum L-type current in the larval muscles is observed in the range of -10 to 0 mV. Vertebrate L-type

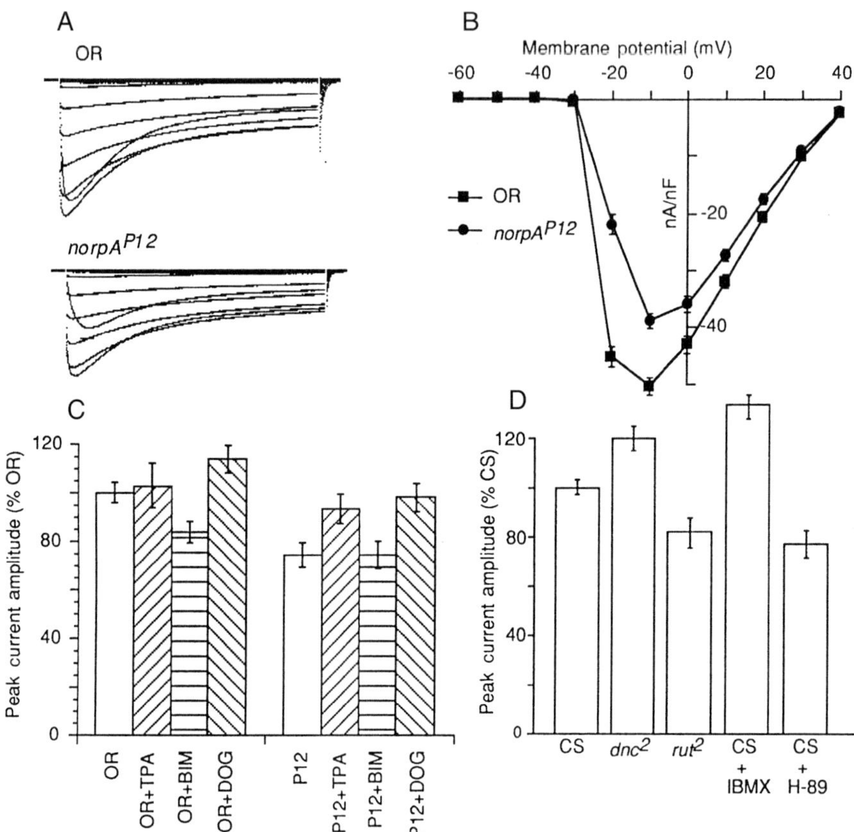

FIG. 7. Modulation of L-type Ca^{2+} channels via phospholipase C–DAG–PKC and cAMP–PKA pathways. (A and B). The *norpA*^{P12} mutation reduces the L-type current. Traces show currents through Ca^{2+} channels measured by using Ba^{2+} as the charge carrier. Current traces and I/V plots are shown in response to voltage steps from a V_h of −40 mV to voltages of −60 to 40 mV in increments of 10 mV. (C) The *norpA*^{P12} mutant current can be rescued by TPA or DOG, whereas the wild-type (OR) current is reduced by BIM. (D) The L-type current is reduced in *rut*^{2} mutants and is enhanced in *dnc*^{2} mutants. IBMX increases the wild-type (CS) current while H-89 decreases the current. A, B, and C were reproduced from Gu and Singh,[73] with permission. D shows unpublished data from Bhattacharya and Singh.

channels are generally blocked by 1,4-dihydropyridines (DHPs, e.g., nifedipine), phenylalkylamines (e.g., verapamil), and benzothiazepines (e.g., diltiazem).[93] The L-type current in *Drosophila* larval muscles has a slightly different pharmacological profile. It is blocked by PN200-110, nifedipine, and nitrendipine (DHPs) and by diltiazem (a benzothiazepine). Another hallmark of the vertebrate L-type current is the agonistic effect of DHPs such as Bay K-8644. Similarly, the DHP-sensitive current in larval muscles shows a dramatic increase in the tail currents in response to Bay K-8644. However, the L-type current in larval muscles of *Drosophila* is much less sensitive to verapamil (a phenylalkylamine) than the L-type current in vertebrates.

The L-type channels in larval body wall muscles also differ from related channels in other tissues of *Drosophila*. Blockade of larval muscle channels by DHPs and their relative insensitivity to verapamil[91] contrast with the resistance of most L-type channels from brain membrane preparations to DHPs and their very high sensitivity to phenylalkylamines.[90] It will be very interesting to investigate the evolutionary relationship and functional significance of the subtypes of L-type channels.

In addition to the L-type current, larval muscles of *Drosophila* show another component of the Ca^{2+} channel current that is related to the vertebrate T-type current. Fewer specific blockers are available for vertebrate T-type channels, amiloride being among the more selective ones.[94–96] In the larval muscles of *Drosophila*, 1 m*M* amiloride almost completely blocks the T-type related current.[91] As in the case of vertebrate T-type Ca^{2+} channels, amiloride-sensitive channels in *Drosophila* are inactivated by holding the membrane at −30 or −40 mV.[73,91] This provides a convenient way to physiologically resolve the L-type current in the larval muscles.

In vertebrate preparations, the T-type current generally activates at low potentials in the range of −70 to −50 mV.[97,98] *Drosophila* Ca^{2+} channels characterized so far from muscles as well as neurons, including amiloride-sensitive Ca^{2+} channels, all activate at higher thresholds.[87–89,91] In addition, the amiloride-sensitive current from *Drosophila* muscles shows slower inactivation than the vertebrate T-type current. This presents an interesting correlation to the slow contraction speed of the larval muscles. It will be instructive to investigate evolutionary and functional significance of this diversity of T-type related channels.

Two Ca^{2+} channel α_1 subunits, Dmca1D[99] and Dmca1A,[100] have been cloned from *Drosophila*. Like the voltage-gated Ca^{2+} channels from vertebrates, *Drosophila* channels show four repeat structures with six transmembrane domains each. Null mutations in the gene *l(2)35Fa*, which codes for the Dmca1D subunit, cause recessive embryonic lethality. A missense mutation in the gene reduces the L-type Ca^{2+} current in larval muscles, leaving the amiloride-sensitive current intact.[101] This mutation identifies the gene coding for the L-type channels in the larval muscles. The gene coding for Dmca1A was earlier identified by a courtship-song mutation *cacophony*.[102] Identification of these two Ca^{2+} channel genes and analysis of their splice variants and differential expression will be very helpful in the role of Ca^{2+} channels in larval muscles and other excitable tissues in *Drosophila*.

A. Modulation of Ca^{2+} Currents

The significance of ion channels as targets of modulation by second messenger system pathways has already been discussed in Section II. The modulation of *Drosophila* Ca^{2+} channels via these pathways assumes considerable importance due to the role of these channels in cellular physiology and pathology in general[103,104] and their role in insect muscle contraction in particular. Mutations that disrupt specific steps in signal transduction pathways help greatly in determining the role of these pathways in the modulation of channels. Among the available mutations that affect second messenger system pathways in *Drosophila,* mutations in cAMP and phospholipase C pathways have been examined for their effect on L-type Ca^{2+} channels.

Mutations in the *norpA* gene greatly reduce phospholipase C-β and have been very helpful in elucidating the role of phospholipase C in visual[105–107] and olfactory[108] transduction in *Drosophila.* The role of phospholipase C-mediated pathways in modulating *Drosophila* L-type channels has now been examined with the help of *norpA* mutations and pharmacological agents that disrupt steps in the phosphoinositide cascade. Expression of the *norpA*-coded phospholipase C, originally found in heads, has subsequently been observed in other regions of the body, including legs, thorax, and abdomen.[109]

It has generally been difficult to assign a specific phospholipase C isozyme to a specific signaling pathway.[109,110] Because of the specificity of genetic mutations, these experiments show that the same phospholipase C isozyme that mediates phototransduction and olfaction via IP3[108,111] also mediates the modulation of L-type Ca^{2+} channels in larval muscles of *Drosophila* via DAG-PKC. *norpA* mutations reduce the L-type current in the larval muscles (Figs. 7A and 7B). Since phospholipase C converts phosphatidylinositol 4,5-bisphosphate (PIP_2) into inositol trisphosphate (IP_3) and DAG (DAG in turn activating PKC), the effect of phospholipase C on the current could be mediated via IP_3, DAG, or both. Application of PKC activators such as phorbol 12-myristate 13-acetate (TPA) (Fig. 7C) and phorbol 12,13-didecanoate (PDD, but not its inactive analog 4αPDD) rescue the current in mutant fibers without significantly affecting the normal current. Bisindolylmaleimide (BIM), an inhibitor of PKC, reduces the current in normal fibers without affecting the mutant current (Fig. 7C). An analog of DAG, sn-1,2-dioctanoyl-glycerol (DOG), increases the current in mutant fibers (Fig. 7C). These experiments suggest that the DHP-sensitive Ca^{2+} channels in *Drosophila* may be modulated by the phospholipase C–DAG–PKC pathway.

Similar experiments have been performed with mutations and drugs that disrupt the cAMP cascade (Bhattacharya and Singh, unpublished).

The *dnc* and the *rut* mutations, discussed earlier, disrupt phosophodiesterase and adenylyl cyclase, respectively, increasing and decreasing the levels of cellular cAMP. The L-type current is increased in *dnc* larvae and reduced in *rut* larvae (Fig. 7D). The effect of the *dnc* mutation can be mimicked by 3-isobutyl-1-methylxanthine (IBMX), an inhibitor of phosphodiesterase (Fig. 7D), and the effect of *rut* can be rescued by forskoline, which activates adenylyl cyclase. These experiments suggest that the L-type current in the larval muscles may be modulated via the cAMP mediated pathway. The role of protein kinase A (PKA) in modulation is suggested by the use of *N*-(2-[(*p*-bromocinnamyl)amino]ethyl)-5-isoquinolinesulfonamide (H-89), which is an inhibitor of PKA. H-89 reduces the current in wild-type muscles to the level of the current in *rut* muscles (Fig. 7D). The cAMP pathway plays an important role in regulating the contraction of insect muscles.[112,113] This provides a context to study the functional significance of the modulation of L-type channels described earlier, and of the T-type channels, which await further investigation.

IV. Conclusions

The larval muscle preparation provides an excellent *in situ* assay system for the studies of ion channels in *Drosophila.* This review demonstrates that the function of the native K^+ and Ca^{2+} channels can be analyzed with the advanced biophysical and pharmacological techniques that have been applied to other systems. The physiological and biophysical properties of ion channels can be analyzed from the single channel to the whole cell level, and the functional roles of different classes of ion channels can be directly elucidated in *Drosophila* muscles because of the available mutants.

Since the first cloning and sequencing of the *Sh* gene,[29–31] a number of genes coding for other families of K^+ channels[32–34,43] have been first identified in *Drosophila,* allowing a reconstruction of their evolutionary relationships.[3] Many of the K^+ channel subunits have been analyzed in exquisite details using heterologous expression systems. However, the subunit stoichiometry, functional modulation, and cellular roles of the native channels require *in situ* characterization in a well-defined *in vivo* system. The *Drosophila* larval preparation is likely to make further contributions in these areas of channel research.

The structure and function of the Ca^{2+} channels in larval muscle are being unraveled. The interaction of inward Ca^{2+} and outward K^+ currents in the control of membrane excitability and Ca^{2+}-dependent cell regulation still awaits further exploration. The modulation and developmental regulation, as well as the gene-dosage dependence and subunit composition of Ca^{2+} channels, remain to be uncharted territory in *Drosophila*. It will be important to determine how the K^+ channels and Ca^{2+} channels are coregulated in this system to meet the functional requirements. The newly discovered peptide regulation of K^+ channel functions in this system opens a door to the hormone- and transmitter-controlled cell functions that are mediated by ion channels and related second messenger systems.[76,78]

There are several important lines of research that can be further explored in this preparation. These include the pre- and postsynaptic roles of the various Ca^{2+} and K^+ channels in the efficacy and plasticity of synaptic transmission[114–116] and the interactions among these channels for regulating the excitation–contraction coupling for the initiation of muscle contraction. Similarly, the functional interaction of ion channels with other categories of membrane-associated and cytoskeletal proteins can be determined in the larval muscle preparation. Different categories of mutations are now available in *Drosophila,* including those affecting proteins involved in cell movement,[116] membrane recycling, and channel and receptor aggregation,[117] to help demonstrate the functional roles of ion channels in a broader context.

References

1. Wu, C.-F., and Ganetzky, B. (1992). Neurogenetic studies of ion channels in *Drosophila*. *In* "Ion Channels" (T. Narahashi, ed.), Vol. 3, pp. 261–314. Plenum Press, New York.
2. Jegla, T., and Salkoff, L. (1994). Molecular evolution of K^+ channels in primitive eukaryotes. *Soc. Gen. Physiol. Ser.* **49,** 213–222.
3. Jan, L. Y., and Jan, Y. N. (1997). Cloned potassium channels from eukaryotes and prokaryotes. *Annu. Rev. Neurosci.* **20,** 91–123.
4. Saito, M., and Wu, C.-F. (1993). Ionic channels in cultured *Drosophila* neurons. *In:* Pichon Y, ed. "Comparative Molecular Neurobiology" (Y. Pichon, ed.), pp. 366–389, Birkhauser, Basel.
5. Schwarz, T. L., Papazian, D. M., Carreto, R. C., Jan, Y. N., and Jan, L. Y. (1990). Immunological characterization of K^+ channel components from *Shaker* locus and differential distribution of splicing variants in *Drosophila*. *Neuron* **2,** 119–127.
6. Mottes, J. R., and Iverson, L. (1995). Tissue-specific alternative splicing of hybrid *Shaker/lacZ* genes correlates with kinetic differences in *Shaker* K^+ currents in vivo. *Neuron* **14,** 613–623.
7. Rogero, O., Hammerle, B., and Tejedor, F. J. (1997). Diverse expression and distribution of *Shaker* potassium channels during the development of the *Drosophila* nervous system. *J. Neurosci.* **17,** 5108–5118.
8. Jan, L. Y., and Jan, Y. N. (1976). Properties of the larval neuromuscular junction in *Drosophila melanogaster*. *J. Physiol.* **262,** 189–214.

9. Ueda, A., and Kidokoro, Y. (1996). Longitudinal body wall muscles are electrically coupled across the segmental boundary in third instar larvae of *Drosophila melanogaster. Invertebr. Neurosci.* **1,** 315–322.
10. Wu, C.-F., Ganetzky, B., Haugland, F. N., and Liu, A.-X. (1983). Potassium currents in *Drosophila:* Different components affected by mutations of two genes. *Science* **220,** 1076–1078.
11. Wu, C.-F., and Haugland, F. N. (1985). Voltage clamp analysis of membrane currents in larval muscle fibers of *Drosophila:* Alteration of potassium currents in *Shaker* mutants. *J. Neurosci.* **5,** 2626–2640.
12. Singh, S., and Wu, C.-F. (1989). Complete separation of four potassium currents in *Drosophila. Neuron* **2,** 1325–1329.
13. Haugland, F. N., and Wu, C.-F. (1990). A voltage-clamp analysis of gene-dosage effects of the *Shaker* locus on larval muscle potassium currents in *Drosophila. J. Neurosci.* **10,** 1357–1371.
14. Zhong, Y., and Wu, C.-F. (1991). Alteration of four identified K$^+$ currents in *Drosophila* muscle by mutations in *eag. Science* **252,** 1562–1564.
15. Hagiwara, S., and Byerly, L. (1981). Calcium channel. *Annu. Rev. Neurol.* **4,** 69–125.
16. Ashcroft, F. M., and Stanfield, P. R. (1982). Calcium and potassium currents in muscle fibers of an insect (*Carausius morosus*). *J. Physiol.* **323,** 93–115.
17. Schwartz, L. M., and Stuhmer, W. (1984). Voltage-dependent sodium channels in an invertebrate striated muscle. *Science* **225,** 523–525.
18. Bate, M. (1993). The mesoderm and its derivatives. *In* "The Development of *Drosophila melanogaster*" (M. Bate and A. M. Arias, eds.), pp. 1013–1090. Cold Spring Harbor Laboratory Press, Cold Spring Harbor, NY.
19. Wei, A., and Salkoff, L. (1986). Occult *Drosophila* calcium channels and twinning of calcium and voltage-activated potassium channels. *Science* **233,** 780–782.
20. Elkins, T., Ganetzky, B., and Wu, C.-F. (1986). A *Drosophila* mutation that eliminates a calcium-dependent potassium current. *Proc. Natl. Acad. Sci. USA* **83,** 8415–8419.
21. Elder, H. Y. (1975). Muscle structure. *In* "Insect Muscle" (P. N. R. Usherwood, ed.), pp. 1–74. Academic Press, New York.
22. Aidley, D. J. (1989). "The Physiology of Excitable Cells." Cambridge University Press, Cambridge.
23. Gho, M., and Mallart, A. (1986). Two distinct calcium-activated potassium currents in larval muscle fibers of *Drosophila melanogaster. Pfueg. Arch.* **407,** 526–533.
24. Hodgkin, A. L., and Huxley, A. F. (1952). Currents carried by sodium and potassium ion through the membrane of the giant axon of *Loligo. J. Physiol.* **116,** 449–472.
25. Kaplan, W. D., and Trout, W. E. (1969). The behavior of four neurological mutants of *Drosophila. Genetics* **61,** 399–409.
26. Salkoff, L., and Wyman, R. J. (1983). Ion currents in *Drosophila* flight muscles. *J. Physiol.* **337,** 687–709.
27. Singh, S., and Wu, C.-F. (1990). Properties of potassium currents and their role in membrane excitability in *Drosophila* larval muscle fibers. *J. Exp. Biol.* **152,** 59–76.
28. Zhong, Y., and Wu, C.-F. (1993). Modulation of different K$^+$ currents in *Drosophila:* A hypothetical role for the eag subunit in multimeric K$^+$ channels. *J. Neurosci.* **13,** 4669–4679.
29. Kamb, A., Iverson, L. E., and Tanouye, M. A. (1987). Molecular characterization of *Shaker,* a *Drosophila* gene that encodes a potassium channel. *Cell* **50,** 405–413.
30. Papazian, D. M., Schwarz, T. L., Tempel, B. L., Jan, Y. N., and Jan, L. Y. (1987). Cloning of genomic and complementary DNA from *Shaker,* a putative potassium channel gene from *Drosophila. Science* **237,** 749–753.

31. Pongs, O., Kecskemethy, N., Muller, R., Krah-Jentgens, I., Baumann, A., Kiltz, H. H., Canal, I., Liamazares, S., and Ferrus, A. (1988). *Shaker* encodes a family of putative potassium channel proteins in the nervous system of *Drosophila. EMBO J.* **7,** 1087–1096.
32. Chouinard, S. W., Wilson, G. F., Schlimgen, A. K., and Ganetzky, B. (1995). A potassium channel beta subunit related to the aldo-keto reductase superfamily is encoded by the *Drosophila Hyperkinetic* locus. *Proc. Natl. Acad. Sci. USA* **92,** 6763–6767.
33. Warmke, J., Drysdale, R., and Ganetzky, B. (1991). A distinct potassium channel polypeptide encoded by the *Drosophila eag* locus. *Science* **252,** 1560–1562.
34. Atkinson, N. S., Robertson, G. A., and Ganetzky, B. (1991). A component of calcium-activated potassium channels encoded by *Drosophila solo* locus. *Science* **253,** 551–555.
35. Kamb, A., Tseng-Crank, J., and Tanouye, M. A. (1988). Multiple products of the *Drosophila Shaker* gene may contribute to potassium channel diversity. *Neuron* **1,** 421–430.
36. Schwarz, T. L., Tempel, B. L., Papazian, D. M., Jan, Y. N., and Jan, L. Y. (1988). Multiple potassium-channel components are produced by alternative splicing at the *Shaker* locus of *Drosophila. Nature* **331,** 137–142.
37. Salkoff, L., and Wyman, R. J. (1981). Genetic modification of potassium channels in *Drosophila Shaker* mutants. *Nature* 293, 228–230.
38. Hoshi, T., Zagotta, W. N., and Aldrich, R. W. (1990). Biophysical and molecular mechanisms of *Shaker* potassium channel inactivation. *Science* **250,** 533–538.
39. Isacoff, E. Y., Jan, Y. N., and Jan, L. Y. (1991). Putative receptor for the cytoplasmic inactivation gate in the *Shaker* K^+ channel. *Nature* **353**(6339), 86–90.
40. Iverson, L. E., Tanouye, M. A., Lester, H. A., Davidson, N., and Rudy, B. (1988). A-type potassium channels expressed from *Shaker* locus cDNA. *Proc. Natl. Acad. Sci. USA* **85,** 5723–5727.
41. Timpe, L. C., Schwarz, T. L., Tempel, B. L., Papazian, D. M., Jan, Y. N., and Jan, L. Y. (1988). Expression of functional K^+ channels from *Shaker* cDNA in Xenopus oocyte. *Nature* **331,** 143–145.
42. Iverson, L. E., and Rudy, B. (1990). The role of the divergent amino and carboxyl domains on the inactivation properties of potassium channels derived from the *Shaker* gene of *Drosophila. J. Neurosci.* **10,** 2903–2916.
43. Butler, A., Wei, A., Baker, K., and Salkoff, L. (1989). A family of putative potassium channel genes in *Drosophila. Science* **243,** 943–947.
44. Wei, A., Covarrubias, M., Butler, A., Baker, K., Pak, M., and Salkoff, L. (1990). K^+ current diversity is produced by an extended gene conserved in *Drosophila* and mouse. *Science* **248,** 599–601.
45. Tsunoda, S., and Salkoff, L. (1995). The major delayed rectifier in both *Drosophila* neurons and muscle is encoded by *Shab. J. Neurosci.* **15,** 5209–5221.
46. Wu, C.-F., Tsai, M.-C., Chen, M.-L., Zhong, Y., Singh, S., and Lee, C. Y. (1989). Actions of dendrotoxin on K^+ channels and neuromuscular transmission in *Drosophila melanogaster,* and its effects in synergy with K^+ channel-specific drugs and mutations. *J. Exp. Biol.* **147,** 21–41.
47. Kraliz, D., and Singh, S. (1997). Selective blockade of the delayed rectifier potassium current by tacrine in *Drosophila. J. Neurobiol.* **32,** 1–10.
48. Kraliz, D., Bhattacharya, A., and Singh, S. (1998). Blockade of the delayed rectifier potassium current in *Drosophila* by quinidine and related compounds. *J. Neurogenet.* **12,** 25–39.
49. Pichon, Y., and Ashcroft, F. M. (1985). Nerve and muscle: Electrical activity. *In* "Comprehensive Insect Physiology, Biochemistry and Pharmacology" (G. A. Kerkut and L. L. Gilbert, eds.), pp. 85–114. Pergamon Press, New York.

50. Aidley, D. J. (1975). Excitation-contraction coupling and mechanical properties. *In* "Insect Muscle" (P. N. R. Usherwood, ed.), pp. 337–356. Academic press, New York.
51. Lagrutta, A., Shen, K., North, R. A., and Adelman, J. P. (1994). Functional differences among alternatively spliced variants of *slowpoke,* a *Drosophila* calcium-activated potassium channel. *J. Biol. Chem.* **269,** 20347–20351.
52. Komatsu, A., Singh, S., Rathe, P., and Wu, C.-F. (1990). Mutational and gene dosage analysis of calcium activated potassium channels in *Drosophila:* Correlation of micro- and macroscopic currents. *Neuron* **4,** 313–321.
53. Gorczyca, M., and Wu, C.-F. (1991). Single-channel K^+ currents in *Drosophila* muscle and their pharmacological block. *J. Membr. Biol.* **121,** 237–248.
54. Zagotta, W. N., Brainard, M. S., and Aldrich, R. W. (1988). Single-channel analysis of four distinct classes of potassium channels in *Drosophila* muscle. *J. Neurosci.* **8,** 4766–4779.
55. Solc, C. K., Zagotta, W. N., and Aldrich, R. W. (1987). Single-channel and genetic analyses reveal two distinct A-type potassium channels in *Drosophila. Science* **236,** 1094–1098.
56. Zhao, M.-L., Sable, E. O., Iverson, L. E., and Wu, C.-F. (1995). Functional expression of *Shaker* K^+ channels in cultured *Drosophila* "Giant" neurons derived from *Sh* cDNA transformants: Distinct properties, distribution, and turnover. *J. Neurosci.* **15**(2), 1406–1418.
57. Lichtinghagen, R., Stocker, M., Wittka, R., Boheim, G., Stuhmer, W., Ferrus, A., and Pongs, O. (1990). Molecular basis of altered excitability in *Shaker* mutants of *Drosophila melanogaster. EMBO J.* **9,** 4399–4407.
58. MacKinnon, R. (1991). Determination of the subunit stoichiometry of a voltage activated potassium channel. *Nature* **350,** 232–235.
59. MacKinnon, R., Aldrich, R. W., and Lee, A. W. (1993). Functional stoichiometry of *Shaker* potassium channel inactivation. *Science* **262,** 757–759.
60. Wu, C.-F., and Chen, M. L. (1995). Co-assembly of potassium channel subunits in *Drosophila:* The combinatorial hypothesis revisited. *Chin. J. Physiol.* **38,** 131–138.
61. Siddiqi, O., and Benzer, S. (1976). Neurophysiological defects in temperature-sensitive paralytic mutants of *Drosophila. Proc. Natl. Acad. Sci. USA* **73,** 3253–3257.
62. Tanouye, M. A., and Wyman, R. J. (1980). Motor outputs of giant nerve fiber in *Drosophila. J. Neurophysiol.* **44,** 405–421.
63. Salkoff, L., and Wyman, R. (1980). Facilitation of membrane electrical excitability in *Drosophila. Proc. Natl. Acad. Sci. USA* **77,** 6216–6220.
64. Elkins, T., and Ganetzky, B. (1988). The roles of potassium currents in *Drosophila* flight muscle. *J. Neurosci.* **8,** 428–434.
65. Wu, C.-F., and Ganetzky, B. (1988). Genetic and pharmacological analysis of potassium channels in *Drosophila. In* "Molecular Basis of Drug and Pesticide Action" (G. G. Lunt, ed.), pp. 311–323. Elsevier, Amsterdam.
66. Stewart, B. A., Atwood, H. L., Renger, J. J., Wang, J., and Wu, C.-F. (1994). Improved stability of *Drosophila* larval neuromuscular preparations in haemolymph-like physiological solutions. *J. Comp. Physiol.* **175,** 179–191.
67. Kandel, E. R., and Schwartz, J. H. (1982). Molecular biology of learning: Modulation of transmitter release. *Science* **218,** 433–443.
68. Byrne, J. H., and Kandel, E. R. (1996). Presynaptic facilitation revisited: State and time dependence. *J. Neurosci.* **16,** 425–435.
69. Levitan, I. B. (1988). Modulation of ion channels in neuron and other cells. *Annu. Rev. Genet.* **11,** 119–136.
70. Zhong, Y., and Wu, C.-F. (1993). Differential modulation of potassium currents by cAMP and its long-term and short-term effects: *Dunce* and *Rutabaga* mutants of *Drosophila. J. Neurogenet.* **9,** 15–27.

71. Chopra, M., and Singh, S. (1994). Developmental temperature selectively regulates a voltage-activated potassium current in *Drosophila*. *J. Neurobiol.* **25,** 119–126.
72. Griffith, L. C., Wang, J., Zhong, Y., Wu, C.-F., and Greenspan, R. J. (1994). Calcium/calmodulin-dependent protein kinase II and potassium channel subunit *eag* similarily affect plasticity in *Drosophila*. *Proc. Natl. Acad. Sci. USA* **91,** 10044–10048.
73. Gu, G.-G., and Singh, S. (1997). Modulation of the dihydropyridine-sensitive calcium channels in *Drosophila* by a phospholipase C-mediated pathway. *J. Neurobiol.* **33,** 265–275.
74. Byers, D., Davis, R., Kiger, J. A., Jr. (1981). Defect in cyclic AMP phosphodiesterase due to the *dunce* mutation of learning in *Drosophila melanogaster*. *Nature* **289,** 79–81.
75. Livingstone, M. S., Sziber, P. P., and Quinn, W. G. (1984). Loss of calcium/calmodulin responsiveness in adenylate cyclase of *rutabaga,* a *Drosophila* learning mutant. *Cell* **37,** 205–215.
76. Zhong, Y. (1995). Mediation of PACAP-like neuropeptide transmission by coactivation of Ras/Raf and cAMP signal transduction pathways in *Drosophila*. *Nature* **375,** 588–592.
77. Zhong, Y., and Pena, L. A. (1995). A novel synaptic transmission mediated by a PACAP-like neuropeptide in *Drosophila*. *Neuron* **14,** 527–536.
78. Guo, H. F., The, I., Hannan, F., Bernards, A., and Zhong, Y. (1997). Requirement of *Drosophila* NF1 for activation of adenylyl cyclase by PACAP38-like neuropeptides. *Science* **276,** 795–798.
79. Robertson, G. A., Warmke, J. M., and Ganetzky, B. (1996). Potassium currents expressed from *Drosophila* and mouse *eag* cDNA in *Xenopus* oocytes. *Neuropharmacology* **35,** 841–850.
80. Chen, M. L., Hoshi, T., and Wu, C.-F. (1996). Heteromultimeric interactions among K^+ channel subunits from *Shaker* and *eag* families in Xenopus oocytes. *Neuron* **17,** 535–542.
81. Wang, J. W., and Wu, C.-F. (1996). In vivo functional role of the *Drosophila Hyperkinetic* β subunit in gating and inactivation of *Shaker* K^+ channels. *Biophys. J.* **71,** 3167–3176.
82. Ikeda, K., and Kaplan, W. D. (1970). Patterned neural activity of a mutant *Drosophila melanogaster*. *Proc. Natl. Acad. Sci. USA* **66,** 765–772.
83. Yao, W.-D., and Wu, C.-F. (1995). *Drosophila* hyperkinetic mutants affecting K channel β subunits after firing pattern and K current properties in cultured "giant" neurons. *Soc. Neurosci.* **21,** 284. [Abstract]
84. Ciorba, M. A., Heinemann, S. H., Weissbach, H., Brot, N., and Hoshi, T. (1997). Modulation of potassium channel function by methionine oxidation and reduction. *Proc. Natl. Acad. Sci. USA* **94,** 9932–9937.
85. Wilson, G. E., Wang, Z., Chouinard, S. W., Griffith, L. C., and Ganetzky, B. (1997). Interactions of the *Drosophila* K^+ channel β subunit, *Hyperkinetic,* with members of the *eag* superfamily. *Soc. Neurosic.* **23,** 309. [Abstract]
86. Suzuki, N., and Kano, M. (1977). Development of action potentials in larval muscle fibers in *Drosophila melanogaster*. *J. Cell Physiol.* **93,** 383–388.
87. Byerly, L., and Leung, H.-T. (1988). Ionic currents of *Drosophila* neurons in embryonic cultures. *J. Neurosci.* **8,** 4379–4393.
88. Leung, H.-T., and Byerly, L. (1991). Characterization of single calcium channels in *Drosophila* embryonic nerve and muscle cells. *J. Neurosci.* **11,** 3047–3059.
89. Saito, M., and Wu, C.-F. (1991). Expression of ion channels and mutational effects in giant *Drosophila* neurons differentiated from cell division-arrested embryonic neuroblasts. *J. Neurosci.* **11,** 2135–2150.
90. Pelzer, S., Barhanin, J., Pauron, D., Trautwein, W., Lazdunski, M., and Pelzer, D. (1989). Diversity and novel pharmacological properties of Ca^{2+} channel in *Drosophila* brain membranes. *EMBO J.* **8,** 2365–2371.
91. Gielow, M. L., Gu, G.-G., and Singh, S. (1995). Resolution and pharmacological analysis of the voltage-dependent calcium channels of *Drosophila* larval muscles. *J. Neurosci.* **15,** 6085–6093.

92. Gu, G.-G., and Singh, S. (1995). Pharmacological analysis of heartbeat in *Drosophila. J. Neurobiol.* **28,** 269–280.
93. Triggle, D. J. (1990). Calcium, calcium channels and calcium channel antagonists. *Can. J. Physiol. Pharm.* **68,** 1474–1481.
94. Tang, C. M., Presser, F., and Morad, M. (1988). Amiloride selectively blocks the low threshold (T) calcium channel. *Science* **240,** 213–215.
95. Hirano, Y., Fozzard, H. A., and January, C. T. (1989). Characteristics of L-type and T-type calcium currents in canine-cardiac-Purkinje cells. *Am. J. Physiol.* **256,** H1478–H1492.
96. Tytgat, J., Vereecke, J., and Carmeliet, E. (1990). Mechanism of cardiac T-type channel blockade by amiloride. *J. Pharm. Exp. Ther.* **254,** 546–551.
97. Tsien, R. W., Lipscombe, D., Madison, D. V., Bley, K. R., and Fox, A. P. (1988). Multiple types of neuronal calcium channels and their selective modulation. *TINS* **11,** 431–437.
98. Hille, B. (1992). "Ionic Channels of Excitable Membranes." Sinauer, Sunderland.
99. Zheng, W., Feng, G., Ren, D., Eberl, D. F., Hannan, F., Dubald, M., and Hall, L. M. (1995). Cloning and characterization of a calcium channel $\alpha 1$ subunit from *Drosophila melanogaster* with similarity to the rat brain type D isoform. *J. Neurosci.* **15**(2), 1132–1143.
100. Smith, L. A., Wang, X., Peixoto, A. A., Neumann, E. K., Hall, L. M., and Hall, J. C. (1996). A *Drosophila* calcium channel $\alpha 1$ subunit gene maps to a genetic locus associated with behaviour visual defects. *J. Neurosci.* **16,** 7868–7879.
101. Hall, L. M., Ren, D., Eberl, D. F., Xu, H., and Chopra, M. (1996). A *Drosophila* mutant identifies an amino acid affecting calcium channel activation rate and peak current. *Soc. Neurosci.* **22,** 6. [Abstract]
102. Kulkarni, S. J., and Hall, J. C. (1987). Behavioral and cytogenetic analysis of the cacophony courtship song mutant and interacting genetic variants in *Drosophila melanogaster. Genetics* **115,** 461–475.
103. Gopalakrishnan, M., and Triggle, D. J. (1990). The regulation of receptors, ion channels, and G proteins in congestive heart failure. *Cardiovasc. Drug Rev.* **8,** 255–302.
104. Ferrante, J., and Triggle, D. J. (1990). Drug and disease-induced regulation of voltage-dependent calcium channels. *Pharmacol. Rev.* **42,** 29–43.
105. Pak, W. L., Ostroy, S. E., Deland, M. C., and Wu, C.-F. (1976). Photoreceptor mutant of *Drosophila:* Is protein involved in intermediate steps of phototransduction? *Science* **194,** 956–959.
106. Yoshioka, T., Inoue, H., and Hotta, Y. (1983). Defective phospholipid metabolism in the retinular cell membrane of *norpA (no receptor potential)* visual transduction mutants of *Drosophila. Biochem. Biophys. Res. Commun.* **111,** 567–573.
107. Bloomquist, B. T., Shortridge, R. D., Schneuwly, S., Perdew, M., Montell, C., Steller, H., Rubin, G. M., and Pak, W. L. (1988). Isolation of a putative phospholipase C gene of *Drosophila, norpA,* and its role in phototransduction. *Cell* **54,** 723–733.
108. Riesgo-Escovar, J., Raha, D., and Carlson, J. R. (1995). Requirement for a phospholipase C in odor response: Overlap between olfaction and vision in *Drosophila. Proc. Natl. Acad. Sci. USA* **92,** 2864–2868.
109. Zhu, L., McKay, R. R., and Shortridge, R. D. (1993). Tissue-specific expression of phospholipase C encoded by the *norpA* gene of *Drosophila melanogaster. J. Biol. Chem.* **268,** 15994–16001.
110. Ross, C. A., MacCumber, M. W., Glatt, C. E., and Snyder, S. H. (1989). Brain phospholipase C isozymes: Differential mRNA localizations by in situ hybridzation. *Proc. Natl. Acad. Sci. USA* **86,** 2923–2927.
111. Hardie, R. C., and Minke, B. (1993). Novel Ca^{2+} channels underlying transduction in *Drosophila* photoreceptors: Implications for phosphoinositide-mediated Ca^{2+} mobilization. *Trends Neurosci.* **16**(9), 371–376.

112. Evans, P. D., and O'Shea, M. (1978). The identification of an octopaminergic neurone and the modulation of a myogenic rhythm in the locust. *J. Exp. Biol.* **73,** 235–260.
113. Evans, P. D., Swales, L. S., and Whim, M. D. (1988). Second messenger systems in insects: An introduction. *In* "Molecular Basis of Drug and Pesticide Action" (G. G. Lunt, ed.). Elsevier, Amsterdam.
114. Zhong, Y., and Wu, C.-F. (1991). Altered synaptic plasticity in *Drosophila* memory mutants with a defective cyclic AMP cascade. *Science* **251,** 198–201.
115. Delgado, R., Latorre, R., and Labarca, P. (1994). *Shaker* mutants lack post-tetanic potentiation at motor end-plates. *Eur. J. Neurol.* **6,** 1160–1166.
116. Wu, C.-F., Renger, J. J., and Engel, J. E. (1998). Activity-dependent functional and developmental plasticity of *Drosophila* neurons. *In* "Advances in Insect Physiology," Vol. 27. Academic Press, London.
117. Tejedor, F. J., Bokhari, A., Rogero, O., Gorczyca, M., Zhang, J., Kim, E. S., and Budnik, V. (1997). Essential role of *dlg* in synaptic clustering of *Shaker* K^+ channels in vivo. *J. Neurosci.* **17,** 152–159.
118. Poulain, C., Ferrus, A., and Mallart, A. (1994). Modulation of type A K^+ current in *Drosophila* larval muscle by internal Ca^{2+}; effects of the overexpression of frequenin. *Pflüg. Arch.* **427,** 71–79.

DEVELOPMENT OF THE ADULT NEUROMUSCULAR SYSTEM

Joyce J. Fernandes and Haig Keshishian

Department of Biology, Yale University, New Haven, Connecticut 06520

I. Introduction
II. The Adult Musculature
  A. Thoracic and Head Musculature
  B. Abdominal Musculature
III. Motor Neurons
  A. Flight Motor Neurons
  B. Jump Motor Neurons
  C. Other Adult Motor Neurons
IV. Formation of Adult Neuromuscular Junctions
  A. Morphology of Adult Junctions
  B. Development of Neuromuscular Junctions
  C. Orthograde Signaling, from Nerve to Muscle
  D. Retrograde Signaling, from Muscle to Nerve
V. Functional Development of the Adult Neuromuscular System
VI. Conclusions and Perspectives
  References

I. Introduction

During the life cycle of *Drosophila* two sets of neuromuscular junctions are formed (Fig. 1). The first is the larval set, which is established during embryogenesis (Fig. 1A). It is initially involved in embryonic contractions and hatching and subsequently in the characteristic larval behaviors of crawling, feeding, and molting. Larval neuromuscular junctions progressively enlarge as the animal grows during its three larval instars. By the third instar, the muscles are 10 times their embryonic lengths, and motor neurons also show a significant expansion in synaptic branching and bouton number.[1] Larval neuromuscular junctions rapidly dedifferentiate into small, bulb-like contacts during early pupal development.[2,3] This marks the onset of development of a second, adult set of neuromuscular junctions with a largely novel set of muscle fibers. The new neuromuscular system serves the specialized functions of the adult, such as flight, walking, feeding, and copulation (reviewed in Ref. 4).

Metamorphosis is thus a profound transition phase in the *Drosophila* life cycle, wherein a relatively simple larval body plan is transformed into

0074-7742/99 $30.00

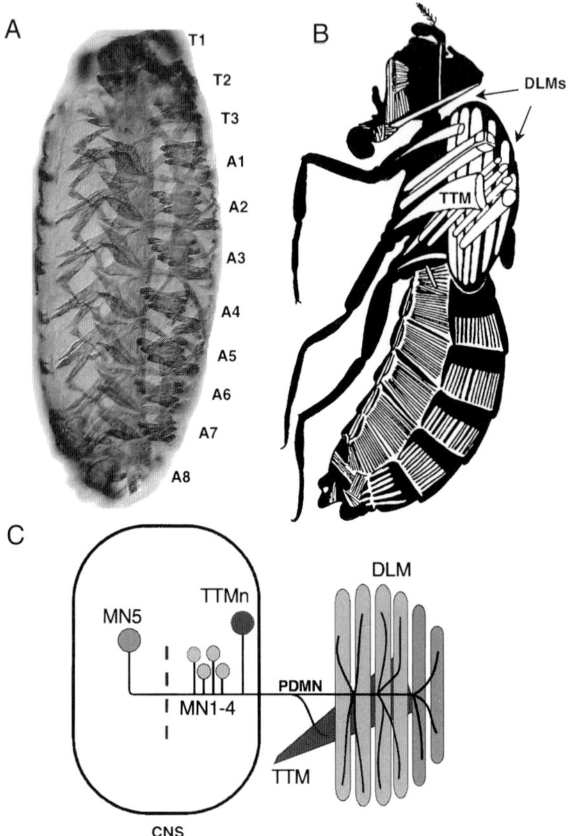

Fig. 1. Motor systems in the embryo and the adult. (A) Musculature of a late stage 16 embryo, when motor neurons have established their initial connections. The embryo is oriented to reveal ventral and lateral musculature in three thoracic (T1–T3) and eight abdominal (A1–A8) segments. Note the segmental uniformity of the pattern, which consists of 30 fibers in each abdominal hemisegment from A1–A7. Muscles labeled with antimyosin. (B) Schematic view of the major muscle groups of the adult (Modified from Miller, 1950). Note that the largest muscles are located in the thorax. Abdominal fibers retain the simpler organization found in the embryo. Other specialized muscles are found in the head and terminal segments. DLMs and the TTM are labeled. Unlabeled fibers in the thorax are DVMs. These run dorsoventrally. (C) Schematic of the motor neurons that innervate DLM flight muscles and the TTM jump muscle. These motor neurons are driven during the light-off escape reflex. PDMN, posterior dorsal mesothoracic nerve; DLM, dorsal longitudinal muscle; TTM, tergotrochanteral muscle; DVM, dorsal longitudinal muscle.

a more complex adult form. The transformation is under hormonal control and involves the programmed degeneration of most larval tissues, together with the generation of adult specific structures.[5] Primordia of most of the adult structures are set aside within the embryo and proliferate during larval life to subsequently take over during metamorphosis.[6,7] For example, the larval epidermis is replaced by adult epidermal cells, which are found as imaginal discs in the thoracic regions and as histoblast nests in the abdomen. Similarly, almost all of the larval musculature is histolyzed, replaced by the proliferation of muscle progenitor cells that are set aside by the end of embryogenesis.[8] The few larval muscles that are retained during pupal development enable head eversion within the first few hours of metamorphosis; others serve as scaffolds for the development of specific adult muscle fibers.[5]

Unlike the musculature, the central nervous system (CNS) is built on a preexisting scaffold of neurons. While many embryonically generated neurons may undergo apoptosis, it is probable that most, if not all, motor neurons survive metamorphosis and are structurally respecified to innervate adult muscle targets.[4,9] The restructuring takes place to accommodate novel, adult-specific motor circuits and behaviors. Much of the remodeling is associated with the dominance of the adult thorax in locomotor function and control, in striking contrast to the larva, where locomotor control is distributed more evenly along the segments.[4] BuDR birth dating has also shown that about 90% of the CNS outside of the brain consists of postembryonically generated neurons, with most of the newly generated cells located in the expanded thoracic neuromeres. Studies in the moth, *Manduca,* have shown that the respecified motor neurons innervate adult muscles located in the same general vicinity of their earlier larval muscle targets.[10]

This review describes the specialized adult motor systems and the morphological development of the musculature and innervation and explains why the study of these motor systems will throw light on some interesting developmental problems that are unique to the adult (see Table 1).

II. The Adult Musculature

One of the major differences between embryonic and adult neuromuscular systems is that the latter is more specialized. The larval neuromuscular pattern is based on an essentially uniform, segmentally repeated array of dorsal, ventral, and lateral muscle fibers, with variations in specific thoracic and abdominal segments.[11,12] The motor neurons are also segmentally repeated, projecting to homologous muscle targets.[13–15] In the adult there

TABLE I

ROLE OF THE NERVE AND OF THE MUSCLE DURING ADULT NEUROMUSCULAR DEVELOPMENT[a]

Events in adult neuromuscular development	Dependence on nerve or muscle
1. Arrival of myoblasts	Autonomous[74]
2. Myoblast proliferation	Nerve[38,76,74]
3. Muscle Patterning	
Muscle identity	Nerve, e.g., MSM[45,38]
Muscle formation	Nerve, e.g., DVM,[74] abdominal muscles[38]
4. Expression of actin isoforms	Autonomous, Act88F in IFMs[74]
	Nerve dependent, 79B in MSM[38]
5. Retraction of larval neuromuscular junctions	Not known
6. Primary branch outgrowth	
Onset	Autonomous[74]
Ordered growth	Muscle[74]
7. Elaboration of higher order branches	Muscle[74]
8. Clustering of glutamate receptors	Not determined

[a] During adult neuromuscular development, the development of innervation and of the musculature takes place in close proximity, a situation that is distinct from the embryo. Thus nerve–muscle interactions can shape the emerging neuromuscular pattern. Some of the key events in neuromuscular formation that occur during the first 24 hr of metamorphosis are listed. The influence of the nerve/muscle is also indicated.

are distinct set muscles in the head, thorax, and abdomen that greatly vary in fiber number and arrangement (reviewed in Ref. 5 and 16). Among these are the specialized pharyngeal muscles involved in feeding behaviors, the dorsal thoracic muscles in flight and wing movements, the ventral thoracic muscles in walking and jump movements, and the abdominal muscles in copulation and oviposition.

A second major difference between embryonic and the adult musculature is in the use of mesodermal founder cells. A specialized founder cell prefigures each embryonic muscle fiber.[11,17] These cells contain all the necessary developmental information to specify muscle fiber identity. The embryonic myoblast population is thus thought to contain at least two cell types: a founder cell population and a fusion-competent population, which fuses with the founders as the muscle fibers develop. Evidence for this hypothesis is supported by mutants of the gene *myoblast city* (*mbc*), where fusion does not occur.[17] In *mbc* embryos, founder cells span the correct attachment sites, receive appropriate motor neuronal innervation, and express myosin. A similar founder cell population has not yet been identified for the adult.

Given the apparent absence of founder cells for adult muscles, other cells have been proposed to help pattern the myoblasts. For example, in the adult abdomen, developing peripheral nerves have been implicated in patterning the myoblasts into separate muscle fibers.[2] Larval muscle fibers are also known to pattern adult muscles. In the thorax, the formation of one set of flight muscles, the dorsal longitudinal muscles (DLMs), uses persistent larval muscle fibers as myoblast fusion targets.[18–20] Myoblasts from the wing imaginal disc fuse into the larval scaffolds, in a manner reminiscent of muscle founders. However, when larval fibers are eliminated by laser ablation, DLM muscle fibers still form, although the final number of fibers becomes highly variable.[21,22] The persistent larval muscles in this case are thought to partition myoblasts and thereby regulate the number of muscle fibers formed. In other insects, the nerve,[23] larval muscle scaffolds,[24,25] and the epidermis[26] have been proposed to control muscle formation.

A. Thoracic and Head Musculature

In keeping with the focus of locomotor control in the thorax, this part of the body has the largest number of muscle fibers.[27] Each hemithorax has about 80 muscle fibers that, based on clonal analyses, are divided into dorsal and ventral sets.[28] These sets correlate well with the association of myoblasts with specific imaginal discs. Imaginal discs are epithelial invaginations that proliferate during larval life and give rise to the thoracic epidermis. Within the folds of the discs are proliferating muscle progenitor cells, also referred to as adepithelial cells. [29] At the onset of metamorphosis, the discs undergo a process of unfolding and expansion.[7] Simultaneously, the associated muscle progenitors are released into the thoracic cavity and assemble over the developing epidermis to begin the formation of the adult musculature (Fig. 2, see color insert).[20] The wing and haltere discs are dorsally located, while the three leg discs are ventral and the associated myoblasts give rise to correspondingly located dorsal and ventral muscle fibers. Clonal analysis reveals lineage restrictions between these muscle fibers as early as embryonic gastrulation, suggesting that sequestration into discs imposes restrictions on the myoblast pools. During pupal development myoblasts from separate dorsal and ventral discs can generate muscle fibers that lie adjacent to each other. For example, the large leg muscles, derived from ventral discs, develop in the proximity of groups of flight muscles that have dorsal disc origins. There does not appear to be any apparent cell mixing, indicating that restrictions continue to operate on the myoblast pools during development.

Much of the dorsal thorax is occupied by the indirect flight muscles (IFMs), consisting of 13 muscle fibers.[5] Six of these comprise the dorsal longitudinal muscles (DLMs), and the remaining seven fibers are grouped into three bundles that make up the dorsoventral muscles (DVMs) [Figs.

1B and 2B (see color insert for Fig. 2B)]. These muscle fibers are fibrillar in structure and have opposing functions. DLMs are the wing depressors, whereas DVMs are the wing elevators. The alternate contraction and relaxation of these muscles flexes and resonates the box-like thorax, generating the wing beats during flight. IFM fibers are asynchronous, where each cycle of contraction and relaxation is not driven by synchronized nerve impulses. Rather, a single burst of nerve activity is sufficient to sustain several rounds of muscle contraction and relaxation. Among their other specializations, these muscles uniquely express a specific form of actin, Act88F.[30] Screens for flightless mutations have revealed a number of loci that regulate development,[31–33] or structural aspects.[34–36] The other group of dorsal thoracic muscles are the direct flight muscles (DFMs), These are 17 in number, express a distinct form of actin (Act79B), and their morphology is referred to as tubular, the same as most other thoracic muscles, including the leg muscles.[5,37] Unlike IFMs, which provide the power for flight, smaller DFMs, located at the base of the wing, are responsible for steering functions of the wing.

Although most of the dorsal musculature originates from the wing disc, muscle progenitors from this disc give rise to muscles with distinct morphologies, composition, and modes of development. For example, among the two groups of IFMs, DLMs use persistent larval muscles as scaffolds whereas DVMs develop *de novo*.[20] DFMs, however, differ widely in morphology and structural components from the DLMs and DVMs, but share the *de novo* mode of development with DVMs.[5] This raises interesting questions about the way in which a single myoblast pool diversifies. Do these differences exist autonomously within the wing disc or do they arise as a result of interactions with the epidermis or perhaps with their innervating neurons? In at least one muscle, the male-specific muscle of the abdomen, it is known that innervation plays a key role in its identity and also determines the expression of 79B actin (see below).[38]

The largest tubular muscle in the thorax is the jump muscle, the tergal depressor of the trochanter (TDT), also known as the tergo-trochantral muscle (TTM, Fig. 1B). This muscle, like most other tubular muscles, expresses the Act79B isoform.[37] The TTM is a ventral mesothoracic muscle and is a key component of the light-off escape response. The light-off stimulus startles the fly, which then jumps and immediately initiates flight.[39] The TTM executes the jump motion. It receives excitation from the giant fiber interneuron, which has its cell body in the brain. The jump response has been a valuable and elegant assay for identifying mutations affecting both synaptic connectivity and neuromuscular development. These screens have identified mutants such as *bendless* (*ben*), *passover* (*pas*), and *nonjumper* (*nj*).[40] Apart from neurophysiological defects in *ben* and *nj*, TTM develop-

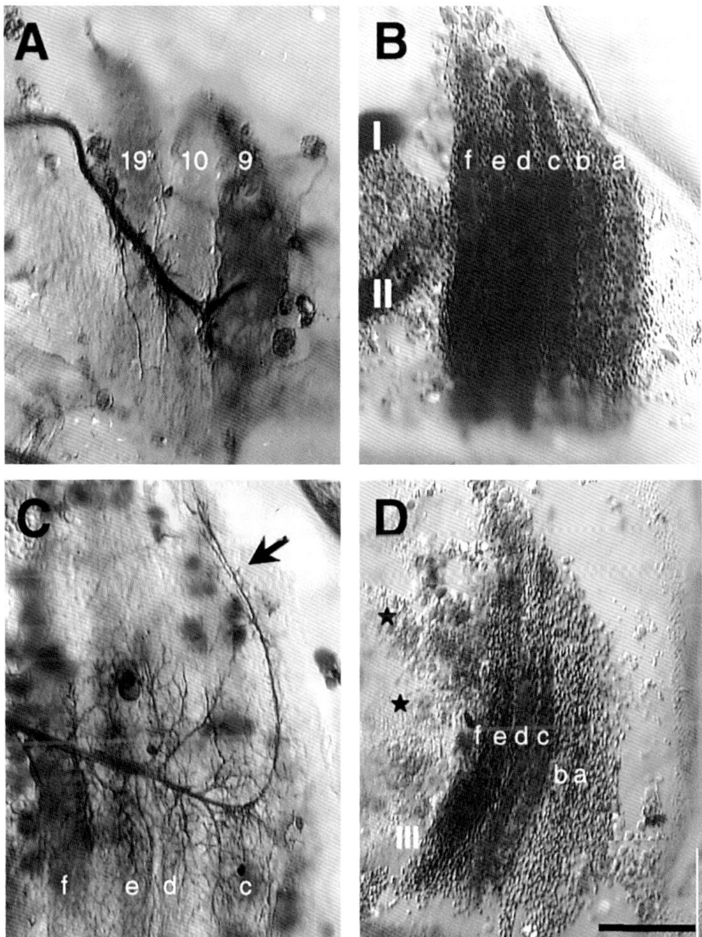

FIG. 2. Neuromuscular development of adult DLM flight muscles, in both normal and experimentally manipulated animals. (A) The three persistent dorsal larval muscle fibers (9,10,19') at 8 hr after puparium formation (APF) showing the dedifferentiated neuromuscular contacts made by the intersegmental nerve. (B) The larval muscle fibers split to give rise to the six DLM fibers as myoblasts fuse. The muscles are shown surrounded by the Twist-positive myoblasts that are initially associated with the wing disc. Also visible are two groups of DVMs, I and II. DVM III is present underneath the developing DLMs and is out of the plane of focus. (C) Effect of larval muscle ablation on nerve development. The dorsal most larval fiber (MF9; see A) was ablated in the larva, thus removing the muscle targets for motor neuron MN5.[58, 59] Shown here at 24 hr APF, a terminal nerve branch that normally innervates DLMs a and b projects (arrow) to neighboring DLMc. Normal motor neuronal contacts have elaborated on DLMs c-f by this time. (D) The distribution of Twist-positive myoblasts surrounding the developing DLMs in an animal where the mesothoracic nerve was cut at the larval stage. Shown here at 24 hr, the myoblast distribution is reduced in comparison to the control depicted in B. Also, the DLM fibers are thinner. Denervation usually prevents DVM formation (*), particularly DVMs I and II. Muscle fibers are labeled with myosin heavy chain LacZ in A and C and by the Act88F-lacZ reporter in B and D and are revealed using x Gal. Nerves are labeled with anti-HRP, whereas myoblasts are labeled with anti-Twist, both revealed with peroxidase cytochemistry. Scale: 100 μm. Adapted from Fernandes and Keshishian.[74]

ment in these mutants is also compromised.[33,41] Studies have shown that apart from the TDT, the IFMs, the DFMs, another leg muscle, the TLM, is also involved in the escape response.[42]

Among the prominent muscles in the head are the pharyngeal dilators involved in feeding, the rostral protractors, which control movements of the proboscis, and the ptilinium retractors, which are used during adult emergence.[5,43] All head muscles are of the tubular fiber type. Studies on the origins of these muscles have indicated that clonally distinct muscles are associated with the eye-antennal disc, the cibarial disc, and labial discs.[44]

B. Abdominal Musculature

In contrast to their thoracic and head counterparts, adult abdominal muscles have a simple organization and bear a closer similarity to larval musculature. There are sets of dorsal, ventral, and lateral muscle fibers in segments A1–A6.[27] A few segment-specific specializations of muscle fibers are also evident in the adult abdomen. These include the male-specific muscle (MSM, also known as muscle of Lawrence, MOL) found in the fifth abdominal segment.[28,45] This muscle is conspicuous due to its large muscle span. It is interesting that formation of the MSM is not under the control of the sex-determination pathway.[46] However, the *fruitless* gene, which is involved in male courtship behavior, also plays a role in MSM development.[47] The terminal segments also show sex specific muscle fiber patterning variations.[27] In addition, there are also specific muscles associated with the ovary and the testes.

Unlike the disc origins of adult thoracic muscle progenitors, abdominal muscles arise from progenitor cells that are closely associated with larval nerves.[2,8] Myoblasts proliferate during larval life and continue to do so after pupariation. Their progeny subsequently assemble over the developing epidermis (the expanding histoblast nests) to form the groups of dorsal, ventral, and lateral muscles. A similar association of myoblasts is also seen in thoracic segments, but their role in muscle fiber development remains unknown.

The entire complement of muscle fibers in each adult abdominal hemisegment thus traces its lineage to six precursor cells in the embryo.[8] The six precursor cells are characterized by a persistent expression of the transcription factor Twist. In the embryo there is a single ventral, two lateral, and three dorsal precursor cells, ultimately generating the corresponding ventral, lateral, and dorsal muscle fiber groups. This relationship has been tested by blocking cell proliferation during larval development with hydroxyurea. The reduced number of nerve associated precursors correlates well with the absence of specific muscle groups in the adult, suggesting a direct clonal relationship.[48]

III. Motor Neurons

Each hemisegment of the *Drosophila* larval body wall is innervated by about 40 motor neurons, which in the abdomen innervate 30 muscle fibers.[49] However, motor neuron identities in the head and thoracic segments remain to be determined. Motor neurons that innervate the adult musculature are believed to be remodeled larval neurons. The persistence of larval motor neurons into metamorphosis to innervate developing adult muscle targets has been studied in several insects, including *Manduca*[10,50–53] and the beetle, *Tenebrio*.[54] The process of motor neuronal remodeling occurs in two steps: the withdrawal of larval processes (Fig. 2A, see color insert) and the elaboration of adult-specific branches (Fig. 2C, see color insert).[55] This is true of both central and peripheral arbors. In the periphery, larval neuromuscular junctions are withdrawn synchronous with the histolysis of musculature. Subsequently, as new muscle fibers arise, the motor neurons elaborate adult-specific processes to innervate the developing fibers. In the CNS, dendritic reorganization takes place to presumably integrate the newly generated interneurons into adult specific neural circuits. In *Drosophila,* it is believed that larval motor neurons are similarly generated.

Both *Drosophila* and *Manduca* undergo complete metamorphosis, resulting in adults that are radically different from their larval forms. Changes in the body plan found in holometabolous insects are associated with corresponding radical changes in neural circuits, both to innervate the new musculature and to generate characteristic adult behaviors. It has been shown that dendritic remodeling in the *Manduca* CNS correlates with behavioral alterations.[56] The process of neuronal and synaptic remodeling is a complex problem. It is integrated with several processes that must occur simultaneously: the reorganization of musculature in the periphery, the reorganization of the CNS, which apart from respecification involves the birth of interneurons, and the selective death of others. Furthermore, this orchestration is timed by changes in hormonal levels that directly or indirectly affect most of the processes taking place during metamorphosis.[57]

A. Flight Motor Neurons

Given the importance and size of the indirect flight muscles, their motor neurons were among the first to be characterized in adult *Drosophila,* using both HRP and cobalt backfills. Five motor neurons (labeled MNs1–5) innervate the 6 DLM fibers (Fig. 1C).[58,59] The dorsal-most pair of DLMs (DLMa and b) is innervated by MN5, while the rest of the fibers are singly innervated. The motor neurons are mesothoracic and are likely derived from larval counterparts, which remain unidentified. The motor neurons innervating the 7 DVM fibers have also been identified. Each of the 7 fibers in a hemithorax are singly innervated. It is of interest that all 13 of the indirect flight muscle fibers are mononeuronally innervated, in contrast to the polyneuronal innervation seen in the embryo and larva. How this change in peripheral connectivity is achieved remains unknown. While the indirect flight muscles are responsible for powering flight, the more subtle wing movements are controlled by the direct flight muscles (DFMs). Motor neurons for a subset of the DFMs have been identified and classified according to morphology.[42] These motor neurons are thought to receive inputs from sensory afferents, including mechanosensory inputs and visual inputs from the compound eyes.

B. Jump Motor Neurons

The TTMn, which innervates the jump muscle (TTM), is another mesothoracic neuron that has been well studied in the context of the light-off escape response and provides a glimpse into the way motor circuitry is organized in the adult.[39] The TTMn receives input from the giant fiber (GF), the command interneuron for the escape reflex. Giant fiber (GF) axons descend to the mesothorax, where at a characteristic bend they make a very fast electrical synapse with the TTMn on each side.[40,60] A GF collateral branch also excites IFMs via the peripherally synapsing interneuron (PSI). The GF–PSI synapse is also electrical, whereas the PSI–DLMn synapses are chemical. In *bithorax* mutants, in which the metathorax (T3) in transformed into a mesothorax (T2), the TTM as well as its motor neuron are duplicated.[61] GF axons synapse with both TTMns. After contacting the mesothoracic TTMn, GF axons continue posteriorly through the transformed metathorax to synapse with the duplicated TTMn, suggesting that cues for the navigation of the GF axons through the metathorax are duplicated in these mutants.[62]

C. Other Adult Motor Neurons

Only a few other adult motor neurons have been characterized. Among these are the pharyngeal motor neurons (PMNs). Using GAL4 driver lines that are specifically expressed in these neurons, a UAS-tetanus toxin construct was directed to the PMNs to block synaptic transmission. As a result, feeding is severely diminished (M. Tissot, personal communication). PMNs, like most of the adult motor neurons, are also likely to be modified larval neurons. Mutations in *fruitless,* a locus shown to play a role in MSM development and in courtship, also indicate a more general role for the locus in the formation of somatic muscle synapses (B. Taylor, personal communication). The function of the male-specific muscle remains unknown. In another mutant called *dissatisfaction,* the females are sterile due to a failure in uterine muscle innervation. Males also show slower copulation because of defects in abdominal curling (B. Taylor, personal communication).

IV. Formation of Adult Neuromuscular Junctions

A. Morphology of Adult Junctions

Some of the earliest studies on fibrillar muscle innervation described the ramification of fine branches over the muscle surface.[63] Electron microscopic observations have shown that the nerve branches penetrate deeply into the muscle fiber, running along the muscle plasma membrane, which is thrown into folds.[64] Deep within the fiber, the axons make synaptic contact with the muscle membrane, where synaptic vesicles are visible. This system of internal synapses is typical for fibrillar muscles[63,65] and differs dramatically from embryonic and larval junctions, which are located on the muscle surface.[66] In case of adult abdominal muscles, the neuromuscular junctions also lie on the muscle surface.[38] However, there are differences in the size of varicosities and degree of branching compared to larval neuromuscular junctions.

B. Development of Neuromuscular Junctions

The formation of a neuromuscular synapse involves a series of coordinated interactions between the motor neuron and the muscle. The most well-studied skeletal synapse, in light of the problem of orthograde and retrograde signaling, is the vertebrate neuromuscular junction (reviewed in Ref. 67). Although the differentiation of synaptic partners may initially be independent, intercellular communication between the two cell types is essential for the initiation, formation, and maintenance of a neuromuscular synapse. It is useful to recapitulate what is known from vertebrate studies to point to possible analogous events in *Drosophila*. Some of the prominent effects of the nerve on vertebrate muscle differentiation include fiber-type specification, the clustering of acetylcholine receptors at synaptic sites, and the insertion of other synaptic membrane-specific proteins.[67] On the other hand, the target muscle releases neurotrophins for nerve survival[68] and affects neurotransmitter release.[69] Interactions between a muscle and its motor neuron continue through the life of an animal, and those that take place during initial formation are often recapitulated during nerve/muscle injury.[67] A major goal in understanding synapse development in *Drosophila* is to define the nature of similar orthograde and retrograde signals. In the *Drosophila* embryo, innervation is important for the clustering of glutamate receptors and for the organization of subsynaptic specializations.[70,71]

At the onset of adult neuromuscular development in *Drosophila,* the larval junctions are withdrawn, although they retain contacts with persistent larval muscles (Fig. 2A, see color insert) or at distinct sites on the epidermis.[2,3] Subsequently, new contacts are elaborated as myogenesis begins. Thus the motor neuron arbor remains in the periphery throughout adult neuromuscular development, where it can contact the epidermis, myoblasts, and in particular, the developing muscle fibers. This scenario is very different from the *Drosophila* embryo, where motor neurons arrive at an already formed array of muscle fibers.[12,11] Adult neuromuscular development therefore presents an advantageous situation for examining the interactions between a differentiating motor neuron and its muscle targets (see Table 1).

C. Orthograde Signaling, from Nerve to Muscle

The association of muscle progenitors with nerves was first described in the primitive dipteran *Simulium*.[72] These nerve-associated cells are precursors of the fibrillar thoracic muscles. Similar cells have also been observed in butterflies, where they are found along nerve sheaths,[25] blowflies,[73] and *Drosophila*.[2,3,8] Thus, the motor neurons have an opportunity to directly influence myoblast proliferation and differentiation. Work done in Lepidoptera indicated for the first time that the nerve might play a trophic role during myogenesis.[23] These studies examined effects of nerve cuts during DLM development. It was shown that there is a critical period for nerve requirement and that nerve cuts made prior to this period result in thinner fibers, whereas late nerve cuts do not alter DLM development. This implicated role of the nerve in myoblast proliferation has subsequently been demonstrated in denervation studies done in *Drosophila*,[38,74] as well as in *Manduca*.[75]

In contrast to this likely trophic/mitogenic role for the nerve during myoblast proliferation, there are some examples of the nerve determining both muscle identity and pattern. The identity of the male-specific muscle found in the fifth abdominal segment of *Drosophila* males depends on the innervating motor neuron.[28,45] These studies showed for the first time that muscle identity was determined not by contributing myoblasts but by extrinsic factors, in this case, the innervating neurons. Subsequent denervation studies have confirmed this conclusion and also showed that the fiber-specific actin79B expression in the MSM was abolished following denervation.[38] Studies on the *fruitless* mutation have revealed that the nerve could be involved in recruiting myoblasts into the male-specific muscle as well as in proliferation.[46,76]

Another example of the nerve affecting muscle pattern comes from denervation studies done in the thorax of *Drosophila* (Fig. 2B, D, see color insert).[74] Mesothoracic denervation affected the formation of one group of indirect flight muscles, the DVMs. The other group of muscles, the DLMs, still develop after being denervated, but the muscle fibers are thinner, presumably due to an effect on myoblast proliferation. It is postulated that this differential effect of denervation reflects the two different modes of myogenesis used by the two related muscles. While DLMs use larval scaffolds for development, DVMs develop *de novo.* In DVM development, innervation is thought to play a role by sustaining myoblast fusion that gives rise to a mature fiber. During DLM development, this function is thought to be carried out at least in part by the larval scaffolds. This latter hypothesis has been tested. Although DLM development still occurs following the ablation of larval scaffolds,[21] muscle fiber development is completely abolished if the nerve is also ablated.[74] This observation suggests that a dependence on innervation may be a general feature of adult myogenesis, even for fibers that use larval scaffolds.

D. Retrograde Signaling, from Muscle to Nerve

Far less is understood about the influence that the developing adult muscle has on motor neuron development. The developing musculature may be involved in the arrival of axons, as well as in the more local process of elaboration of neuromuscular endings. In vertebrates, the sites of neuromuscular junction elaboration are thought to be determined by the muscle surface, probably by extracellular matrix molecules.[67] During metamorphosis in *Drosophila,* some aspects of adult-specific motor neuron branching are dependent on the muscle fiber. By ablating one of the larval scaffolds that give rise to DLM fibers,[21,22] it is possible to create situations where the targets of motor neuron MN5, DLM a and b, develop in a delayed fashion (Fig. 2C). Under these conditions of transient target loss, it was found that while the primary motor neuron branches of MN5 did emerge, the appearance of higher order branches was stalled until the correct muscle fibers formed. This occurred even though adjacent, nontarget DLM fibers were within reach of the arbor of MN5.[74] Thus, muscle fiber target specificity for motor neurons exists, as has been demonstrated in the embryo.[77] This suggests that surface features of the developing muscle fiber are used as cues for the morphological development of motor neuron arbors, likely involving retrograde signaling mechanisms.

V. Functional Development of the Adult Neuromuscular System

The development of electrical excitability and ion channel expression in the adult musculature has been examined most extensively for DLM fibers. Three voltage-activated ion currents have been identified in the muscle, including two K^+ currents and a Ca^{2+} current.[78] The first current to appear is the fast transient K^+ current (I_A), mediated by channels encoded by the *Shaker* gene. This current appears between 60 and 70 hr after pupariation (APF) and is eliminated in *Sh*^{KS133} mutants. Following the appearance of this current, the delayed rectifier current becomes apparent, which can be studied in isolation in the mutant *Sh*^{KS133}. The third current is a voltage-activated Ca^{2+} current and is the last to develop. This current is abolished in mutations at the *slowpoke* (*slo*) locus.[79,80] In addition to these voltage-activated currents, the largest transmembrane current is actually the glutamatergic synaptic current, evoked by motor neuronal stimulation, resulting in a fast depolarization of the muscle membrane.[81] The sequence in which these currents develop resembles that later described for the embryo,[82] albeit over a longer period of time in the pupa. It should be noted that a role for activity in shaping adult motor neuronal arbors has not been demonstrated, in contrast to the embryo.[83,84]

VI. Conclusions and Perspectives

Most of the research in neuromuscular development in *Drosophila* has focused on the embryonic and larval stages.[49,85] However, the adult system provides specific scenarios not found in the embryo, which make for intriguing developmental strategies. One of the most important is the potential for extensive two-way signaling between the innervating motor neuron and the developing muscle. The large size of the muscles and nerves also makes manipulations easier. Another novel aspect of the adult neuromuscular system is that the motor neurons are probably modified larval neurons, and their remodeling provides an excellent system for studying morphological plasticity. Furthermore, the presence of specialized motor systems in the adult provides an opportunity to investigate the common threads that underlie development of these neuromuscular systems and to address the basis of the differences.

Much remains to be done to gain a cellular and molecular understanding of the mechanisms underlying adult neuromuscular development. For example, although adult muscle development has been studied as early as the 1950s, we still do not know whether precursor cells pattern the musculature, and the details of neuromuscular development remain cursorily understood, in contrast to the embryonic situation. Nonetheless, muscle patterning is influenced by nerves and by persistent larval muscles, mechanisms that are distinct to the adult. It also remains to be seen if the distinct adult modes of muscle development use the same molecules that are expressed during myogenesis and innervation in the embryo. The transcription factor Twist, expressed in muscle progenitors in the embryo, is also expressed in adult myoblasts,[8] suggesting that downstream molecules might be similar to those found in the embryo. However, some downstream genes may be novel to the adult, as has been shown for Erect Wing.[86] Another area that remains to be explored are the molecular mechanisms that govern motor neuronal remodeling, which might throw light on genes important for plasticity.

The amenability of *Drosophila* to molecular genetic approaches is one of its greatest strengths. Genes specific to the adult neuromuscular system have been previously identified using screens for adult-specific behaviors such as flight, walking, jumping and eclosion.[31,32,39] These need to be exploited further, in light of studies that demonstrate nerve–muscle interactions during neuromuscular development. The large number of molecules already known to play a role during embryonic and larval neuromuscular development can serve as a starting point, which will throw light on the mechanisms that operate during adult development. The characterization of motor neuronal identities and their remodeling will provide important insights into neuronal plasticity. Finally, novel genetic expression and mosaic methods (such as GAL4 and FLP/FRT),[87,88] as well as vital labeling techniques (utilizing green fluorescent protein),[89] will make it possible to clarify the signaling events that occur during myogenesis and synapse formation and to test hypotheses about the development of the adult neuromuscular junction.

Acknowledgments

We thank members of the Keshishian laboratory for fruitful discussions during the preparation of this review and Eric Gallela for providing the embryo illustrated in Fig. 1A. We also thank Drs. B. Taylor, M. Tissot, Y. A. Sun, and R.J. Wyman for communicating results prior to publication. This work was supported by grants from the NIH, NSF, and NASA.

References

1. Keshishian, H., Chiba, A., Chang, T. N., Halfon, M. S., Harkins, E. W., Jarecki, J., Wang, L., Anderson, M., Cash, S., Halpern, M. E., and Johnasen, J. (1993). Cellular mechanisms governing synaptic development in *Drosophila melanogaster*. *J. Neurobiol.* **24,** 757–787.
2. Currie, D., and Bate, M. (1991). The development of adult abdominal muscles in *Drosophila:* Myoblasts express Twist and are associated with nerves. *Development* **113,** 91–102.
3. Fernandes, J., and VijayRaghavan, K. (1993). Development of innervation to the indirect flight muscles of *Drosophila. Development.* **118,** 123–139.
4. Truman, J. W., Taylor, B. J., and Awad, T. A. (1993). Formation of the adult nervous system. *In* "The Development of *Drosophila melanogaster*" (M. Bate and A. Martinez-Arias, eds.). Cold Spring Harbor Press, Cold Spring Harbor, NY.
5. Crossley, C. A. (1978). The morphology and development of the *Drosophila* muscular system. *In* "The Genetics and Biology of *Drosophila*" (M. Ashburner and T. F. R. Wright, eds.), pp. 499–560. Academic Press, New York.
6. Cohen, S. (1993). Imaginal disc development. *In* "The Development of *Drosophila melanogaster*" (M. Bate and A. Martinez-Arias, eds.), pp. 747–841. Cold Spring Harbor Laboratory Press, Cold Spring Harbor, NY.
7. Fristrom, D., and Fristrom, J. W. (1993). The metamorphic development of the adult epidermis. *In* "The Development of *Drosophila melanogaster*" (M. Bate and A. Martinez-Arias, eds.), pp. 843–897. Cold Spring Harbor Laboratory Press, Cold Spring Harbor, NY.
8. Bate, M., Rushton, E., and Currie, D. A. (1991). Cells with persistent *twist* expression are the embryonic precursors of adult muscles in *Drosophila. Development* **113,** 79–89.
9. Truman, J. W., and Bate, M. (1988). Spatial and temporal patterns of neurogenesis in the CNS of *Drosophila melanogaster. Dev. Biol.* **125,** 146–157.
10. Thorn, R., and Truman, J. W. (1989). Sex-specific neuronal respecification during the metamorphosis of the genital segments of the tabacco hornworm moth, *Manduca sexta. J. Comp. Neurol.* **284,** 489–503.
11. Bate, M. (1990). The embryonic development of larval muscles in *Drosophila. Development* **110,** 791–804.
12. Johansen, J., Halpern, M. E., and Keshishian, H. (1989). Axonal guidance and the development of muscle fiber-specific innervation in *Drosophila* embryos. *J. Neurosci.* **9,** 4318–4332.
13. Sink, H., and Whitington, P. M. (1991). Location and connectivity of abdominal motor neurons in the embryo and larva of *Drosophila melanogaster. J. Neurobiol.* **22,** 298–311.
14. Halpern, M. E., Chiba, A., Johansen, J., and Keshishian, H. (1991). Growth cone behavior underlying the development of stereotypic synaptic connections in *Drosophila* embryos. *J. Neurosci.* **11,** 3227–3238.

15. Bate, M. (1993). The mesoderm and its derivatives. *In* "The Development of *Drosophila melanogaster*" (M. Bate and A. Martinez-Arias, eds.), pp. 1013–1090. Cold Spring Harbor Laboratory Press, Cold Spring Harbor, NY.
17. Rushton, E., Drysdale, R., Abmayr, S. M., Michelson, A. M., and Bate, M. (1995). Mutations in a novel gene, *myoblast city,* provide evidence in support of the founder cell hypothesis for *Drosophila* muscle development. *Development* **121,** 1979–1988.
18. Shatoury, H. E. (1956). Developmental interactions in the development of the imaginal muscles of *Drosophila. J. Embryol. Exp. Morphol.* **4,** 228–239.
19. Costello, W. J., and Wyman, R. J. (1986). Development of an indirect flight muscle in a muscle-specific mutant of *Drosophila melanogaster. Dev. Biol.* **118,** 247–258.
20. Fernandes, J., Bate, M., and VijayRaghavan, K. (1991). Development of the indirect flight muscles of *Drosophila. Development* **113,** 67–77.
21. Farrell, E. R., Fernandes, J. J., and Keshishian, H. (1996). Muscle organizers in *Drosophila:* The role of persistent larval muscle fibers in adult flight muscle development. *Dev. Biol.* **176,** 220–229.
22. Fernandes, J., and Keshishian, H. (1996). Patterning the dorsal longitudinal muscles of *Drosophila:* Insights from the ablation of larval scaffolds. *Development* **122,** 3755–3763.
23. Neusch, H. (1985). Control of muscle development. *In* "Comprehensive Insect Biochemistry Physiology and Pharmacology" (G. A. Kerkut and L. I. Gilbert, Eds.), pp. 425–452. Pergamon Press, Oxford.
24. Smith, W. A., and Velzing, E. H. (1986). Fusion of myocytes with larval muscle remnants during flight muscle development in the Colorado beetle. *Tissue Cell* **18,** 469–478.
25. Cifuentes-Diaz, C. (1989). Mode of formation of the flight muscles in the butterfly. *Peiris brassicae. Tissue Cell* **21,** 875–889.
26. Williams, G. J. A., and Caveney, S. (1980). A gradient of morphogenetic information involved in muscle patterning. *J. Embryol. Exp. Morphol.* **58,** 35–61.
27. Miller, A. (1950). The internal anatomy and histology of the imago *Drosophila melanogaster. In* "Biology of *Drosophila,*" pp. 420–534. Hafner Publishing, New York.
28. Lawrence, P. A., and Johnston, P. (1984). The genetic specification of pattern in a *Drosophila* muscle. Cell **29,** 493–503.
29. Reed, C. T., Murphey, C., and Fristrom, D. (1975). The ultrastructure of the differentiating pupal leg of *Drosophila melanogaster. Wilhem Roux's Arch. Dev. Biol.* **179,** 285–302.
30. Fyrberg, E. (1984). Actin genes of *Drosophila. In* "Oxford Surveys of Eukaryotic Genes," pp. 61–86.
31. Deak, I. I. (1977). Mutations of *Drosophila melanogaster* that affect muscles. *J. Embryol. Exp. Morphol.* **40,** 35–63.
32. Deak, I. I., Bellamy, P. R., Bienz, M., Dubois, Y., Fenner, E., Gollin, M., Rahmal, A., Ramp, T., Reinhardt, C. A., and Cotton, B. (1982). Mutations affecting the indirect flight muscles of *Drosophila melanogaster. J. Embryol. Exp. Morphol.* **69,** 61–81.
33. de la Pompa, J. L., Garcia, J. R., and Ferrus, A. (1989). Genetic analysis of muscle development in *Drosophila melanogaster. Dev. Biol.* **131,** 439–454.
34. Mogami, K., and Hotta, Y. (1981). Isolation of *Drosophila* flightless mutants which affect myofibrillar proteins of insect flight muscle. *Mol. Gen. Genet.* **183,** 409–417.
35. Bernstein, S. I., O'Donnell, P. T., and Cripps, R. M. (1993). Molecular genetic analysis of muscle development, structure and function in *Drosophila. Int. Rev. Cytol.* **143,** 63–152.
36. Cripps, R. M., Ball, E., Stark, M., Lawn, A., and Sparrow, J. C. (1994). Recovery of dominant autosomal flightless mutants of *Drosophila melanogaster* and identification of a new gene required for normal muscle structure and function. *Genetics* **137,** 151–164.
37. Courchesne-Smith, C. L., and Tobin, S. L. (1989). Tissue-specific expression of the 79B Actin gene during *Drosophila* development. *Dev. Biol.* **133,** 313–321.

38. Currie, D., and Bate, M. (1995). Innervation is essential for the development and differentiation of a sex-specific adult muscle in *Drosophila. Development* **121,** 2549–2557.
39. Wyman, R. J., Thomas, J. B., Salkoff, L., and Costello, W. (1985). The *Drosophila* thorax as a model system for neurogenetics. *In* "Model Neural Networks and Behaviour" (A. I. Stevenson, ed.). Plenum, New York.
40. Thomas, J. B., and Wyman, R. J. (1984). Mutations altering synaptic connectivity between identified neurons in *Drosophila. J. Neurosci.* **4,** 530–538.
41. Edgecomb, R. S., Ghetti, C., and Schneiderman, A. M. (1993). *Bendless* alters thoracic musculature in *Drosophila. J. Neurogenet.* **8,** 201–219.
42. Trimarchi, J. T., and Schneiderman, A. M. (1994). The motor neurons innervating the direct flight muscles of *Drosophila melanogaster* are morphologically specialised. *J. Comp. Neurol.* **340,** 427–443.
43. Kimura, K.-I., Shimozawa, T., and Tanimura, T. (1986). Muscle degeneration in the posteclosion development of a *Drosophila* mutant, abnormal proboscis extension reflex C, (aperC). *Dev. Biol.* **117,** 194–203.
44. VijayRaghavan, K., and Pinto, L. (1984). The cell lineage of the muscles of the *Drosophila* head. *J. Embryol. Exp. Morphol.* **85,** 285–294.
45. Lawrence, P. A., and Johnston, P. (1986). The muscle patterns of a segment of *Drosophila* may be determined by neurons and not by contributing myoblasts. *Cell.* **45,** 505–513.
46. Gailey, D. A., Taylor, B. J., and Hall, J. C. (1991). Elements of the *fruitless* locus regulate development of the muscle of Lawrence, a male-specific structure in the abdomen of *Drosophila melanogaster* adults. *Development* **113,** 879–890.
47. Taylor, B. J. (1992). Differentiation of a male specific muscle in *Drosophila melanogaster* does not require the sex-determining genes, *doublesex* and *intersex. Genetics* **15,** 275–296.
48. Broadie, K. S., and Bate, M. (1991). The formation of adult muscles in *Drosophila:* Ablation of identified muscle precursor cells. *Development* **113,** 103–118.
49. Keshishian, H., Broadie, K., Chiba, A., and Bate, M. (1996). The *Drosophia* neuromuscular junction: A model system for studying synaptic development and function. *Annu. Rev. Neurosci.* **19,** 545–575.
50. Casaday, G. B., and Camhi, J. M. (1976). Metamorphosis of flight motor neurons in the moth *Manduca sexta. J. Comp. Physiol.* **112,** 143–158.
51. Kent, K. S., and Levine, R. B. (1988). Neural control of leg movements in a metamorphic insect: Persistence of larval leg motor neurons to innervate the adult legs of *Manduca sexta. J. Comp. Neurol.* **276,** 30–43.
52. Levine, R. B., and Truman, J. W. (1985). Dendritic reorganization of abdominal motor neurons during metamorphosis of the moth, *Manduca sexta. J. Neurosci.* **5,** 2424–2431.
53. Weeks, J. C., and Truman, J. W. (1986). Hormonally mediated reprogramming of muscles and motor neurons during the larva-pupal transformation of the tobacco hornworm, *Manduca sexta. J. Exp. Biol.* **125,** 1–13.
54. Breidbach, O. (1987). The fate of persisting thoracic neurons during metamorphosis of the meal beetle *Tenebrio molitor. Roux's Arch. Dev. Biol.* **196,** 93–100.
55. Levine, R. B., and Truman, J. W. (1982). Metamorphosis of the insect nervous system: Changes in the morphology and synaptic interactions of identified cells. *Nature* **299,** 250–252.
56. Levine, R. B., and Weeks, J. C. (1990). Hormonally mediated changes in simple reflex circuits during metamorphosis in the moth, *Manduca sexta. J. Neurobiol.* **21,** 1022–1036.
57. Levine, R. B., Morton, D. B., and Restifo, L. L. (1995). Remodeling of the insect nervous system. *Curr. Opin. Neurobiol.* **5,** 28–35.
58. Ikeda, K., and Koenig, J. H. (1988). Morphological identification of the motor neurons innervating the dorsal longitudinal muscle of *Drosophila melanogaster. J. Comp. Neurol.* **273,** 436–444.

59. Coggshall, J. C. (1978). Neurons associated with the dorsal longitudinal flight muscles of *Drosophila melanogaster. J. Comp. Neurol.* **177,** 707–720.
60. Tanouye, M. A., and Wyman, R. J. (1980). Motor outputs of the giant nerve fiber in *Drosophila. J. Neurophysiol.* **44,** 405–421.
61. Schneiderman, A. M., Tao, M. L., and Wyman, R. J. (1993). Duplication of the escape response neural pathway by mutation of the *Bithorax*-Complex. *Dev. Biol.* **157,** 455–473.
62. Thomas, J. B., and Wyman, R. J. (1984). Duplicated neural structure in *Bithorax* flies. *Dev. Biol.* **102,** 531–533.
63. Tiegs, O. W. (1955). The flight muscles of insects. *Phil. Trans. Roy. Soc. Lond. B* **238,** 221–348.
64. Shafiq, S. A. (1964). An electron microscopic study of the innervation and sarcoplasmic reticulum of the fibrillar flight muscle of *Drosophila melanogaster. Quart. J. Micr. Sci.* **105,** 1–6.
65. Smith, D. S., (1960). Innervation of the fibrillar flight muscle of an insect: *Tenebrio molitor* (Coleoptera). *J. Biophys. Biochem. Cytol.* **8,** 447–466.
66. Johansen, J., Halpern, M. E., Johansen, K. M., and Keshishian, H. (1989). Sterotypic morphologyh of glutamatergic synapses on identified muscle cells of *Drosophila* larvae. *J. Neurosci.* **9,** 710–725.
67. Hall, Z., and Sanes, J. (1993). Synaptic structure and development: The neuromuscular junction. *Cell/Neuron* **72–10**(Suppl.), 99–121.
68. Funakoshi, H., Belluardo, N., Arenas, E., Yamamoto, Y., Casabona, A., Persson, H., and Ibanez, C. F. (1995). Muscle-derived Neurotrophin-4 as an activity-dependent trophic signal for adult motor neurons. *Science* **268,** 1495–1499.
69. Xie, Z.-P., and Poo, M.-M. (1986). Initial events in the formation of neuromuscular synapse: Rapid induction of actetylcholine release from embryonic neuron. *Proc. Natl. Acad. Sci. USA* **83,** 7069–7073.
70. Budnik, V. (1996). Synapse maturation and structural plasticity at *Drosophila* neuromuscular junctions. *Curr. Opin. Neurobiol.* **6,** 858–867.
71. Zito, K., Fetter, R. D., Goodman, C. S., and Isacoff, E. Y. (1997). Synaptic clustering of Fasciclin II and Shaker: Essential targetting sequences and role of Dlg. *Neuron* **19,** 1007–1016.
72. Hinton, H. E., (1995). Origin of indirect flight muscles in primitive flies. *Nature* **183,** 557–558.
73. Peristianis G. C., and Gregory, D. W. (1971). Early stages of flight muscle development in the blowfly, *Lucila cuprina:* A light and electron microscopic study. *J. Insect Physiol.* **17,** 1005–1022.
74. Fernandes, J. J., and Keshishian, H. (1998). Nerve-muscle interactions during indirect flight muscle development in *Drosophila. Development* **125,** 1769–1779.
75. Consoulas, C., and Levine, R. B. (1997). Accumulation and proliferation of adult leg muscle precursors in *Manduca* are dependent on innervation. *J. Neurobiol.* **32,** 531–553.
76. Taylor, B., and Knitte, L. M. (1995). Sex-specific differentiation of a male-specific muscle, the Muscle of Lawrence, is abnormal in hydroxy-urea and in *fruitless* male flies. *Development* 121, 3079–3088.
77. Chiba, A., Hing, H., Cash, S., and Keshishian, H. (1993). Growth cone choices of *Drosophila* motoneurons in response to muscle fiber mismatch. *J. Neurosci.* **13,** 714–732.
78. Salkoff, L., and Wyman, R. J. (1983). Ion channels in *Drosophila* muscle. *TINS* **6,** 128–133.
79. Elkins, T. B., Ganetzky, B., and Wu, C.-F. (1986). A gene affecting a calcium-dependent potassium current in *Drosophila. Proc. Natl. Acad. Sci. USA* **83,** 8415–8419.
80. Elkins, T. B., and Ganetzky, B. (1988). The roles of potassium currents in *Drosophila* flight muscle. *J. Neurosci.* **8,** 428–434.

81. Salkoff, L., and Wyman, R. J. (1983). Ion currents in *Drosophila* flight muscle. *J. Physiol.* **337,** 687–709.
82. Broadie, K. S., and Bate, M. (1993). Development of the embryonic neuromuscular synapse of *Drosophila melanogaster. J. Neurosci.* **13,** 144–166.
83. Jarecki, J., and Keshishian, H. (1995). Role of neural activity during synaptogenesis in *Drosophila. J. Neurosci.* **15,** 8177–9190.
84. Budnik, V., Zhong, Y., and Wu, C.-F. (1990). Morphological plasticity of motor axons in *Drosophila mutants with altered excitability. J. Neurosci.* **10,** 3754–3786.
85. Broadie, K. S., and Bate, M. (1995). Wiring by fly: The neuromuscular system of the *Drosophila* embryo. *Neuron* **15,** 513–525.
86. DeSimone, S., Coelho, C., Roy, S., VijayRaghavan, K., and White, K. (1996). ERECT WING, the *Drosophila* member of a family of DNA binding proteins is required in imaginal myoblasts for flight muscle development. *Development* **122,** 31–39.
87. Brand, A. H., Manoukian, A. S., and Perrimon, N. (1994). Ectopic expression in *Drosophila. In "Drosophila melanogaster:* Practical Uses in Cell and Molecular Biology'' (L. S. B. Goldstein and E. A. Fyrberg, eds.), pp. 635–654. Academic Press, San Diego.
88. Xu, T., and Rubin, G. M. (1993). Analysis of genetic mosaics in developing and adult *Drosophila* tissues. *Development* **117,** 1223–1237.
89. Brand, A. (1995). GFP in *Drosophila. Trends Genet.* **11,** 247–256.

CONTROLLING THE MOTOR NEURON

James R. Trimarchi, Ping Jin, and Rodney K. Murphey

Department of Biology, Morrill Science Center, University of Massachusetts, Amherst, Massachusetts 01003

I. Introduction
II. The Three Reflex Circuits
A. Reflex Circuit 1: The Flight-Related Reflex Circuit and Its Modification by *shaking-B*[2]
B. Reflex Circuit 2: The Leg Resistance Reflex Circuit and Its Disruption by *Glued*[1]
C. Reflex Circuit 3: The Hair Plate Reflex Circuit and Its Disruption by *unsteady*
III. Future Analysis Using Intracellular Recordings from Central Neurons
IV. Using the GAL4/UAS System to Characterize Neural Circuits Underlying Behavior
V. Conclusions
References

I. Introduction

Creative behavioral and anatomical screens of *Drosophila* mutants have identified many genes necessary for proper neural development and function. In contrast, relatively few neural circuits suitable for analyzing the role these gene products play in circuit assembly and function have been characterized.[1–3] The paucity of experimentally amenable neural circuits in *Drosophila* and the belief that electrophysiological techniques are difficult to apply to the *Drosophila* central nervous system (CNS) have hindered the determination of links between genes, neural function, and behavior. The bulk of the reviews in this volume describe studies utilizing the neuromuscular junction to investigate the link between genes and synaptic properties or synaptic development. These reviews illustrate the effectiveness of the neuromuscular junction as a model system; however, some differences undoubtedly exist between the neuromuscular junction and central synapses. Exploring these differences should enhance our understanding of the molecular mechanisms underlying synaptic development and operation. Our goal has been to characterize a set of neural circuits that exhibit diverse properties and are, therefore, useful for analyzing mechanisms by which gene products establish central synaptic properties underlying behavior. This review describes three sensory-to-motor reflexes and their associated central circuits and synapses. The uniquely identifiable neurons

0074-7742/99 $30.00

comprising these reflex circuits offer an excellent opportunity for investigating detailed molecular mechanisms underlying the development and operation of the CNS.

Some of the genetic approaches used to characterize these reflex circuits have been used to investigate the substrate underlying other behaviors in *Drosophila.* In particular the GAL4/UAS system[4] has become the ''technique of choice'' for identifying neural substrates participating in such behaviors as olfactory jump,[5] olfactory learning,[6] and sexual behavior.[7–9] Many of these studies, however, exhibit distinct shortcomings. First, the neural circuits underlying these behaviors have not been characterized and most likely employ almost all of the neurons that constitute the fly's nervous system. For example, the neural circuit participating in olfactory conditioning includes neurons of the antennal glomeruli, mushroom bodies, and central brain, in addition to descending neurons coordinating walking and all of the leg motor neurons and proprioceptive sensory afferents. In fact, the neural circuits underlying most of the behaviors presently studied genetically are so inclusive and complicated that investigators have not attempted to characterize their specific neural components. A second consideration is that most of the GAL4 lines used for these behavioral studies express GAL4 in hundreds, if not thousands, of neurons that may or may not participate in the given behavior.[5–9] Third, analysis of the behavior and the performance of the nervous system occurs on the second, minute, and often hour time scale. During these extremely long times information flows rapidly throughout many neural circuits within the nervous system and thereby obscures the precise genetic manipulations, making it extremely difficult to resolve the effect of particular genetic alterations on neural function at a mechanistic level.

In contrast, the reflex neural circuits that we have characterized are composed of relatively few neurons that are uniquely identified. For example, the hair plate reflex circuit employs only eight sensory neurons that exhibit synaptic connections with a single identified motorneuron (see Section II,C). Because of the simplicity of the three reflex circuits, we can analyze the function of gene products in establishing both dendritic and axonal anatomy in conjunction with their function in shaping the physiology of central neurons and synapses. Importantly, the GAL4 lines with which we manipulate gene expression are also extremely selective and thereby promote expression in very limited subsets of sensory neurons within the reflex neural circuits. Finally, the employment of electrophysiological techniques to monitor nervous system performance allows an analysis of the activity of individual motor neurons in the millisecond time scale. These properties enhance our ability to investigate the genetic mechanisms underlying the assembly and function of the nervous system.

II. The Three Reflex Circuits

We have identified three sensory-to-motor reflex circuits in *Drosophila* by employing traditional neuroanatomical and neurophysiological techniques in combination with modern genetic approaches.[1,2,10] This merger of techniques has proven very effective for characterizing neural circuits underlying simple behaviors. Each of the neural circuits we have characterized in *Drosophila* has been extensively studied in larger insects. We have drawn from this body of information to assist in the design and interpretation of the genetic manipulations we employ and expedite the characterization of the reflex circuits. Not surprisingly, these genetic manipulations have revealed that the reflex circuits in *Drosophila* are remarkably similar to those of larger insects.

The three reflex circuits we have characterized all follow the same general theme; sensory neurons synapsing with and controlling the activity of a motor neuron. In each reflex, a subset of sensory neurons serves as the input to the circuit. These sensory neurons can be consistently and precisely stimulated under experimental conditions. As a monitor of reflex circuit function, the activity of specific motor neurons can be reliably recorded using either electromyograms from their associated muscles or, more directly, by using intracellular recording techniques from their cell bodies. Together with dye injection techniques for visualizing the structure of identified neurons, we can assess both the anatomical and physiological integrity of neurons and synapses comprising the three reflex circuits.

A. Reflex Circuit 1: The Flight-Related Reflex Circuit and Its Modification by *shaking-B*[2]

In this neural circuit, sensory input controls the activity of the first basalar motor neuron, B1MN, through prominent electrical synapses (Fig. 1).[2] The B1MN innervates a small direct flight muscle (B1) that adjusts wing movements, resulting in steering and course corrective maneuvers. During each wing beat cycle, B1MN exhibits one action potential phase locked to the ventral-most extent of the wing beat path. Unilateral shifts in the phase of activation of B1MN relative to the wing beat cycle result in turning.[11 11] The precise timing of activation of B1MN is regulated, in part, by sensory neurons that monitor aspects of flight.[2,11–14] In particular, the B1MN receives direct electrical synapses from sensory neurons of the ipsilateral haltere.[2,15] Halteres are reduced metathoracic wings that serve as gyroscopic sensory organs during flight.[16] The ~200 sensory neurons

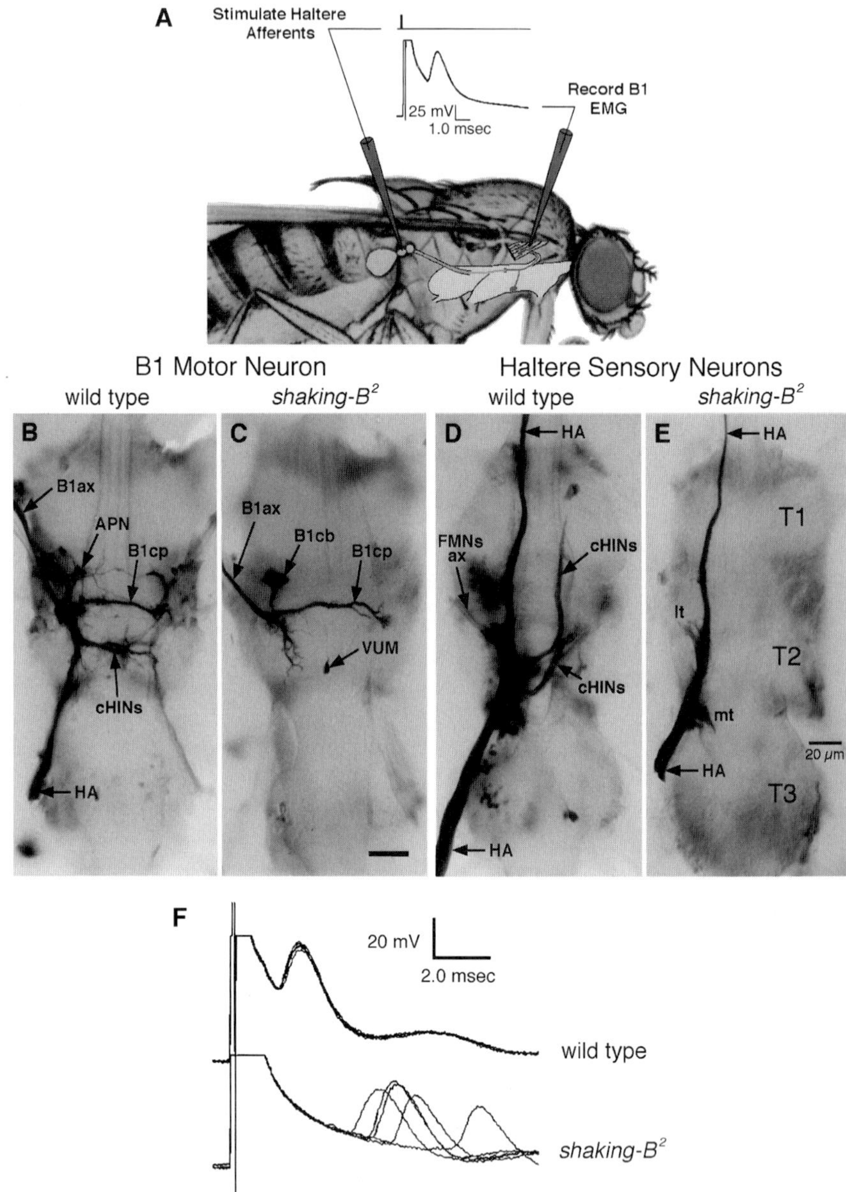

FIG. 1. The anatomy and physiology of the flight-related reflex circuit and its disruption by *shaking-B*^{2}. (A) Schematic of the preparation employed to record the reflex. Haltere sensory neurons are stimulated with a brief (20 μsec) depolarization and the evoked response of B1MN is recorded as electromyograms from the B1 muscle.[2] In wild-type flies (B and D), dye passes in both directions between haltere afferents and B1MN indicative of the presence of

associated with each haltere monitor flight parameters such as torque and acceleration.[12,15–18] Haltere sensory neurons project axons into the central nervous system of the fly where they convey flight information to neurons coordinating flight (Figs. 1D and 1E).[2,15,16]

Anatomical, physiological, and genetic evidence demonstrate that electrical synapses exist between haltere afferents and B1MN in *Drosophila*.[2] Electrical synapses are composed of gap junctions that make the cytoplasm contiguous between the synapsing neurons, thereby allowing the passage of ions and small molecular weight dyes between the neurons.[19–21] This unique feature of electrical synapses was used to demonstrate that the dye neurobiotin readily passes in both directions between haltere afferents and B1MN (Figs. 1B and 1D), and therefore suggests the presence of electrical synapses.

Electrophysiological results support these anatomical findings. Strong stimulation of haltere afferents reliably evokes electromyograms from the B1 muscle that occur at invariant short latencies (1.7 ± 0.16 msec) (Fig. 1F). Intracellular recordings from the B1MN cell body further demonstrate that haltere sensory neurons evoke an action potential from B1MN following a short delay (<1.7 msec). These short latencies from stimulation to the evoked response are consistent with the presence of electrical synapses between haltere afferents and B1MN and corroborate the observed dye coupling between these neurons. Together these data suggest that haltere sensory neurons exhibit prominent electrical synapses with B1MN that are responsible, in part, for controlling the timing of activation of B1MN characteristically observed during flight.[2,11,13–15]

---

electrical synapses. Dye coupling is also observed from B1MN (B) and haltere afferents (D) to a select set of interneurons (APN and cHINs). In contrast, in *shaking-B*^{2} flies (C and E), no dye coupling is observed between haltere afferents and B1MN or any other central neurons indicative of the absence of electrical synapses. The morphology of B1MN in *shaking-B*^{2} flies was not different from that observed in wild-type flies upon using a dye that does pass through electrical synapses. In C the ventral unpaired median cell that also innervates the B1 muscle is visible. (**F**) Repetitive stimulation of haltere afferents (80 V; 1 Hz) in wild-type flies reliably evokes a response from the B1 muscle at an invariant latency indicative of the presence of electrical synapses (five responses overlaid to demonstrate consistency). Similar stimulation in *shaking-B*^{2} flies evokes responses from the B1 muscle that occur at longer and more variable latencies indicative of the absence of electrical synapses (five traces overlaid). (APN, anterior projecting interneurons; B1ax, B1MN axon in peripheral nerve; B1cb, B1MN cell body; B1cp, B1MN neurite extending across midline; cHIINs, contralaterally projecting haltere interneurons; FMNsax, flight motor neurons axons including the axon of B1MN; HA, haltere afferents; lt, lateral tuft of the haltere afferent projection; mt, medial tuft of the haltere afferent projection; T1,T2, and T3, pro-, meso-, and metathoracic neuromeres, respectively; VUM, ventral unpaired median cell).

We found that the electrical synapses between haltere afferents and B1MN are genetically abolished by the *shaking-B²* mutation (also known as *shaking-B(neural)*), resulting in abnormal timing of activation of B1MN.[2] The *shaking-B²* mutation results in a stop codon replacing a leucine in the open reading frame and behaves as a viable genetic null[22,23] and, interestingly, *shaking-B²* flies are unable to fly.[24] The *shaking-B²* mutation eliminates the dye coupling observed between haltere afferents and B1MN (Figs. 1C and 1E). Mutant *shaking-B²* flies also exhibit concurrent physiological defects consistent with the loss of gap junctions. Strong stimulation of haltere afferents in *shaking-B²* flies evokes action potentials from B1MN and electromyograms from the B1 muscle that occur at variable latencies longer that those observed in wild-type flies (Fig. 1F). These longer and more variable latencies suggest that the electrical synapses between haltere afferents and B1MN are eliminated by the *shaking-B²* mutation and substantiate the absence of the dye coupling observed between these neurons. It seems likely that a *shaking-B²* fly's inability to fly arises, in part, from aberrant control of B1MN due to the absence of electrical synapses between haltere afferents and B1MN.

Complementary results have been obtained in larger flies and suggest that electrical synapses between haltere afferents and B1MN exist in other Diptera. In *Calliphora,* dye coupling between haltere afferents and B1MN has been observed using two different tracers: cobalt[25,26] and neurobiotin.[15] Haltere afferent stimulation evokes biphasic synaptic potentials in the dendrite of B1MN.[15] The early phase occurs at a short and invariable latency, is calcium independent, and does not fatigue upon repetitive stimulation. These physiological characteristics, in conjunction with dye coupling, suggest that electrical synapses exist between haltere afferents and B1MN in other flies. Utilizing electrical synapses to control the precise timing of B1MN activation during flight may be a neural feature common to all Diptera.

In addition to haltere afferents, many other sensory afferents exhibit dye coupling with central neurons, indicating that electrical synapses between sensory neurons and central neurons are perhaps more prevalent that previously believed. For example, neurobiotin readily passes between coxal hair plate sensory neurons (see Section II,C) and several interneurons. In each case, Shaking-B is required for dye coupling between sensory neurons and central neurons. Mutations in the *shaking-B* gene also eliminate electrical synapses between central neurons in the giant fiber circuit of *Drosophila.*[23,27–30] Although our findings do not distinguish between a role for the Shaking-B protein in the gap junction complex and its role in the construction or maintenance of the gap junction complex, it is clear that most, if not all, central electrical synapses in *Drosophila* require Shaking-B.

In summary, the flight-related reflex is composed of haltere sensory neurons that utilize direct electrical synapses to control the activity of a flight motor neuron.[2] These electrical synapses participate in the precise timing of activation of B1MN during flight.[2,11,13–15] The *shaking-B*[2] mutation abolishes these electrical synapses, resulting in abnormal control of B1MN activation and an inability to fly. Analysis of the anatomy and physiology of the identified neurons in this reflex circuit is enhancing our understanding of genetic mechanisms participating in the establishment and operation of electrical synapses and gap junctions.

B. Reflex Circuit 2: The Leg Resistance Reflex Circuit and Its Disruption by *Glued*[1]

In this neural circuit, sensory input controls the activity of the extensor tibia motor neurons through chemical synapses. In insects including *Drosophila*, only two motor neurons innervate the tibial extensor muscle: the slow extensor tibia motor neuron (SETI) and the fast extensor motor neuron (FETI).[31,32] As its name implies, contraction of the extensor tibia muscle extends the tibia by increasing the femur–tibia joint angle.[33] A variety of behaviors, including walking, jumping, and grooming, require coordinated movements of this leg joint and, therefore, proper activation of the extensor motor neurons of the tibia. The simplicity of the innervation of the extensor muscle and the richness of leg sensory innervation have made the control of this leg joint the focus of a great deal of attention. This historical background facilitates genetic analysis of the circuit.

In larger insects, which are amenable to standard electrophysiological recording methods, the activation of SETI and FETI has been extensively studied and demonstrated to be regulated, in part, by sensory neurons that monitor aspects of leg position and movement.[31,34] One of the crucial sensory organs at this joint is the femoral chordotonal organ (FECO), which monitors joint position, velocity, and acceleration of the tibia.[35,36] Studies in the locust and stick insect have revealed that the FECO is composed of a functionally heterogeneous population of sensory neurons that can be classified according to their physiology and axon projection within the CNS.[34–38] Flexion of the tibia activates FECO sensory neurons, which, in turn, excite the motor neurons to the extensor muscle, which resists the imposed flexion of the tibia; hence the name "resistance reflex."[31,34] Although both FETI and SETI receive synaptic input from FECO sensory neurons, synapses with SETI primarily mediate a resistance reflex.[31] The neural circuitry underlying the activation of SETI during the resistance reflex has been studied extensively in the locust and stick insect. A subset of FECO sensory neurons exhibit monosynaptic chemical connections with SETI and FETI as well as synapses with local nonspiking interneurons, spiking interneurons, and intersegmental interneurons, all combining to regulate joint position.[31,39,40]

Drosophila exhibits a resistance reflex that is remarkably similar to that studied in larger insects. In particular, the resistance reflex in *Drosophila* is mediated mainly by synaptic connections between FECO sensory neurons and SETI (Fig. 2). In *Drosophila*, rhythmic flexion of the tibia evokes muscle junctional potentials from the motor neurons innervating the extensor tibia muscle.[10] The electromyograms usually exhibit junctional potentials of only one size (Fig. 2C); however, muscle junctional potentials of a larger size are also occasionally observed (Fig. 2A). These junctional potentials of two distinct amplitudes result from activity in the two excitatory motor neurons (SETI and FETI) innervating the extensor of the tibia muscle. We determined that the larger junctional potentials arise from FETI by showing that a junction potential of the same size was evoked by stimulation of the giant fiber circuit.[10,32] The FETI motor neuron was identified unequivocally by retrograde staining from the muscle (Fig. 2B, see Ref. 32). Since this large junction potential is seldom recorded during passive flexion of the tibia, it was concluded that the SETI is the predominant motor neuron activated during the resistance reflex in *Drosophila.* This is consistent with findings in other insects where SETI has been demonstrated to be the primary motor neuron active during the reflex.[31] So far we have been unable to stain SETI anatomically by the retrograde filling method and its exact structure remains unclear in *Drosophila.*

In *Drosophila* the FECO is the largest sensory organ in the leg[41] and, as in larger insects, it mediates the resistance reflex in *Drosophila.* This was demonstrated by targeting tetanus toxin light chain to subsets of FECO sensory neurons and showing that the reflex was blocked. Two independent

---

FIG. 2. The anatomy and physiology of the resistance reflex circuit and its disruption by *Glued*^{1}. (A) Schematic of the preparation used to record the resistance reflex. The recording electrode is placed in the extensor tibia muscle (located in the femur) and the tibia was moved in sinusoidal movements with a small probe attached to a speaker. Each flexion evokes a burst of muscle potentials in the tibia extensor muscle indicative of activation of the slow extensor motor neuron (SETI) with occasional activation of the fast extensory of the tibia (FETI). (B) Structure of the FETI in the second thoracic segment stained by retrograde labeling from the muscle (taken from Trimarchi and Schneiderman[32]). We have been unable to label the SETI motor neuron in isolation for technical reasons. (C) Typical recordings from a wild-type specimen in response to leg movement. In this record the FETI does not respond to the movement. (D and E) Expression patterns of two GAL4 lines visualized by anti-β-gal staining. (D1) The P[GAL4]-50Y exhibits GAL4 expression in a small cluster

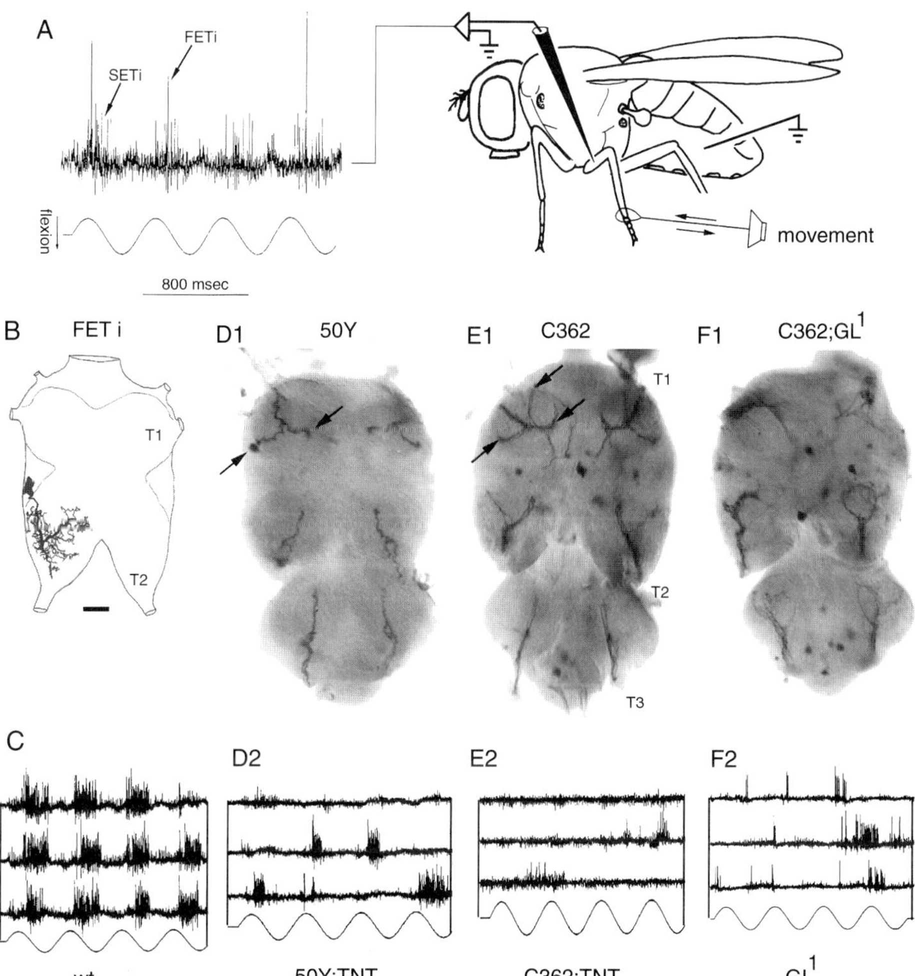

of sensory neurons in the FECO of each leg and projects axon collaterals to two parts of the leg neuropil (arrows). (E1) The c362 insertion is expressed in a larger subset of FECO sensory neurons, which exhibits three axon collaterals that terminate in the leg neuropil (arrows). (D2) Targeted expression of tetanus toxin to subsets of FECO sensory axons labeled by the P[Gal4]-50Y insert resulted in a significantly reduced resistance reflex. (E2) Targeted expression of tetanus toxin to subsets of FECO sensory axons labeled by the P[Gal4]-C362 insert resulted in a significantly reduced resistance reflex. (F1) FECO sensory neuron axon projections are disrupted in the *Glued¹* mutant as revealed by the P[Gal4]-c362. Instead of exhibiting three distinct axon branches as observed in wild type flies (E1), FECO sensory neurons in *Glued¹* flies have numerous disorientated branches. (F2) The resistance reflex was disrupted in the *Glued¹* mutant as indicated by the presence of very few muscle potentials evoked by each.

GAL4 lines that shared expression in subsets of FECO afferents (Figs. 2D1 and 2E1) were used to express tetanus toxin and yielded similar physiological and behavioral results. Targeted expression of tetanus toxin in FECO sensory neurons blocked chemical synaptic communication between FECO afferents and SETI and blocked the resistance reflex (Figs. 2D2 and 2E2). Targeted expression of an inactive form of tetanus toxin in FECO afferents did not block the reflex and served as a control. Two conclusions can be draw from these genetic manipulations. First, FECO sensory neurons are necessary for the resistance reflex, consistent with data in large insects.[34] Second, because tetanus toxin cleaves synaptobrevin, a synaptic vesicle protein required for evoked neurotransmitter release[5] chemical synapses must exist between these FECO sensory neurons and central neurons. In larger insects some of this input to the SETI occurs through monosynaptic chemical connections[39] and our results suggest this is the case in *Drosophila* as well. The behavior of flies expressing tetanus toxin in subsets of FECO afferents was consistent with the reflex blockade. Their posture was abnormal, the legs were often abnormally splayed, and the femur–tibia joint was flexed in an unusual position. In addition, the flies appeared clumsy because the legs became tangled during walking. These behavioral abnormalities are consistent with disruption of the resistance reflex.

The identified neurons comprising the resistance reflex offer a unique opportunity to analyze the physiological effects of axon pathfinding errors. Although many genes that participate in axon outgrowth and pathfinding have been identified, the functional consequences of altered neuronal morphology have been largely unexplored. It was found that the *Glued*^{1} mutation, which disrupts a necessary component of the retrograde cytoplasmic motor, results in abnormal FECO axon pathfinding and in a disrupted resistance reflex.[1] *Glued*^{1} is a dominant-negative mutation in the P150 component of the dynactin complex[42] and acts as a poison subunit within the retrograde motor complex. In *Glued*^{1} flies, FECO sensory neurons arborize abnormally within the CNS (Fig. 2F1). Since the abnormal branching was still confined to the proper proprioceptive neuropil layer it was possible that appropriate synapses between these sensory neurons and SETI might still exist. We demonstrated, however, that the resistance reflex was abnormal in *Glued*^{1} flies. It is likely that the disrupted anatomical projections of FECO afferents lead to miswiring of the resistance reflex circuit and the concurrent physiological defects. These observations suggest that the precise structure of sensory neuron projections is essential for proper function and highlight the utility of the resistance reflex for bridging the gap between molecular, genetic, and physiological inquires.

In summary, the leg resistance reflex is composed of FECO sensory neurons that utilize chemical synapses to control the activity of a leg motor neuron. The control of the activity of SETI through chemical synapses in the resistance reflex is in contrast to the control of B1MN by direct electrical synapses in the flight-related reflex described earlier and therefore offers a unique opportunity to compare genetic mechanisms underlying the construction and function of central chemical and electrical synapses. Analysis of the anatomy and physiology of the identified neurons in this reflex has already enhanced our understanding of the role cytoplasmic motors play in axon outgrowth and pathfinding[1] and the involvement of calcium/calmodulin dependent protein kinase II in a simple form of learning.[10] Further employment of anatomical and physiological methods for analyzing the resistance reflex will enhance our understanding of genetic mechanisms participating in the development and operation of central circuits and synapses.

C. Reflex Circuit 3: The Hair Plate Reflex Circuit and Its Disruption by *unsteady*

In this neural circuit, sensory input controls the activity of the prothoracic leg motor neuron 29, 29MN through prominent chemical synapses (Fig. 3). The 29MN innervates an extrinsic muscle of the prothoracic leg (muscle 29, see Ref. 33). Muscle 29 functions as a remotor and abductor of the coxa. Based on analogy with other insects, muscle 29 in *Drosophila* is presumed to be active during several behaviors, including walking, front leg reaching and grooming. The activation of leg muscles, including muscle 29, is regulated, in part, by proprioceptive sensory neurons that monitor joint angles of the leg.[43–54] In particular, the 29MN receives chemical synapses from sensory neurons associated with a hair plate on the ipsilateral coxa, coxal hair plate eight (CXHP8) (Fig. 3). CXHP8 is composed of a cluster of eight small hairs (sensilla tricodea; ~15 μm in length) located on the proximal anterior rim of the prothoracic coxa.[55–58] An analogous hair plate has been studied in stick insect (BF2[43,51–53] and CXHPV[46,47]) locust (CXHP1[44,45,49,51]), and *Manduca* (AHP[59]). This sensory structure monitors the position of the thoracic–coxal joint and its hairs become deflected when the leg is abducted or promoted.[43–53,58] The eight sensory neurons associated with CXHP8 of *Drosophila* project axons into the central nervous system where they convey joint angle and coxal motion information to neurons coordinating leg movements (Fig. 3D).[57,58] In other insects, sensory input to leg motor neurons from hair plate sensory neurons is, in part, monosynaptic chemical,[45,48–50,52–54] and findings in *Drosophila* are consistent with a prominent chemical connection between CXHP8 afferents and 29MN.

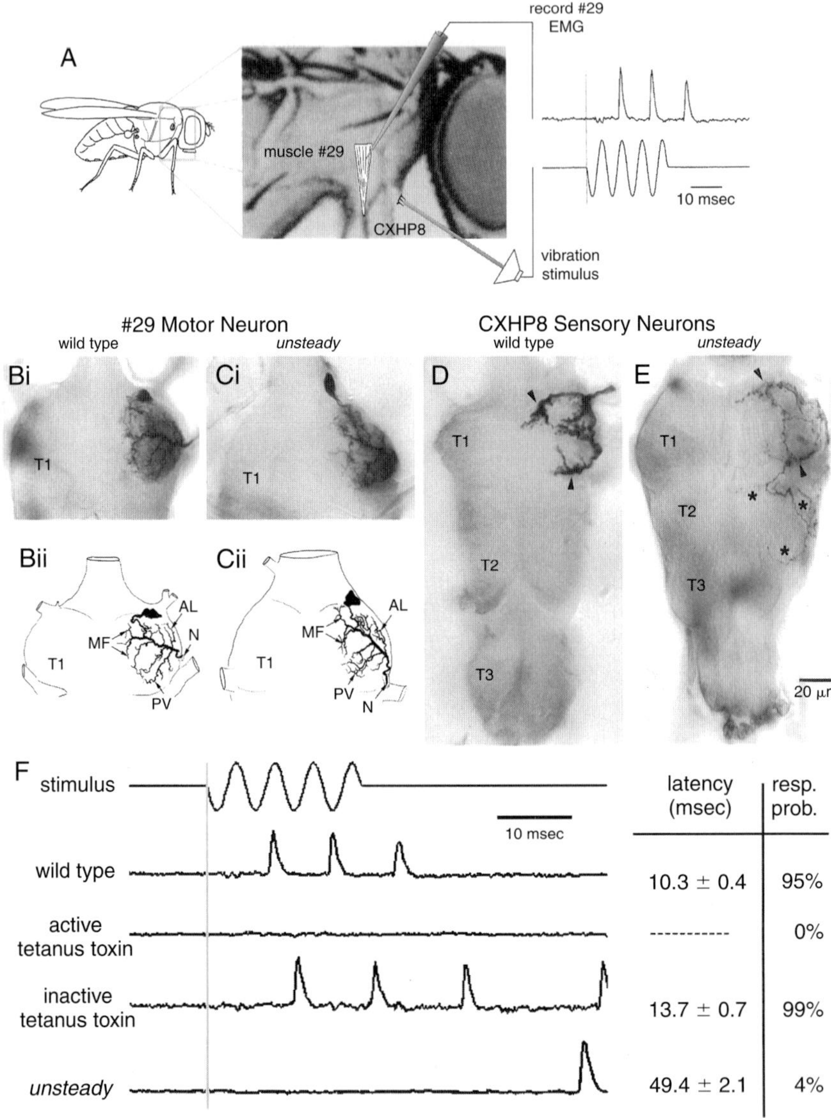

FIG. 3. The anatomy and physiology of the hair plate reflex circuit and its disruption by *unsteady*. (A) A schematic of the preparation employed to record the reflex. CXHP8 sensory neurons selectively stimulated with a glass probe vibrated with a 200-Hz sine wave evoke two to four muscle potentials from 29MN that can be recorded as electromyograms from muscle 29. (B and C) The *unsteady* mutation does not disrupt the dendritic projections of 29MN. The projections of 29MN in both wild-type (Bi) and unsteady (Ci) flies are confined to the ipsilateral prothoracic leg neuromere (T1). Several prominent branches of 29MN are consistently observed in both wild-type (Bii) and *unsteady* (Cii) flies (arrowheads). (D and E) The unsteady mutation disrupts the morphometrics of the thoracic and abdominal ganglia,

Anatomical, physiological, and genetic evidence in *Drosophila* corroborate the conclusion that chemical synapses exist between CXHP8 afferents and 29MN. Although the axonal branches of hair plate sensory neurons overlap the dendritic branches of 29MN (Figs. 3B and 3D), dye coupling is not observed between these neurons, suggesting that direct electrical synapses do not exist between these neurons. Mechanical stimulation of the hair plate and direct electrical stimulation of the hair plate sensory neurons, however, reliably evoke a response from 29MN recorded as electromyograms from muscle 29 (>90% of stimuli evoke a response). The latency from the onset of the mechanical stimulus to the first evoked response from muscle 29 is 10.3 ± 0.4 msec (Fig. 3F). These latencies are similar to those observed for the activation of motor neurons by monosynaptic connections from hair plate afferents in other insects and are consistent with the activation of 29MN, in part, by monosynaptic chemical connections from CXHP8 afferents.[45,48–50,52–54]

It was demonstrated in *Drosophila*, that chemical synapses between CXHP8 afferents and central neurons are necessary for proper function of the hair plate reflex. Using the GAL4/UAS system to target tetanus toxin expression[5] to CXHP8 sensory neurons, chemical synaptic communication between CXHP8 afferents and central neurons was prohibited and the reflex was blocked. Five GAL4 lines were employed to express tetanus toxin in subsets of sensory neurons and their participation in the hair plate reflex circuit was tested. Two lines, c17 and c161,[60] exhibited prominent GAL4 expression in hair plate afferents, and targeting tetanus toxin expression

---

resulting from abnormal condensation of the thoracic neuromeres (T1–T3). In contrast to B and C, where only the prothoracic ganglia are shown, in D and E the entire thoracic and abdominal ganglia are shown. The unsteady mutation also results in ectopic projections from CXHP8 sensory neurons axons. In wild-type flies (D), the eight hair plate sensory neurons exhibit axonal projections that are confined to the prothoracic and arborize around the rind of the prothoracic leg neuromere. In ~45% of unsteady flies (E), hair plate sensory neurons exhibit abnormal (*) projections to posterior neuromeres in addition to arborizing around the rind of the prothoracic leg neuromere. (F) The hair plate reflex is blocked by targeted expression of tetanus toxin to CXHP8 sensory neurons using the c17 GAL4 line that expresses in hair plate afferents. In wild-type flies, mechanical stimulation of hair plate afferents reliably (95% response probability; resp. prob.) evokes a response from the B1 muscle that occurred with an average latency of 10.3 msec from the onset of the stimulus. The GAL4-driven expression of active tetanus toxin in CXHP8 sensory neurons blocks the hair plate reflex (0% response probability). Expression of inactive tetanus toxin in CXHP sensory neurons does not block the reflex and serves as a control. Only 4% of mechanical stimuli evoke a response from *unsteady* flies and the latency to the response is abnormally long (> 40 msec), demonstrating that in addition to resulting in neuroanatomical abnormalities, the *unsteady* mutation also results in physiological defects.

with these lines blocked the reflex (Fig. 3F). Expression of a molecularly altered and inactive form of tetanus toxin[5] in hair plate afferents using c17 and c161 did not block the reflex and served as a control (Fig. 3F). These results suggest that CXHP8 sensory neurons exhibit prominent chemical synapses that participate in the activation of 29MN and, similar to larger insects, it is likely that a portion of this activation occurs through monosynaptic chemical connections.

The other three GAL4 lines, 480, 50Y, and c362,[1] were used as controls to direct tetanus toxin expression to sensory neurons other than hair plate afferents. Targeting tetanus toxin expression with these lines did not block the hair plate reflex. This observation is noteworthy because targeting expression of tetanus toxin to FECO afferents using 50Y and c362 results in a blocked resistance reflex in sibling flies that exhibited a normal hair plate reflex (see Section II,B). We were able to use the GAL4/UAS system to completely block the resistance reflex circuit without disrupting the physiology of the related hair plate reflex circuit. These experiments demonstrate the selectivity of the GAL4 lines we employed, the specificity of the circuits underlying the leg reflexes, and the precision of the physiological assays we have developed. By using the appropriate GAL4 lines, specific neural circuits can be precisely disrupted without altering functionally related circuits.

Interestingly, flies expressing tetanus toxin in CXHP8 afferents, and therefore a blocked hair plate reflex, executed abnormal behaviors reminiscent of those performed by locust and stick insects with surgically lesioned hair plates,[43–45,47,52,53] further suggesting that the GAL4 lines we used to target tetanus toxin expression are relatively specific. *Drosophila* exhibiting a blocked hair plate reflex displayed postural defects, often fully extend one leg for abnormally long times, and have difficulty maintaining leg joint angles while grooming with other legs. Tetanus toxin expressing flies also walk clumsily, extending the legs abnormally, often resulting in the legs becoming tangled. In particular, flies expressing tetanus toxin in CXHP8 have difficulty releasing their tarsi from the substrate at the end of the stance phase of the walking cycle. These behavioral abnormalities are characteristic of those performed by larger insect that have undergone a lesion of coxal hair plates[43–45,47,52,53] and are consistent with the disruption of the hair plate reflex that we demonstrated physiologically.

The hair plate reflex, with its remarkably simple underlying neural circuit, offers an excellent opportunity to investigate the anatomical and physiological consequences of mutations isolated from behavioral screens. The grooming mutant, *unsteady,* results in specific anatomical and physiological defects in neurons comprising the hair plate reflex. The *unsteady* mutation was originally isolated in a grooming screen, in which mutagenized flies were coated with a fine dust and mutant flies were identified as those remaining dusty after a 2-hr grooming period.[61] None of the *unsteady* flies were able to remove the dust during the grooming period while 96% of wild-type flies successfully removed the dust. The dust-laden *unsteady* flies attempted to groom, however, and spent a similar length of time trying to remove the dust as did wild-type flies. The uncoordinated grooming behavior *unsteady* flies prevented the removal of the dust and suggested that central neural circuits might be abnormal.

The anatomy of sensory neurons and the physiology of the hair plate reflex circuit are disrupted in *unsteady* flies. Anterograde staining of CXHP8 sensory neurons revealed disrupted projections within the prothoracic neuromere and ectopic projections to ipsilateral mesothoracic and metathoracic neuromeres (Fig. 3E). Despite disruptions in sensory neuron axonal projections, the morphology of the dendrites of 29MN was relatively unaltered (Fig. 3C). These results demonstrate that *unsteady* is required for proper axon pathfinding or branching but may not be required for the branching of dendrites.

Although sensory neuron axons exhibited disrupted projections, axonal arbors of CXHP8 sensory neurons and dendrites of 29MN overlap and could potentially establish functional connections. Mechanical stimulation of CXHP8 afferents in *unsteady* flies, however, was virtually unable (4% response probability) to evoke a response from 29MN recorded as electromyograms from muscle 29 (compare to 95% response probability in wild-type flies). Responses that were evoked in *unsteady* flies occurred at abnormally long latencies (49.4 ± 2.1 msec) in contrast to the short latency (10.3 ± 0.4 msec) responses observed in wild-type flies (Fig. 3F). To determine if this physiological defect resulted from a peripheral mechanosensory transduction defect or from a more central defect, CXHP8 sensory neurons were directly stimulated electrically, thereby circumventing the mechanosensory apparatus and directly activating the sensory neurons. Electrical stimulation of CXHP8 afferents in *unsteady* flies yielded results similar to those obtained from mechanical stimulation, failing to activate 29MN (6% response probability) or resulting in abnormally long response latencies (13.7 ± 1.6 msec). In contrast, electrical stimulation of CXHP8 afferents in wild type flies reliably (75% response probability) evoked electromyograms from muscle 29 that occurred at short latencies (9.4 ± 1.4 msec). These observations suggest that in addition to the anatomical defects in axonal branching, the *unsteady* mutation may also disrupt the physiology of the neurons or synaptic connections in the hair plate reflex circuit.

The gene disrupted by the *unsteady* mutation appears to be previously uncharacterized. Recombination tests with deficiencies delimited the mutation to the 13A2–13A5 region of the X chromosome.[62] Several mutations within this region, including *ether-a-go-go*,[62] *irregular optic chiasm A*,[63] and *rutabaga*,[64,65] fully complement the *unsteady* phenotype. The hair plate reflex, with its identified sensory and motor neurons and physiological assay, should prove effective for identifying novel genes that participate in neural development and function.

III. Future Analysis Using Intracellular Recordings from Central Neurons

The application of intracellular recording to identified central neurons in *Drosophila* will provide an enhanced characterization of the neurons underlying the three reflexes described in this review and should facilitate our understanding of the effects mutations have on the physiological properties of central neurons and central synapses. During the course of characterizing the neural substrate of these three reflexes a technique was developed for reliably recording intracellularly from neurons in the thoracic ganglia of *Drosophila*. The procedure devised draws heavily on the advancements forged by Ikeda and colleagues during their intracellular investigations of flight motor neurons in *Drosophila*.[66–68] Using sharp microelectrodes and stabilizing the ganglia against endogenous movements, the activity of identified central neurons can be reliably and repeatedly recorded.[2] For example, the motor neurons innervating the dorsal longitudinal muscles (DLMns)[69,70] exhibit ~20-mV action potentials and ~3-mV inhibitory potentials (Fig. 4A). These inhibitory potentials are presumed to arise from contralateral DLMns.[68] In addition, synaptic potentials evoked by haltere afferents can be recorded from a variety of flight motor neurons, including B1MN (Fig. 4B), second basalar motor neuron, first pterale I motor neuron, and third pterale III motor neuron.[71] Intracellular recordings from a variety of leg motor neurons are also possible, including FETI,[32] 29MN, and tergotrochanteral motor neurons.[72] In addition, intracellular impalements and recordings were obtained from a variety of unidentified interneurons (Fig. 4C).

The application of intracellular recording techniques to flight motor neurons has been instrumental in determining that the *shaking-B*^{2} mutation disrupts central synapses between haltere afferents and B1MN rather than altering peripheral synapses between B1MN and the B1 muscle.[2] In addition, intracellular recording techniques have been employed to demonstrate that, unlike many of the direct flight motor neurons, the DLMs do not receive synaptic input from haltere afferents. The application of intracellular recording to identified central neurons in *Drosophila* will greatly improve the characterization of central synapses between identified neurons and enhance investigations linking genes to neuronal and synaptic properties.

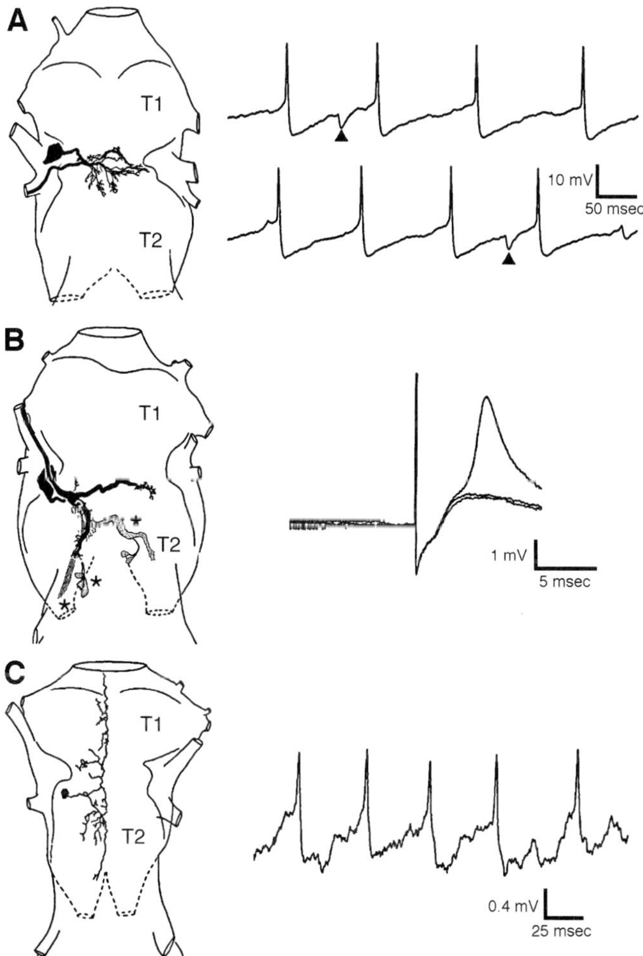

FIG. 4. Intracellular recordings and dye injections of several central neurons. (A) Action potentials and synaptic potentials (arrowhead) can be recorded from the cell bodies of DLM motor neurons. (B) Stimulation of haltere afferents (arrowhead) evokes synaptic potentials and action potentials from B1MN that can be recorded from the large B1MN cell body. Dye injected intracellularly into B1MN passed through electrical synapses to haltere afferents and several interneurons (stippled; *). (C) Action potentials can also be recorded intracellularly from the small cell bodies of interneurons.

IV. Using the GAL4/UAS System to Characterize Neural Circuits Underlying Behavior

Blocking the resistance reflex and the hair plate reflex by targeting expression of tetanus toxin to subsets of sensory neurons demonstrates the utility of the GAL4/UAS system for dissecting neural circuits underlying specific behaviors. Indeed the GAL4/UAS system has become the ''technique of choice'' for determining the neural substrate underlying *Drosophila* behavior.[5–9] The GAL4/UAS technique, however, should be used with caution and there are several issues that should be addressed before interpreting results obtained by using this system to genetically manipulate particular behaviors. Most importantly, the expression pattern of the GAL4 line must be very selective, ideally being restricted to a few identified neurons. If GAL4 expression is too widespread, then targeted expression of tetanus toxin or any other gene product will undoubtedly result in the altered function of many neural circuits and the disruption of several behaviors. Interpreting either the neural circuit underlying particular behaviors or the molecular mechanism by which a gene acts from such nonselective GAL4 experiments is extremely difficult. We guarded against nonspecific tetanus toxin expression by testing if flies that exhibited a block of a particular reflex could successfully execute other reflexes and behaviors that require coordinated leg movements.[1] We found that flies expressing tetanus toxin in FECO afferents and therefore exhibiting a blocked resistance reflex could nevertheless successfully perform a scratch reflex in response to deflection of hairs at the base of the wing.[73] Moreover, it was physiologically determined that the hair plate reflex was unaltered in flies expressing tetanus toxin in FECO sensory neurons and exhibiting a blocked resistance reflex (Fig. 3) (see Section II,C). These results suggest that the GAL4 lines employed are extremely selective and that specific neural circuits can be precisely disrupted without altering other related circuits. Care should be taken to identify and use GAL4 lines that exhibit selective expression, thereby limiting the number of neurons and neural circuits genetically manipulated.

None of the GAL4 lines we[1] or others[5–9] have employed are ideally selective and therefore multiple GAL4 lines that exhibit different but overlapping expression should be used. For example, we employed two GAL4 lines to target tetanus toxin expression to FECO afferents (Fig. 2). Although both lines are highly selective and predominantly express in FECO sensory neurons, each exhibits expression in some neurons other than FECO sensory neurons. Using either GAL4 line to target tetanus toxin expression resulted in a blocked resistance reflex. Therefore, it is the overlapping expression of tetanus toxin in the FECO sensory neurons that is responsible for the blocked reflex and not the spurious expression in the few other sensory neurons that the two GAL4 lines do not share in common. Another example of the selective use of the GAL4/UAS system to genetically manipulate specific neural circuits was given earlier (see Section II,C). By utilizing multiple GAL4 lines that overlap in their expression, one can identify the relevant neurons responsible for the behavioral or physiological results.

Proper use of the GAL4/UAS system also requires that the developmental profile of GAL4 expression be determined. Knowledge of when and where the genetic manipulations are taking place is necessary for interpreting the results. For example, the two GAL4 lines used to target tetanus toxin to FECO afferents begin expressing in late pupal stages. It is, therefore, unlikely that the block of the resistance reflex observed in the adult fly results from changes to the nervous system during development. In fact, the axonal projections of FECO sensory neurons expressing tetanus toxin are not different from the projection of FECO sensory neurons in wild-type flies.[1] This observation suggests that the block of the resistance reflex obtained by targeted expression of tetanus toxin does not result from a disruption in the anatomical connectivity during development but rather arises from the acute elimination of chemical synaptic transmission between FECO sensory neurons and SETI in the adult. Characterizing the developmental profile of GAL4 expression is essential for determining if results obtained in the adult fly arise from acute genetic manipulations or from disruptions during development that subsequently lead to abnormal neural function in the adult.

The GAL4/UAS system is a powerful genetic tool for dissecting neural circuits underlying behavior, provided that (1) the GAL4 lines employed are selective, (2) multiple lines exhibiting overlapping expression are used, and (3) the developmental expression patterns are determined. Considering these caveats can facilitate the interpretation of the results obtained from using the GAL4 system and enhance our understanding of molecular–genetic mechanisms underlying neural function and behavior.

V. Conclusions

The three reflexes we have identified add to the short list of neural circuits amenable to physiological experimentation in *Drosophila*.[1–3] During each reflex behavior, an identified set of sensory neurons controls the activity of a specific motor neuron. The simple neural circuits underlying these reflexes offer the opportunity to analyze both the anatomy and the physiology of uniquely identified neurons and their central synaptic connections. In particular, the flight-related reflex circuit should facilitate investigations of gene products necessary for the establishment, membrane insertion, and modulation of gap junctions[2]; and the resistance reflex circuit will allow the identification of gene products necessary for short-term circuit dynamics[10]; the hair plate reflex circuit will promote the investigation of gene products that function at central chemical synapses. The functional assays that have been developed can also be used to probe the physiological consequences of a wide variety of mutants, including learning and memory mutants, courtship mutants, circadian rhythm mutants, and bang sensitive mutants. Our goal is to employ these reflex circuits to uncover mechanisms by which gene products shape aspects of neural function underlying behavior.

Acknowledgments

Special thanks to Dr. Randall Phillis who keeps us honest about genetics and to Phyllis Caruccio, Suman Reddy, and Michael Getzinger who were very helpful with histology and computer imaging. Supported by NIH grants to RKM (NS15571) and NIH-NRSA to JRT (NA09700).

References

1. Reddy, S., Jin, P., Trimarchi, J. R., Caruccio, P., Phillis, R., and Murphey, R. K. (1997). Mutant molecular motors disrupt neural circuits in *Drosophila. J. Neurobiol.* **33,** 711–723.
2. Trimarchi, J. R., and Murphey, R. K. (1997). The *shaking-B*[2] mutation disrupts electrical synapses in a flight circuit in adult *Drosophila. J. Neurosci.* **17,** 4700–4710.
3. Wyman, R. J., Thomas, J. B., Salkoff, L., and King, D. G. (1984). The *Drosophila* giant fiber system. *In* "Neural Mechanisms of Startle Behavior" (R. C. Eaton, ed.), pp. 133–161. Plenum, New York.
4. Brand, A. H., and Perrimon, N. (1993). Targeted gene expression as a means of altering cell fates and generating dominant phenotypes. *Development* **118,** 401–415.
5. Sweeney, S. T., Broadie, K., Keanne, J., Niemann, H., and O'Kane, C. J. (1995). Targeted expression of tetanus toxin light chain in *Drosophila* specifically eliminates synaptic transmission and causes behavioral defects. *Neuron* **14,** 341–351.
6. Connolly, J. B., Roberts, I. J. H., Armstrong, J. D., Kaiser, K., Forte, Tully, T., and O'Kane, C. J. (1996). Associative learning disrupted by impaired Gs signalling in *Drosophila* mushroom bodies. *Science* **274,** 2104–2107.
7. Ferveur, J.-F., Störtkuhl, K. F., Stocker, R. F., and Greenspan, R.J. (1995). Genetic feminization of brain structures and changed sexual orientation in male *Drosophila. Science* **267,** 902–905.
8. Nakayama, S., Kaiser, K., and Aigaki, T. (1997). Ectopic expression of sex-peptide in a variety of tissues in *Drosophila* females using the P[GAL4] enhancer-trap system. *Mol. Gen. Genet.* **254,** 449–455.
9. O'dell, K. M. C., Armstrong, J. D., Yang, M. Y., and Kaiser, K. (1995). Functional dissection of the *Drosophila* mushroom bodies by selective feminization of genetically defined subcompartments. *Neuron* **15,** 55–61.

10. Jin, P. (1998). "Neural Plasticity and Its Molecular Mechanisms in *Drosophila.*" Ph.D. thesis, University of Massachusetts.
11. Egelhaaf, M. (1989). Dynamic properties of two control systems underlying visually guided turning in house-flies. *J. Comp. Physiol. A* **161,** 777–783.
12. Heide, G. (1983). Neural mechanisms of flight control in Diptera. *In* "BIONA Report" (W. Nachtigall, ed.), pp. 35–52. Akademie der Wissenschaften und der Literatur Mainz, New York.
13. Heide, G., and Götz, K. G. (1996). Optomotor control of course and altitude in *Drosophila melanogaster* is correlated with distinct activities of at least three pairs of flight steering muscles. *J. Exp. Biol.* **199,** 1711–1726.
14. Tu, M. S., and Dickinson, M. H. (1996). The control of wing kinematics by two steering muscles of the blowfly (*Calliphora vicina*). *J. Comp. Physiol. A* **178,** 719–730.
15. Fayyazuddin, A., and Dickinson, M. H. (1996). Haltere afferents provide direct, electrotonic input to a steering motor neuron in the blowfly, *Calliphora. J. Neurosci.* **16,** 5225–5232.
16. Pringle, J. W. S. (1948). The gyroscopic mechanism of the halteres of Diptera. *Phil. Trans. Roy. Soc. Lond. Ser. B* **233,** 347–381.
17. Cole, E. S., and Palka, J. (1982). The pattern of campaniform sensilla on the wing and haltere of *Drosophila melanogaster* and several of its homeotic mutants. *J. Embryol. Exp. Morphol.* **71,** 41–61.
18. Chan, W. P., and Dickinson, M. H. (1996). Position-specific central projections of mechanosensory neurons on the haltere of the blowfly, *Calliphora vicina. J. Comp. Neurol.* **369,** 405–418.
19. Zimmermann, H. (1993). Electrical synapses. *In* "Synaptic Transmission: Cellular and Molecular Basis" (H. Zimmerman, ed.), pp. 120–123. Oxford Univ. Press, New York.
20. Strausfeld, N. J., and Bassemir, U. K. (1983). Cobalt-coupled neurons of a giant fiber system of Diptera. *J. Neurocytol.* **12,** 971–991.
21. Pereda, A. E., Bell, T. D, and Faber, D. S. (1995). Retrograde synaptic communication via gap junctions coupling auditory afferents to the Mauthner cell. *J. Neurosci.* **15,** 5943–5955.
22. Crompton, D., Todman, M., Wilkins, M., Ji, S., and Davies, J. A. (1995). Essential and neural transcripts from the *Drosophila shaking-B* locus are differentially expressed in the embryonic mesoderm and pupal nervous system. *Dev. Biol.* **170,** 142–158.
23. Krishnan, S. N., Frei, E., Swain, G. P., and Wyman, R. J. (1993). *Passover:* A gene required for synaptic connectivity in the giant fiber system of *Drosophila. Cell* **73,** 967–977.
24. Homyk, T. Jr., Szidonya, J., and Suzuki, D. T. (1980). Behavioral mutants of *Drosophila melanogaster.* III. Isolation and mapping of mutations by direct visual observations of behavioral phenotypes. *Mol. Genet* **177,** 553–565.
25. Hausen, K., Hengstenberg, R., and Wiegand, T. (1988). Flight control circuits in the nervous system of the fly: convergence of visual and mechanosensory pathways onto motoneurons of steering muscles. Sense organs between environment and behavior. *In* "Proceedings of the 16th Göttingen Neurobiology Conference" (N. Elsner and F. G. Barth, eds.), p. 130. Thieme, New York.
26. Hengstenberg, R., Hausen, K., and Hengstenberg, B. (1988). Cobalt pathways from haltere mechanoreceptors to inter- and motoneurons controlling head posture and flight steering in the blowfly Calliphora. Sense organs between environment and behavior. *In* "Proceedings of the 16th Göttingen Neurobiology Conference" (N. Elsner and F. G. Barth, eds.), p. 129. Thieme, New York.
27. Baird, D. H., Schalet, A. P., and Wyman, R. J. (1990). The *Passover* locus in *Drosophila melanogaster:* Complex complementation and different effects on the giant fiber neural pathway. *Genetics* **126,** 1045–1059.

28. Phelan, P., Nakagawa, M., Wilkin, M. B., Moffat, K. G., O'Kane, C. J., Davies, J. A., and Bacon, J. B. (1996). Mutations in shaking-B prevent electrical synapse formation in the *Drosophila* giant fiber system. *J. Neurosci.* **16,** 1101–1113.
29. Sun, Y.-A., and Wyman, R. J. (1996). *Passover* eliminates gap junctional communication between neurons of the giant fiber system in *Drosophila. J. Neurobiol.* **30,** 340–348.
30. Thomas, J. B., and Wyman, R. J. (1984). Mutations altering synaptic connectivity between identified neurons in *Drosophila. J. Neurosci.* **4,** 530–538.
31. Bässler, U. (1993). The femur-tibia control system of stick insects: A model system for the study of the neural basis of joint control. *Brain Res. Rev.* **18,** 207–226.
32. Trimarchi, J. R., and Schneiderman, A. M. (1993). Giant fiber activation of an intrinsic muscle in the mesothoracic leg of *Drosophila melanogaster. J. Exp. Biol.* **177,** 149–167.
33. Miller, A. (1994). The internal anatomy and histology of the imago of *Drosophila melanogaster. In* "Biology of *Drosophila* (M. Demerec, ed.), pp. 420–535. Cold Spring Harbor Laboratory Press, Cold Spring Harbor, NY.
34. Field, L. H., and Pflüger, H. J. (1989). The femoral chordotonal organ: A bifunctional orthopteran (*locusta migratoria*) sense organ. *Comp. Biochem. Physiol. A* **93,** 729–743.
35. Burns, M. D. (1974). Structure and physiology of the locust femoral chordotonal organ. *J. Insect Physiol.* **20,** 1319–1339.
36. Matheson, T. (1998). Response and location of neurons in the locust metathoracic femoral chordotonal organ. *J. Comp. Physiol. A* **166,** 915–927.
37. Hofmann, T., Koch, U. T., and Bässler, U. (1985). Physiology of the femoral chordotonal organ in the stick insect, *Cuniculina impigra. J. Exp. Biol.* **114,** 207–223.
38. Hofmann, T., and Kock, U. T. (1985). Acceleration receptors in the femoral chordotonal organ of the stick insect, *Cuniculina impigra. J. Exp. Biol.* **114,** 225–237.
39. Burrows, M. (1987). Parallel processing of proprioceptive signals by spiking local interneurons and motor neurons in the locust. *J. Neurosci.* **7,** 1064–1080.
40. Burrows, M., Laurent, G. J., and Field, L. H. (1988). Proprioceptive inputs to nonspiking local interneurons contribute to local reflexes of a locust hind leg. *J. Neurosci.* **8,** 3085–3093.
41. Shanbhag, S. R., Singh, K., and Singh, R. N. (1992). Ultrastructure of the femoral chordotonal organs and their novel synaptic organization in the legs of *Drosophila melanogaster* Meigen (Diptera: Drosophilidae). *J. Insect Morphol. Embryol.* **21,** 311–322.
42. Waterman-Storer, C. M., and Hozbaur, E. L. F. (1996). The product of the *Drosophila* gene, *Glued,* is a functional homologue of the $p150^{glued}$ component of the vertebrate dynactin complex. *J. Biol. Chem.* **271,** 1153–1159.
43. Bässler, U. (1977). Sensory control of leg movements in the stick insect *Carausius morosus. Biol. Cybern.* **25,** 61–72.
44. Bräunig, P., and Hustert, R. (1985). Actions and interactions of proprioceptors of the locust hind leg coxo-trochanteral joint. I. Afferent responses in relation to joint position and movement. *J. Comp. Physiol.* **157,** 73–82.
45. Bräunig, P., and Hustert, R. (1985). Actions and interactions of proprioceptors of the locust hind leg coxo-trochanteral joint. II. Influence on the motor system. *J. Comp. Physiol.* **157,** 83–89.
46. Büschges, A., and Schmitz, J. (1991). Nonspiking pathways antagonize the resistance reflex in the thoraco-coxal joint of stick insects. *J. Neurobiol.* **22,** 224–237.
47. Cruse, H., Dean, J., and Suilmann, M. (1984). The contributions of diverse sense organs to the control of leg movements by a walking insect. *J. Comp. Physiol.* **154,** 695–705.
48. Graham, D., and Wendler, G. (1981). The reflex behavior and innervation of the tergo-coxal retractor muscles of the stick insect *Carausius morosus. J. Comp. Physiol.* **143,** 81–91.
49. Kuenzi, F., and Burrows, M. (1995). Central connections of sensory neurons from hair plate proprioceptors in the thoraco-coxal joint of the locust. *J. Exp. Biol.* **198,** 1589–1601.

50. Pearson, K. G., Wong, R. K. S., and Fourtner, C. R. (1976). Connexions between hair-plate afferents and motoneurones in the cockroach leg. *J. Exp. Biol.* **64,** 251–266.
51. Pflüger, H. J., Bräunig, P., and Hustert, R. (1981). Distribution and specific central projections of mechanoreceptors in the thorax and proximal leg joints of locusts. II. The external mechanoreceptors: Hair plates and tactile hairs. *Cell Tissue Res.* **216,** 79–96.
52. Schmitz, J. (1986). The depressor trochanteris motorneurones and their role in coxo-trochanteral feedback loop in the stick insect *Carausius morosus. Biol. Cybern.* **55,** 25–34.
53. Schmitz, J. (1986). Properties of the feedback system controlling the coxa-trochanter joint in the stick insect *Carausius morosus. Biol. Cybern.* **55,** 35–42.
54. Wong, R. K. S., and Pearson, K. G. (1976). Properties of the trochanteral hair plate and its function in the control of walking in the cockroach. *J. Exp. Biol.* **64,** 233–249.
55. Hodgkin, N. M., and Bryant, P. J. (1978). Scanning electron microscopy of the adult of *Drosophila melanogaster. In* "The Genetics and Biology of *Drosophila*" (M. A. Ashburner and T. R. F. Wright, eds.), pp. 337–340. Academic Press, New York.
56. Kankel, D. R., Ferrs, A., Garen, S. H., Harte, P. J, and Lewis, P. E. (1980). The structure and development of the nervous system. *In* "The Genetics and Biology of *Drosophila*" (M. A. Ashburner and T. R. F. Wright, eds.), pp. 295–363. Academic Press, New York.
57. Merritt, D. J., and Murphey, R. K. (1992). Projections of leg proprioceptors within the CNS of the fly *Phormia* in relation to the generalized insect ganglion. *J. Comp. Neurol.* **322,** 16–34
58. Murphey, R. K., Possidente, G., Pollack, G., and Merritt, D. J. (1989). Modality specific axonal projections in the CNS of the flies *Phormia* and *Drosophila. J. Comp. Neurol.* **290,** 185–200.
59. Kent, K. S., and Griffin, L. M. (1980). Sensory organs of the thoracic legs of the moth *Manduca sexta. Cell Tissue Res.* **259,** 209–223.
60. Smith, S. A., and Shepherd, D. (1996). Central afferent projections of proprioceptive sensory neurons in *Drosophila* revealed with the enhancer-trap technique. *J. Comp. Neurol.* **364,** 311–323.
61. Phillis, R. W., Bramlage, A. T., Wotus, C., Whittaker, A., Gramates, L. S., Seppala, D., Farahanchi, F., Caruccio, P., and Murphey, R. K. (1993). Isolation of mutations affecting neural circuitry required for grooming behavior in *Drosophila melanogaster. Genetics* **133,** 581–592.
62. Drysdale, R., Warmke, J., Kreber, R., and Ganetzky, B. (1991). Molecular characterization of *eag:* A gene affecting potassium channels in *Drosophila melanogaster. Genetics* **127,** 497–505.
63. Boschert, U., Ramos, R. G. P., Tix, S., Technau, G. M., and Fischbach, K.-F. (1990). Genetic and developmental analysis of irreC, a genetic function required for optic chiasm formation in *Drosophila. J. Neurogenet.* **6,** 153–171.
64. Byers, D., Davis, R. L., and Kiger, J. A. (1981). Defect in cyclic AMP phosphodiesterase due to the dunce mutation of learning in *Drosophila melanogaster. Nature* **289,** 79–81.
65. Qui, Y., Chan, C. N., Malone, T., Richter, L., Beckendorf, S. K., and Davis, R. L. (1991). Characterization of the memory gene dunce of *Drosophila melanogaster. J. Mol. Biol.* **222,** 553–565.
66. Ikeda, K., and Kaplan, W. D. (1970). Patterned neural activity of mutant *Drosophila melanogaster. Proc. Natl. Acad. Sci. USA* **66,** 765–772.
67. Ikeda, K., and Kaplan, W. D. (1974). Neurophysiological genetics in *Drosophila melanogaster. Am. Zool.* **14,** 1055–1066.
68. Koenig, J. H., and Ikeda, K. (1983). Characterization of the intracellularly recorded response of identified flight motor neurons in *Drosophila. J. Comp. Physiol.* **150,** 295–303.
69. Coggshall, J. C. (1978). Neurons associated with the dorsal longitudinal flight muscles of *Drosophila melanogaster. J. Comp. Neurol.* **177,** 707–719.

70. Ikeda, K., and Koenig, J. H. (1988). Morphological identification of motor neurons innervating the dorsal longitudinal flight muscle of *Drosophila melanogaster. J. Comp. Neurol.* **273,** 436–444.
71. Trimarchi, J. R., and Schneiderman, A. M. (1994). The motor neurons innervating the direct flight muscles of *Drosophila melanogaster* are morphologically specialized. *J. Comp. Neurol.* **340,** 427–443.
72. Swain, G. P., Wyman, R. J., and Egger, M. D. (1990). A deficiency chromosome in *Drosophila* alters neuritic projections in an identified motorneuron. *Brain Res.* **535,** 147–150.
73. Vandervorst, P., and Ghysen, A. (1980). Genetic control of sensory connection in *Drosophila. Nature* **286,** 65–67.

APPENDIX
LARVAL BODY WALL MUSCLES

A montage of confocally derived images of a phalloidin-FITC stained body wall muscle preparation of a third instar larva. The ventral midline (VM) is marked at the far left, the dorsal midline is to the right, and the anterior is to the top. In this preparation, hemisegment A2 is intact and the most internal muscles are numbered. The asterisks near muscle 7 denote that muscles 16 and 17 terminate at the segmental boundary, even though it appears here that they continue into the anterior segment. In A3, muscles 1–7 were dissected away, exposing intermediate level muscles. In A4, the numbered muscles of A3 were removed, revealing those closest to the cuticle. The semitransparent nature of the muscles in this figure demonstrates how the different layers are organized with respect to one another. The numerical terminology used in this figure is based on Crossley (A. Chiba, Ref. 23). Other investigators prefer to use a more descriptive nomenclature which can be seen in the review by M. Bate *et al.* (Fig. 1), where it is cross-referenced to the numerical terminology.

A few body wall muscles have been numbered differently by various authors over the last two decades. In 1978, Crossley described only 27 of the 30 muscles in each hemisegment, numbering them 1–23 and 26–29. He did not include muscles numbered as 24, 25, and 30. In this figure, one can see that muscle 25 is located deep at the intersegmental boundary in the ventral region. Muscle 30 is found between 14 and 15, and has often been mentioned in the literature as 14.2. Some authors have interchanged muscles 14 and 30. Muscle 24 is a deep lateral muscle positioned between 23 and 18. Crossley labeled 24 as 19 in his work, but did not describe what is now referred to as 19, which lies between 11 and 20 (see color insert on following page). Other authors have also interchanged 11 and 19 partly as a result of this confusion.

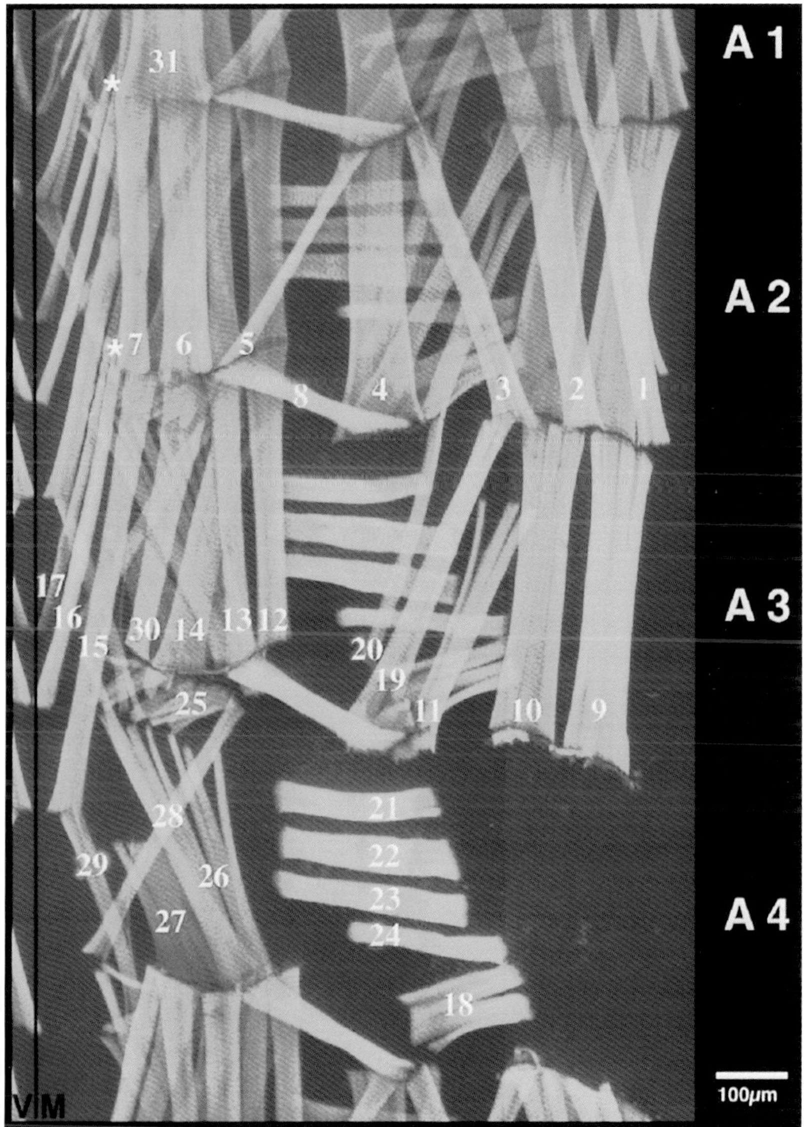
A 1
A 2
A 3
A 4
31
7
6
5
8
4
3
2
1
17
16
15
30
14
13
12
20
19
11
10
9
25
21
22
23
24
28
29
26
27
18
VM
100μm

INDEX

A

Abdomen, musculature, 227–228
abrupt, see clueless
aCC, *see* Anterior corner cell
AChR, *see* Nicotinic acetylcholine receptor
Act79b, 226, 232
Act88f, 226
Actin, 172, 226
α-Actinin, 18
Action potential, 245–246
Activation, 191
Active zones, *see* Vesicle recycling; Neuromuscular junction, presynaptic specialization
Activity-dependent structural plasticity, 127–131
ada, 177
Adaptin, 18, 174
Adaptor proteins, 174
Adenylyl cyclase, 124
Adepithelial cell, 225
Adult muscle precursors
  lineages and significance of founder cell gene expression, 33, 35–37
  segregation, 31
Afferent, *see* Haltere afferents; Proprioceptive sensory afferents
Aggregation; *see also* Clustering
  acetylcholine receptors, 99, 231
  glutamate receptors, 54–55
  glycine receptors, 99
Agrin, receptor clustering role, 99–100
ala, 107, 121, 130
Aldehyde fixative, 169
Aldose reductase, 209
Amiloride, 211
γ-Amino-butyric acid, 52
4-Amino-pyridine, 168, 196
amnesiac, 124
Amorph, 200
A1 channel, 200
4-AP, *see* 4-Amino-pyridine
4aPDD, 212
Amphiphysin, 172, 175, 181
Anchoring, 39
Aneural muscle, 104
ANT, *see* Anterior nerve tract
Antennal glomeruli, 242
Antennapedia, 182
Anterior commissure, 6
Anterior corner cell, muscle cell interactions, 5
Anterior nerve tract, 6
Anterograde staining, 255
Antimorph, 200
ap, muscle specification role, 37
AP-2, 149, 172, 174
Ap180, 172, 174
Aplysia, 107, 127
Apoptosis, 228
APs, *see* Adult muscle precursors
apterous, 35, 37
Argiotoxin, 120
ATPase, 142
Attraction, synaptic, 14–16
Augmentation
  gene mutation studies, 121, 123
  mechanism, 120
Auxilin, 173–174
Axolemma, 164
Axon ending, *see* Bouton
Axon pathways
  Drosophila embryo, 6–7
  pathfinding
    choice point
      hierarchy, 11–12
      mutual independence, 12–13
    fasciculation control, 13
    molecules, 10–11

B

Ball-and-chain domain, 196
bang-sensitive, 123

Basal lamina, 71
Basic helix-loop-helix, 27, 39
Bay K-8644, 211; *see also* 1,4-D: hydropyridine
beaten path, 8, 12–13
Behavior, *see specific behaviors*
Benzothiazepines, 210–211
bHLH, *see* Basic helix-loop-helix
BIM, *see* Bisindolylmaleimide
Biogenic amine; *see* Octopamine
Bipolar neuron, 144
Bisindolylmaleimide, 212
bithorax, 229
BK channels, 198
Blastoderm, 29
blow, see blown fuse
Blowfly, 231
blown fuse, 38
Body wall muscle, *Drosophila*
  adult versus larval studies, 193
  attachment formation, 39–40
  comparison to higher vertebrates, 25–26
  founder cells
    fusion, 38–39
    migration, 39
    model of myogenesis, 31–33
    progenitor cell lineages and gene expression significance, 33, 35–37
    segregation, 30–31
  origins of muscle-forming cells from embryonic mesoderm, 27–30
  preparation for electrophysiologic characterization, 192–193
  types of muscle, 25, 94, 265
B1MN, *see* B1 motor neuron
B1 motor neuron 243, 245–246, 250, 256
Botulinum toxin, 145–146
Bouton, *see also* Subsynaptic reticulum
  development and differentiation, 98, 127
  plasticity, 104
  type I
    neurotransmitters, 95
    subsets, 95
  type II, 95–97
  type III, 97
  ultrastructure, 94, 96
N-[2-(p-Bromocinnamyl)aminoethyl]-5-isoquinolinesulfonamide, 213
BuDR, 223
Butterfly, 231

C

Ca^{2+} ionophore, *see also* Calcium ionophore
cacophony, 211
Cadherin, 11, 13
Caenorhabditis elegans, 140
caki, functions, 103
Calcineurin, 170, 175
Calcium action potential, 202
Calcium channel
  calcium concentration and transmission amplitude, 59
  currents in larval body wall muscle
    functions, 209
    L-type current, 209–212
    modulation by second messengers, 212–213
    T-type current, 211
  currents in synaptogenesis, 48
  neurotransmitter release role, 139–140
  pharmacology, 209–211
  subunit structure, 211
Calcium dependence
  synaptic vesicle recycling, 169–170
  synaptic vesicle release, 59, 121, 140, 142, 148–154
Calcium ionophore, *see specific ionophores*
Calcium sensor, 59
Calcium-free saline, 170, 194, 196
Calcium/calmodulin-dependent protein kinase II
  long-term potentiation role, 121
  modulation of potassium channels, 206
  presynaptic plasticity role, 107, 121
  structural plasticity role, 130
Calliphora, 246
CaM kinase II, *see* Calcium/calmodulin-dependent protein kinase II
CaMGUK, 103
cAMP, *see* Cyclic AMP
CAMs, *see* Cell adhesion molecules
Capacitance, 171, 180–181
CasK, 103
Cation channel, *see specific channels*
Cell adhesion molecules, 13, 15, 105, 107
Cell adhesion mutants, 130–131
Cell-attached patch, 50–52

Cell recognition, 1, 12–17, 102
Cellularization, 145
Central brain, 242
Central circuit, 242, 251
Central nervous system, 6, 10, 12, 223, 228, 241
Central neuron, 242–246, 253, 256–259
Central synapse, 241, 246, 251
cGMP, *see* Cyclic GMP
Channel opening, 198
Chaperone, 152
Chapsyn-110/PSD93
Chemical synapse, 248, 250–253
Choice point
  hierarchy, 11–12
  mutual independence, 12–13
Cibarial disc, 227
Cinchonine, 196
Cisternae, 168–169
Classical genetic, 142
clathrin heavy chain, 174, 181
Clathrin, 168–169, 172–173
Clathrin-associated AP-2 adaptor complex, 172, 174
Clear cored vesicle, 85, 87
Clostridial neurotoxin, 145
clueless, 12
Clustering
  glutamate receptors, 99–100, 103–104, 120
  synaptic proteins, 100–103
CNS, *see* Central nervous system
Coated pit, 168–169, 172
Coated vesicle, 168–169, 172
Cobalt backfill, 229
Coiled coil, Chapter 6, Figure 1 (color)
Collapsin, 16
Collared pits, 166, 168, 171
comatose, 144, 147
commissureless, 12, 17
Complexin, 140, 154
comt, see comatose
Conditional mutant, 166, 177, 181
Conditioning, olfactory, 242
Conductance, *see specific channels*
Confocal microscopy, 73–74, 81, 144, 171
Connectin
  mutagenesis screening for genes, 8
  synaptic attraction role, 15
Constitutive secretion, 139–140, 142–154, 156
Copulation, 221
Coracle, 102
COS7 cell, 101–102, 109
Cotransmitter, 84
Coupling coefficient, 48
Courtship behavior, 123, 211, 227
Coverslip mounting, *Drosophila* embryos for electrophysiological assay, 62–63
Coxa, *see* CXHP8; Hair plate reflex curcuit
Coxal hair plate, 246, 251, 253–255
Crayfish, 86
CREB, *see* Cyclic AMP response element binding protein
CRIPT, 102
Csp, *see* Cysteine string protein
c-syb, see Synaptobrevin
Curare, 180
Current clamp, 143
Current density, 48
Current injection, 19, 204
CXHP8, role in hair plate reflex curcuit, 251, 253–255
Cyclic AMP, 125, 130, 206, 212–213
  modulation of calcium channels, 212–213
  modulation of potassium channels, 206
Cyclic AMP-dependent protein kinase, 121, 125, 132, 213
Cyclic AMP response binding element, 108, 132–133
Cyclic GMP, 207
Cyclic GMP-dependent protein kinase, 206
Cysteine string protein, function in neurotransmitter release, 152–153
Cytoplasm cytoskeleton, 14, 17–18

D

DAG, *see* diacyl glycerol
Dead time, vesicle recycling assay, 179–180
Decapentaplegic, myogenesis role, 29–30
Deep Orange, 173, 176
Delayed rectifier, *see* Voltage-activated non-inactivating outward potassium current
delilah, 39
Dendrite, 251
Dendritic reorganization, 10
Dendritic spine, 78–79
Dendrotoxin, 196
Dense body, 86–89

Dense cored vesicle, 85–86, 96
Density, *see* Postsynaptic densities
Depolarization, 193–194, 233
Depression, 120–121, 177–178
Descending neuron, 242
Desensitization, 51–52
DFM, *see* Direct flight muscle
DG1uR, *see* Glutamate receptor
DHP, *see,* 1, 4-Dihydopyridine
Diacyl glycerol, 127, 212
1,4-Dihydopyridine, 126, 210–212
DiI, 5
Diltiazem, 210–211
Dipteran, 79, 246
Direct flight muscle, 226, 229
Discs-large protein
  domains, 101–102, 109
  fasciclin II interactions, 109–110
  receptor clustering role, 100–102, 109–110
  structural plasticity role, 105–106
  synaptic function studies of mutants, 103
dissatisfaction, 230
Dissection, *Drosophila* embryos for electrophysiological assay, 62–63, 192
DLAR, 8, 11
DLG, *see* Discs-large protein
DLM, *see* Dorsal longitudinal muscle
DlnsR, *see Drosophila* insulin receptor
DmcaIA, 211
DmcaID, 211
Dmef-2, muscle differentiation role, 38, 40
dMerlin, 102
dMoesin, 102
dnaJ, 152, Chapter 7, Figure 1 (color)
dnc, see dunce
Docking, 150–152
Doc2, 149
DOG, *see* sn-1,2-Dioctanoyl-glycerol
Dominant-negative, 182
Dorsal longitudinal fibers
  development, 225, 231–233
  function, 226
  intracellular recordings, 144
  motor neurons, 229
Dorsal longitudinal muscle, 144, 225–226, 229, 232–233, 256
Dorsal midline, 6, 265
Dorsal protein, 27
Dorsal thoracic muscle, 225–226
Dorsoventral muscle
  development, 232
  functions, 225–226
  motor neurons, 229
Dosage compensation, 200
Double mutant, 130, 178
Dpp, *see* Decapetaplegic
DPTP, 8, 11
Drosophila embryo
  axon pathways, 6–7
  cellular organization of neuromuscular system, 3–5
  motor neurons, 6
  synaptic target selections, 7
Drosophila insulin receptor, 85, 126
D-semaphorin II, synaptic inhibition role, 16
dunce, 106–107, 121, 206, 213
  presynaptic plasticity of mutants, 106, 121
  structural plasticity role, 130
DVM, *see* Dorsoventral muscle
Dye coupling, 47–48, 245–246
Dye injection, 5
Dynactin, 250
Dynamin, 18, 165–167, 169, 171–172, 174, 181
Dynein, 18

E

eag, see ether-a-go-go
Early endosome, 164, 170
Eclosion, 234
Ectoderm, 6, 30
Ectopic expression, 9
EJC, *see* Excitatory junctional current
EJP, *see* Excitatory junctional potential
Electrical coupling, 47–48
Electrical synapse, 250
Electrogenesis, 53
Electromyogram, 246, 248, 253, 255
Electrophysiology, 61, 140, 142–144, 173, 241
Electroretinogram, 144, 147, 151, 165
ELISA, *see* Enzyme-linked immunosorbent assay
Embryogenesis, 48, 58, 221
Endocytic vesicles, 164, 167, 169–171
Endocytosis, 163, 167, 173

Endoplasmic reticulum, 139
Endosome, 168
engrailed, 30
Enhancer trap, 156
en passant synapse, 80
Ensemble average, 198–199
Enzyme-linked immunosorbent assay, 109
Epidermis, 223, 225, 227
Epitope tag, 102
Eps15, 172, 175
ER, *see* Endoplasmic reticulum
Erect Wing, adult expression, 234
ERG, *see* Electroretinogram
Escape response, 226–227
ESP, *see* Evoked synaptic potential
EST, *see* Expressed sequence tag
ether-a-go-go
  mutation analysis, 207
  potassium channel subunit encoding, 206
  structure of protein, 206–207
N-Ethylmaleimide sensitive fusion protein
  mutational analysis, 145–148
  neurotransmitter release role, 140, 142
eve, see even-skipped
even-skipped, mesoderm division role, 29
Evoked response, 120, 150–155, 244, 245, 251
Evoked synaptic potential
  measurement, 143
Excitability, 202, 204, 207
Excitatory junctional current, 143
Excitatory junctional potential
  evoked potentials, 96
  mEJP, 82, 104–105
  mutant studies, 126, 131
  vesicle recycling, 178
Excitatory neurotransmitter, *see specific neurotransmitters*
Exocyst, 140
Exocytosis, 69, 178–180
Expanded, 102
Expressed sequence tag, 154
Extensor tibia motorneuron, 247–248, 250
Eye-antennal disc, 227

F

Facilitation, 61, 95, 120–121, 123, 127
  long-term, 132
  paired-pulse, 61
  short-term
    development, 61
    gene mutation studies, 121, 123
    mechanism, 120
Fasciclin
  fasciclin I, 131–132
  fasciclin II
    discs-large protein interactions, 109–110
    expression and postsynaptic density dimensions, 83, 107
    long-term plasticity role, 132
    plasticity role, 108
    retrograde signaling, 105
    synapse stabilization, 107–108
    synaptic attraction role, 15, 109
  fasciclin III, 15, 18, 54, 98
  fasciculation control, 13
  structural plasticity roles, 131
Fast calcium-activated potassium current, 48–49, 193–194, 196, 198, 201–202, 206–207, 233
Fast extensor of the tibia motorneuron, 247–248, 250
Fast motor neuron, 86
Fatigue, 59–60
FEMO, *see* Femoral chordotonal organ
Femoral chordotonal organ, role in leg resistance reflex circuit, 247–248, 250
Femur, 247
FETI, *see* Fast extensor of the tibia motorneuron
Fibrillar muscle, 226
Filopodia, 17–18, 71, 74
Fission, *see* Vesicle recycling
Flickering, 198
Flight, 144, 147, 228–229
Flight muscle, 166, 193, 202, 225
Flight-related reflex circuit
  abolishment by *shaking-B*^{2} mutation, 245–246
  B1 motor neuron activation, 243, 245–246
  haltere sensory neuron role, 243, 245–246
  synapses, 243, 245
FLP/FRT, 234
FM1-43, 144, 170–171, 173, 177–180

FMRFamide, excitatory actions on body wall muscles, 126
Forskoline, 213
Founder cell
  fusion, 38–39
  migration, 39
  model of myogenesis, 31–33, 224
  progenitor cell lineages and gene expression significance, 33, 35–37
  segregation, 30–31
Founder cell model, 30–33, 35–36, 224
Founder myoblast, 30–33, 35–36, 38–40, 224
14-3-3 protein, 121
Frazzled, 16
Freeze fracture, 82
Frequenin, structural plasticity role, 130
Frog neuromuscular junction, 164–165, 171, 179–180
Frog, *see* Frog neuronuscular junction; *Xenopus* oocyte
fruitless, 227, 230, 232
Fusion-competent cell, 32–33

G

GABA, *see* γ-Amino-butyric acid
Gain-of-function, 152
GAL4/UAS system, behavioral circuit analysis
  developmental profiling of GAL4 expression, 259
  GAL4 enhancer trap, 103, 229, 234
  limitations and precautions, 241–242, 258–259
  tetanus toxin targeting in reflex circuits, 253–254, 256, 258
Gap junction, 243, 246
Garland gland, 145
Gastrulation, 29
Gene dosage, 200–202
Genetic dissection, 7–10
Genetic null, 245
Genetic screen, 7–8
Gephyrin, receptor clustering role, 99–100
GFP, *see* Green fluorescent protein
Giant fiber circuit, 229
Giant fiber neuron, 229
Giant fiber pathway, 229
GKAP, 102
Glia, 150
Glial sheath, 71
Glued[1], abolishment of leg resistance reflex circuit, 250
GluR, *see* Glutamate receptor
Glutamate, 52, 79, 84, 94, 120, 143
Glutamate receptor
  cation specificity, 50
  classes, 50–51, 99, 120
  clustering, 54–55, 98–101, 103–104
  distribution in myotubule membrane during development, 52–53
  ionotropic versus metabotropic, 120
  maturation of postsynaptic currents, 50–53
  muscle-specific genes, 50
  postsynaptic densities, 81
  synaptic transmission maturation, 57–59
Glycine receptor, clustering at inhibitory synapses, 99
Glycophorin, 102
Goldfish retinal bipolar synapses, 171
G-protein, 124–126
Grasshopper, 88
Green fluorescent protein, 5, 234
GRIP, 104
Grooming, 247, 251, 254–255
Grooming screen, 254–255
groovin, 39
Growth cone
  motility control, interfacing specific recognition, 17–18
  synaptic transmission maturation, 57–58
Growth cone stage, junctional aggregate development, 71–74
Growth factors, 164
Gsα, 121, 125
GTPase, 125, 165
GTP-binding protein, 152
Guanylate-kinase-like domain, 100–103
GUK domain, *see* Guanylate-kinase-like domain
Gyroscopic sensory organ, 243

H

Hair plate, 246, 251, 253–255
Hair plate reflex curcuit
  CXHP8 sensory neurons, 251, 253–255
  disruption in *unsteady,* 254–255

motor neuron activation, 251, 253–255
tetanus toxin targeting, 253–254, 256
Half-time, membrane internalization in synapes, 179
Haltere afferents, 243, 245–246, 256
Haltere disc, 225
Haltere sensory neuron, role in flight-related reflex circuit, 243, 245–246
Hatching, 77, 104
Head eversion, 223
Head, musculature, 225–227
Heart, 209
Heat shock, 166
H-89, *see* *N*-[2-(p-Bromocinnamyl)aminoethyl]-5-isoquinolinesulfonamide
Hemidesmosome, 83, 87
Hemisegment, 25, 27, 94, 227–228
Hemolymph like saline, 202
Heteroallelic combination, 150
Heterologous expression, 196, 213
Heteromultimeric, 125
Heterophilic interaction, 16
Heterozygote, 201
Hippocampus, 127, 179
Histoblast nest, 223, 227
Histolysis, 193
Hk, see Hyperkinetic
HL3, *see* Hemolymph-like saline
Holding potential, 196, 201
Holometabolous, 228
Homeobox, 32
Homeodomain, *see* LIM homeodomain protein
Homer, 104
Homomultimeric channels, 207
Homophilic, 53, 131
Homozygote, 152, 153, 201
HOOK, domain, 100–103
Hook, 173, 176
Horseradish peroxidase, 165, 229
Hotspots, 172
HRP, *see* Horseradish peroxidase
α-HRP, 79
Hrs2, 154
Hsc70, 173
Human, 1
Hydroxyurea, 227
Hyperexcitable mutant, 106, 123, 128–130
Hyperkinetic
mutation analysis, 207
potassium channel subunit encoding, 206–207
structure of protein, 207, 209
Hyperpolarization, 202
Hypomorph, 83, 103, 108–109, 149, 151

I

I_A, *see* Voltage-activated transient outward potassium current
IBMX, *see* 3-Isobutyl-1-methylxanthine
I_{CF}, *see* Fast calcium-activated potassium current
I_{CS}, *see* Slow calcium-activated potassium current
IFM, *see* Indirect flight muscle
I_K, *see* Voltage-activated non-inactivating outward potassium current
Imaginal disc, 223
Immunoflourescence microscopy, high resolution observation of developing neuromuscular system, 5
in situ hybridization, 192
Inactivation, channels, 194, 196
Indirect flight muscle, 225–226
Induction, 53–57
Inhibition, synaptic, 16
Inhibitory input, 52
Inhibitory potential, 256
Inositol trisphosphate, 127, 212
inscuteable, 34
Insect muscle, 209, 212
Instar, 80–81, 98, 107, 221
Insulin-like peptide, 82–83, 85, 94
Integrins, 18, 39
Intermediate membrane, 164, 170–171
Internalization, collared pits, 168
Interneuron, 246
Intersegmental nerve, 6, 10, 77
Intracellular recording, 62–63, 143, 256
Inward current, 196
Ion transport, *see specific channels*
Ionic currents, *see specific channels*
Ionophore, *see specific ionophores*
Ionotropic, *see* Glutamate receptor
Iontophoresis, 52
IP3, *see* Inositol trisphosphate
irregular optic chiasm A, 255

ISN, *see* Intersegmental nerve
ISNt, *see* Anterior nerve tract
3-Isobutyl-1-methylxanthine, 213
Isometric contractions, 193, 202
Isopotential, 192
Isradipine, 210

J

Jump muscle, *see* Tergo-trochantral muscle
Jumping, 226–227, 229, 247
Junctional aggregate, development
  growth cone stage, 71–74
  prevaricosity stage, 74
  varicosity stage, 74–75, 77–78
Junctional inhibitory input, 50
Junctional potential, 248

K

KD channel, 200
Kinesin, 18
Kinetics, 198, 204–205, 208
"Kiss and run" mechanism, 171
knockout, 35–36
ko, see knockout
Kr, see Kruppel
Kruppel, muscle specification role, 35–37
KST channel, 200

L

Labial disc, 227
Lamina, 74, 80, 88
Lamprey, 182
Late bloomer, 18
Latency, 245–246, 253, 255
Lateral inhibition, 28, 30–31
Latrotoxin, 153
Learning, 119, 121, 127
Leg disc, 225
Leg resistance reflex circuit
  abolishment by *Glued*[1] mutation, 250
  femoral chordotonal organ role, 247–248, 250
  motor neuron activation, 247–248, 250
leonardo, short-term plasticity effects of mutations, 121
Lepidoptera, 231
lethal of scute, muscle differentiation role, 30–31
Leucine-rich repeat, 16
Leukokinin I-like peptide, 94, 126
Light-induced response, 226
Light microscropy, 71, 85
LIM homeodomain protein, 35
linotte, 121
Lipophilic dye, 5
Locust, 88, 120, 126
Long-term Potentiation, 121, 127
Long-term augmentation, development, 61
Long-term facilitation, CREB role, *see* Facilitation
Loss-of-function, 8, 167
  mutant, partial, 149–150
LRR, *see* Leucine-rich repeat
l'sc, see lethal of scute
LT muscles, 37
LTF, *see* Long-term facilitation
LTP, *see* Long-term potentiation
l(2)35Fα, 211
L-type calcium channels, *see* Calcium channel
L-type current, *see* Calcium channel
Lucifer yellow, 5

M

Macroscopic current, 200
MADS box, 38
MAGUK, *see* Membrane-associated guanylate kinase
Male-specific muscle, development, 227, 230, 232
Manduca, 80, 82–83, 86, 88, 223, 228, 231
MAPK, *see* Mitogen-activated protein kinase
mbc, see myoblast city
Mechanical stimulus, 251–252
Mechanosensory, 251–252
mef-2, see myocyte-enhancer factor 2
Membrane compartment, 164, 167–170
Membrane internalization, 165
Membrane patches, *see* Patch clamp
Membrane-associated guanylate kinase, 100–104, 110
Memory, 119, 121, 127
Mesoderm, 28–29, 31, 224
mESP, *see* Miniature excitatory synaptic potential
Metabotropic, *see* Glutamate receptor

Metamorphosis, muscle development, 221, 223
Microelectrode, 256
Microscopic current, 198
Microtubule, 165
Microtubule bundling, 25–27
Midgut, 145
Migration, precursors, 38–40
Miniature excitatory synaptic potential, 143, 145, 151, 153
Mitochondria, 86–87
Mitogen-activated protein kinase, 125
Modulation, 60–61, 212
MOL, *see* Muscle of Lawrence
Molecular motor, 165
Monosynaptic, 253
Mosaic, 234
Motor neuron
  adult structure, 223, 228–229
  axonogenesis, 10–13
  flight motor neurons, 228–229
  jump motor neurons, 229
  muscle development role
    orthograde signaling, 231–232
    retrograde signaling, 232–233
  overview in *Drosophila,* 6, 228
  pharyngeal motor neurons, 229–230
  sensory-to-motor reflexes, *see* Reflex neural circuits
  synaptic target recognition
    accuracy of targeting, 14
    dynamism of target cell biology, 17
    molecules, 11, 13
    synaptic attraction, 14–16
    synaptic inhibition, 16
Mouse, 131, 151–152
MSM, *see* Male-specific muscle
Munc13, 140, 145, 149, 154
Munc18, 140
Muscle contraction, 226
Muscle founder, 30–33, 35–36
Muscle of Lawrence, *see* Male-specific muscle
Muscle progenitor, 25–27, 31
Mushroom bodies, 242
Mutagenesis screen, neuromuscular development genes, 7–8, 65
Myoblast fusion, 3, 25, 30–33, 38–40, 46, 224
myoblast city
  mutant analysis, 224
  retrograde induction of presynaptic specialization, 55, 57
Myocyte, 200
myocyte-enhancer factor-2, retrograde induction of presynaptic specialization, 55–57, 104
MyoD, muscle differentiation role, 37–38
Myogenesis
  origins of muscle-forming cells from embryonic mesoderm, 27–30
  pathway
    conservation, 27
    general pathway, 37–40
  specification, 30–37
Myosin, 18
Myotubule, comparison between *Drosophila* and higher vertebrates, 26–27

N

nap, see no action potential
nau, see nautilus
nautilus, muscle differentiation role, 37–38
NCAM, *see* Neural cell adhesion molecule
Nerve entry point, 70, 72, 88
Netrin B, synaptic attraction role, 15–16
Neural cell adhesion molecule, structural plasticity role, 131
Neurexin, protein–protein interactions, 102–103
Neurobiotin, 245–246
Neurofibromatosis type 1, 125, 206
Neurogenesis, 6
Neurogenic gene, 30–31
Neuroglian, 9
Neurohemal synapse, 88, 97
Neurohormone, 126
Neuroligin, 102
Neuromere, 223
Neuromodulation, 70–71, 78, 88–89, 96, 124–126
Neuromuscular junction
  development, overview, 221, 230–231
  electrophysiological assays in *Drosophila* embryos, 61–64
  history of study, 2–3
  junctional aggregate development
    growth cone stage, 71–74

Neuromuscular junction (*continued*)
prevaricosity stage, 74
varicosity stage, 74–75, 77–78
morphology in adults, 230
nerve types, 70–71
neuropeptides, 124–127
plasticity, 104–110, 119–121, 123–133
postsynaptic densities
adhesion, 80
diameters, 81–82
fasciclin II expression and dimensions, 83
glutamate receptors, 81
type I terminals, 83
type II terminals, 83
type III terminals, 83–85
ultrastructure, 81
presynaptic specializations
active zones, 171–172
T bar, 56, 86, 95, 144, 146
dense bodies, 86–89
microscopy, 85
type I terminals, 86–87
type II terminals, 87–88
vesicle types, 85–86
subsynaptic reticulum
developmental changes, 79–80
distribution in neuromuscular junctions, 78
fixation, 80
function in type I terminals, 78–79
structure, 79
synaptogenesis, 46–50
Neuropeptide, *see specific neuropeptides*
Neuropil, 10
Neurosecretory, *see* Neurotransmitter release
Neurotransmitter release
calcium role, 139–140, 148–150
functional protein identification, 153–154
measuring of vesicle release
biochemical fractionation, 144–145
dorsal longitudinal fibers, intracellular recordings, 144
electrophysiological potentials, 143
electroretinograms in adults, 144
patch-clamp, 143–146
transmission electron microscopy, 144
voltage clamp, 143
SNARE hypothesis
mutational analysis of SNARE complex, 145–148, 151
overview, 140, 142
synaptotagmin regulation, 148–150
synaptic proteins, homology between species, 140–141
Neurotrophins, 164
NF1, *see* Neurofibromatosis type 1
Nicotinic acetylcholine receptor, clustering, 99, 231
Nifedipine, 210–211
Nitrendipine, 210
nj, see nonjumper
NMDA receptor, 103–104
no action potential, 106, 128–130
nonjumper, 226
norpA, 212
Notch signaling, myogenesis, 30, 33, 35
Notch, 30, 33
n-Sec1, 140
NSF, *see N*-Ethylmaleimide sensitive fusion protein
n-syb, see Synaptobrevin
Null mutation, 54–55, 107, 149
Numb, blocking of Notch signaling in myogenesis, 33, 35

O

Octopamine, 94, 96, 125–126
muscle distribution, 125
receptors and signal transduction, 125–126
release by type II endings, 96–97, 125
Octopamine/tyramine receptor, 126
OcTyR, *see* Octopamine/tyramine receptor
Olfactory jump, 241
Olfactory learning, 123, 241
On/off transients, 144, 151, 165
Oocyte, *see Xenopus* oocyte
Orthograde signaling, 230–232
Outward current, 196, 198
Oviposition, 224
Oxidation-reduction, 209
Oxidative metabolism, 86–87

P

PACAP, *see* Pituitary adenylyl cyclase-activated polypeptide

Pair-rule gene, 28
Paired-pulse facilitation, development, *see* Facilitation
para, see paralytic
paralytic
  neural induction of postsynaptic specialization, 54–55
  structural plasticity role, 128
pas, see passover
Passive electrical properties, 192
Passive response, 49–50
passover, 226
Patch-clamp, neuromuscular junction assays in *Drosophila* embryos, 63–64
Pathfinding, *see* Axon pathways
PDD, *see* Phorbol 12,13-didecanoate
PDZ domain, 18, 100–104, 109–110
Peak current, 196
Penetratin, 182
Perforated patch, 64
Peristalsis, 58
Permissive temperature, *see* Temperature-sensitive
Persistent *twist* cells, 71
Pharmacological profile, *see specific drugs and channels*
Pharyngeal dilator, 227
Pharyngeal motorneuron, 229
Pharyngeal muscle, 227, 229
Phasic, 86
Phenylalkylamines, 210–211
Phorbol 12,13-didecanoate, 212
Phorbol 12-myristate, 13-acetate, 212
Phosphatases, 8, 9, 170, 172
Phosphatidyl inositol bisphosphate, 127, 212
Phosphodiesterase, 106, 206, 213
Phosphoinosotide cascade, 172
Phospholipase C, signal transduction at neuromuscular junction, 126–127
Phospholipid, *see specific phospholipids*
Phosphoprotein, 207
Phosphorylation, 148
Phosphotyrosine kinase, *see specific kinases*
Photoreceptor, 212
PIP2, *see* Phosphatidyl inositol bisphosphate
Pits, 168, 171, 181
Pituitary adenylyl cyclase-activated polypeptide
  depolarization of larval body wall muscles, 124
  modulation of potassium channels, 206
  NF1 mutation and defective response, 125
  receptors, 124
  signal transduction pathways, 124–125
PKA, *see* cAMP-dependent protein kinase
PKC, *see* Protein kinase C
PKG, *see* cGMP-dependent protein kinase
Plasma membrane, 139, 164, 168–169
Plasmalemma, 170
Plasticity
  neuromuscular junction structural plasticity, 104–110
  secondary messenger systems at neuromuscular junction, 119–121, 123–133
Plateau, 194
PLC, *see* Phospholipase C
Pleiotropic effect, 166–167
PN200-110, *see* Isradipine
PNT, *see* Posterior nerve tract
Polyphosphoinositide, 149
p115, 154
Pore-forming region, 196
Posterior commissure, 6
Posterior nerve tract, 6
Postsynaptic densities, *see* Neuromuscular junction
Postsynaptic receptor field, 54
Posttetanic potentiation
  development, 61
  mechanism, 120–121
Postural defect, 250
Potassium, *see* Potassium channel
Potassium channel
  currents in larval body wall muscle
    calcium-activated currents
      fast I_{CF}, 196, 198
      slow I_{CS}, 196, 198
    roles in muscle membrane excitability, 202
    single channel currents, 198, 200
    voltage-activated currents
      delayed I_K, 193–194, 196
      pharmacological profiles of channels, 196–197
      transient I_A, 193–194, 196, 233

Potassium channel (*continued*)
currents in neuromuscular junction synaptogenesis, 48–50
presynaptic plasticity of mutants, 106
second messenger modulation
calcium, 204, 206
cyclic AMP, 206
neuropeptides, 206, 214
Shab channel
gene, 196, 200
pharmacological profile, 196–197
Shaker channel
clustering, 101, 109–110
gene discovery, 194, 213
gene dosage dependence of current, 200–201
pharmacological profile, 196–197
short-term plasticity effects of mutations, 121, 123
structure, 194, 196
structural plasticity role, 128–130
subunits, functional overview, 206–207, 209
pox meso, 30
PPF, *see* Paired-pulse facilitation
Prepulse, 194
Prevaricosity stage, junctional aggregate development, 74
Primordia, 223
Proctolin, 88, 94, 126
Progenitor cell, 33, 35–37
Propagation, 79
Proprioception, 251, 253–255
Proprioceptive sensory afferents, 251, 254–255
pros, see prospero
prospero, neural induction of postsynaptic specialization, 53–54, 98
Protein kinase C, 175, 206
Protein 4.1, 102
Prothoracic leg motorneuron, 247–248, 250
PSD95/SAP90, 100–103, 106
Ptilinium retractor, 227
PTP, *see* Posttetanic potentiation
Pupariation, 227, 233
P[LacZ], 32

Q

Quantal content, 143
Quinidine, 196
Quinine, 196

R

Rab, 151–152
Rab3A
mutational analysis, 151–152
neurotransmitter release role, 140, 142
Rabphilin, 140, 149
Raf, 125
Rapsyn, receptor clustering role, 99–100
Ras, 125, 151, 206
RasGAP, *see* Ras GTPase activating protein
Ras GTPase activating protein, 125
Ras opposite protein, 151–152
Ras2, 151
Rectification, 49, 204
Rectifier, 193, 198
Reflex neural circuits
applications of mutant analysis, 260
flight-related reflex circuit
abolishment by *shaking-B*^{2} mutation, 245–246
B1 motor neuron activation, 243, 245–246
haltere sensory neuron role, 243, 245–246
synapses, 243, 245
hair plate reflex curcuit
CXHP8 sensory neurons, 251, 253–255
disruption in *unsteady,* 254–255
motor neuron activation, 251, 253–255
tetanus toxin targeting, 253–254, 256
identification techniques, 241–243, 259
intracellular recordings from central neurons, 256
leg resistance reflex circuit
abolishment by *Glued*^{1} mutation, 250
femoral chordotonal organ role, 247–248, 250
motor neuron activation, 247–248, 250
Regenerative action potential, 202
Regulated secretion, *see* Neurotransmitter release
Repetitive stimulation, 153
Repolarization, 202
Repulsion, 16
Reserve pool, 167
Resistance reflex, 247–248, 250

Restrictive temperature, *see* Temperature-sensitive
Reticulospinal synapses, 172
Retinula cell synapses, 168, 171
Retrograde signaling, 55–57, 104–105, 230–233
Retrograde staining, 248
Reversal potential, 50–51
Reverse genetic, 142
Rho subfamily small G proteins, 18
rolling stone, founder fusion role, 38–39
rop
  mutational analysis, 151
  overexpression phenotypes, 152
rop, *see* Ras Opposite
rost, see rolling stone
Rostral protractor, 227
roundabout, 12
RP neurons, 7, 10, 12, 14–16, 57, 74
rut, see rutabaga
rutabaga
  presynaptic plasticity of mutants, 106–107, 121
  structural plasticity role, 130
  rut-encoded adenylyl cyclase, 125, 127, 206
Rvs167, 172

S

S59, muscle specification role, 32, 35, 37
Sacl, 172
Saline, 196, 202
SAPAP, 102
SAP97, 109
SAP102, 106
Sarcolemma, 83
Scanning electron microscopy, 71, 73–74
SEC1, function in neurotransmitter release, 151–152
Sec8, 145, 154
Sec10, 154
Second messenger, *see specific messengers*
Secretion, *see* Neurotransmitter release
Secretory complex *see* Neurotransmitter release
Sed5, *see* syntaxin
Segmental nerve, 6–7, 10, 12–13
sei, see seizure
seizure, 128
SEM, *see* Scanning electron microscopy
Semaphorin, 16
Sensilla trichodea, 251
Sensory neuron, *see specific neurons*
Sensory-to-motor reflex, 241–243, 245–248, 250–251, 253–260
Septate junctions, 102
SETI, *see* Slow extensor of the tibia motorneuron
7S complex, 142, 147–148, 151, 154
Sexual behavior, 241
Sh, see Shaker
SH3 domain, 100–103
Shab channel, *see* Potassium channel
Shaker channel, *see* Potassium channel
Shaking-B^2, abolishment of flight-related reflex circuit, 245–246
shi, see shibire
shibire mutants
  calcium dependence of vesicle recycling, 169–170
  direct and indirect effects, 181–182
  dynamin function in vesicle fission, 165–166
  pleiotropic phenotypes, 166–167
  synapse recovery studies
    caveats, 167–168
    intermediate compartments in vesicle-depleted nerve terminals, 168–169
  temperature sensitive mutants, 167, 173, 177–178
Short-term facilitation, *see* Facilitation
Signal transduction mutant, 130
Simulium, 231
Single channel, 206
Skeletal muscle, *see* Body wall muscle
slo, see slowpoke
sloppy paired, mesoderm division role, 29–30
Slow calcium-activated potassium current, 48–49, 193–194, 196, 198, 206–207
Slow extensor of the tibia motorneuron, 247–248, 250
Slow motor neuron, 86
slowpoke
  gene dosage dependence of current, 201–202
  I_{CF} subunit encoding, 198, 233
slp, see sloppy-paired
SN, *see* Segmental nerve

sn-1, 2-Dioctanoyl-glycerol, 212
Snail, early mesoderm cell expression, 27–28
SNAP, isoforms, 140, 147–149
α-SNAP
  mutational analysis, 145–146, 148
  neurotransmitter release role, 140, 142
SNAP-25
  mutational analysis, 145–146
  neurotransmitter release role, 140, 142, 150
SNARE hypothesis, *see* Neurotransmitter release
Snt, *see* Posterior nerve tract
Sodium channel, 54, 79, 192, 196; *see also paralytic*
Soluble NSF attachment protein, *see* SNAP
Somatic muscle, 31
Spectrin/actin cytoskeleton, 102
Spike, 247
Splicing, 196
Spontaneous release, *see* Neurotransmitter release
sr, see stripe
SSR, *see* Subsynaptic reticulum
Steady-state input resistance, 64
STF, *see* Short-term facilitation
Stick insect, 251
still life, protein localization at type I boutons, 103
Stoichiometry, channel subunits, 213
Stoned, 175–176, 181
Stretch-activated channel, 49, 200
stripe, 39
Strontium, 209
Styryl dye, 144, 170
Subsynaptic reticulum
  developmental changes, 79–80
  distribution in neuromuscular junctions, 78
  fixation, 80
  function in type I terminals, 78–79
  plasticity, 105–106
  structure, 79
Subunit, potassium channels, 206–207, 209
Suction electrode, 62, 64, 143
Supercontraction, 202
SV2, 154
Synapse maturation, definition, 93
Synapsins, function in neurotransmitter release, 153
Synaptic bouton, 55–57, 108–109, 120, 127
Synaptic cleft, 139
Synaptic domain, 53–54, 98
Synaptic potential, *see* Evoked synaptic potential
Synaptic target recognition
  accuracy of targeting, 14
  dynamism of target cell biology, 17
  interfacing to synaptogenesis initiation, 18
  molecules, 11, 13
  synaptic attraction, 14–16
  synaptic inhibition, 16
Synaptic target selection, 7, 10
Synaptic transmission, 45–46, 57–60, 140
Synaptic vesicle, *see* Neurotransmitter release; Vesicle recycling
Synaptobrevin
  mutational analysis, 145–146
  neurotransmitter release role, 140, 142
Synaptogenesis
  definition, 45
  delayed specializations, 46
  electrical property maturation in embryonic myotubules, 46–50, 97–98
  electrophysiological assays in *Drosophila* embryos, 61–64
  glutamate receptors, 50–53
  ion channels, 46, 48–50
  neural induction of postsynaptic specialization, 53–55
  neuromuscular junctions in *Drosophila,* advantages for study, 46, 64–65
  retrograde induction of presynaptic specialization, 55–57
  synaptic modulation property maturation, 60–61
  synaptic transmission maturation, 57–60
Synaptoids, 87–88
Synaptojanin, 172
Synaptophysin, 140, 154
Synaptoporin, 154
Synaptosome, 166
Synaptotagmin
  C2 domain, 149–150
  calcium regulation of SNARE function, 148–150

domains, 149
genes, 149
mutational analysis, 149
vesicle recycling role, 172
Syncytial precursor, 28, 32
Syntaxin
mutational analysis, 145–146
neurotransmitter release role, 140, 142, 151–152
syt, see Synaptotagmin
syx, see Syntaxin

T

Tacrine, 196
Tail current, 196
Tannic acid, 81
Target membrane SNAP receptor, 142, 145–146, 151, 154
Target muscle, 4–5, 10, 12–16, 58, 63, 71, 93–94, 97–99, 231
Target recognition, *see* Synaptic target recognition
Target selection, *see* Synaptic target selection
Targeting, *see* Synaptic target recognition
T bar, *see* Neuromuscular junctions, presynaptic specializations, active zones
Tβ H, *see* Tyramine-β-hydroxylase
TDT, *see* Tergal depressor of the trochanter
TEA, *see* Tetraethylammonium
TEM, *see* Transmission electron microscropy
temperature-induced-paralysis E, 128–129
Temperature-sensitive, 166, 171
Tenebrio, 228
Tergal depressor of the trochanter, 226–227
Tergo-trochantral muscle
function, 226–227
motor neurons, 229
Tetanic stimulation, 165
Tetanus toxin, targeting in hair plate reflex curcuit, 253–254, 256
Tetraethylammonium, 202
Tetrodotoxin, 54
TGFβ, *see* Transforming growth factor β
Thoracic ganglia, 126, 244–245, 248–249, 252–253, 256–257
Thorax, musculature, 225–227
Threshold, 202
Tibia, 247
Tibial extensor muscle, 247–248, 250
Time constant, 171, 208
tinman, 28
tipE, see temperature-induced-paralysis E
TN, *see* Transverse nerve
toll, mutagenesis screening for genes, 8, 16
Tonic, 86
TPA, *see* Phorbol 12-myristate, 13-acetate
Tracers, 177
Trafficking, vesicles, 140, 142, 146
Transforming growth factor β, 29
Transgene, 55, 108, 142, 156
Transgenic *Drosophila,* loss of function studies, 8–10
Transmission electron microscropy, 85, 144
Transmission failure, 59–60
Transmission fidelity, 45–46, 59–60
Transporter, *see specific transporters*
Transverse nerve, 6
t-SNARE, *see* Target membrane SNAP receptor
tSXV motif, 101, 109
TTM, *see* Tergo-trochantral muscle
TTX, *see* Tetrodotoxin
T-type calcium channel, *see* Calcium channel
T-type current, *see* Calcium channel
Tubular fiber, 226
β1 Tubulin, 39
20S complex, 142, 148
29MN, 251, 253, 255
twi, see twist
twist, 28–29, 31, 38, 55, 98
Twist
adult expression, 234
early mesoderm cell expression, 27–29, 31, 98
retrograde induction of presynaptic specialization, 55, 57
Type I axon, 70, 77, 94
Type I bouton, *see* Bouton
Type I terminal, 70, 77, 78, 81, 83, 85–86
Type Ib bouton, *see* Bouton
Type Is bouton, *see* Bouton

Type II axon, 70, 77, 88, 95–97
Type II bouton, *see* Bouton
Type II terminal, 70, 74, 78–79, 83, 86–88
Type III axon, 70, 97
Type III bouton, *see* Bouton
Type III terminal, 70, 78, 83–86
Tyramine-β-hydroxylase, 96, 126
Tyrosine kinase, *see specific kinases*

U

UAS, *see* GAL4/UAS system
Uncoupling, 48
Unitary current, 51
unsteady, disruption of hair plate reflex curcuit, 254–255
Uptake, dye, 178–179
Uso1, 154

V

Vacuole, 147
VAMP, *see* Synaptobrevin
VA1, 35–37, 39
VAP, *see* Ventral adult precursors
Vap33, 154
Varicosity stage, junctional aggregate development, 74–75, 77–78
VA2, 35–37, 39
Ventral adult precursors, 35, 37, 39
Ventral ganglion, 143
Ventral midline, 6, 265
Ventral nerve cord, 62–64
Ventral thoracic muscle, 224
Ventral unpaired median motorneuron, 7
Verapamil, 210–211
Vertebrate, calcium channels, 209, 211
Vesicle budding, *see* Neurotransmitter release
Vesicle collapse, *see* Neurotransmitter release
Vesicle fusion, *see* Neurotransmitter release
Vesicle lumen, *see* Neurotransmitter release
Vesicle recycling
  active zones, 171–172
  assays
    capacitance measurements, 180–181
    fluorescent dye assays
      dead time measurement, 179–180
      uptake, 173, 178–179
    lethal mutants, 173, 177
    synaptic depression during high frequency stimulation, 177–178
  endocytosis, 163–164
  functional proteins
    identification, 164
    inhibition, 182
    roles, 172–176
  pathway types, 170–172
  shibire mutant analysis
    calcium dependence of vesicle recycling, 169–170
    direct and indirect effects, 181–182
    dynamin function in vesicle fission, 165–166
    pleiotropic phenotypes, 166–167
    synapse recovery studies
      caveats, 167–168
      intermediate compartments in vesicle-depleted nerve terminals, 168–169
    temperature sensitive mutants, 167, 173, 177–178
Vesicle SNAP receptors, 142
Vinculin, 18
Vitelline membrane, 62
Voltage-activated non-inactivating outward potassium current, 48, 193–194, 196, 198, 207
Voltage-activated transient outward potassium current, 48–50, 53, 103, 193–194, 196, 198, 200–202, 207
Voltage-clamp, 47–48, 103, 143, 193–194, 200
Voltage-gated channel, 46, 48–50, 124, 139, 191
Voltage pulse, 196
v-SNARE, *see* Vesicle SNAP receptors
VUM, *see* Ventral unpaired median motorneuron

W

Walking, 221, 247, 254
Whole cell patch-clamp, *see* Patch-clamp
Wing beat, 226, 243

Wing disc, 225
Wingless, myogenesis role, 29–30
Wnt, 30
W7, 206–207

X

Xenopus oocyte, 120, 198

Y

Yeast two-hybrid, 101
Yeast, 139

Z

Zinc-finger, 27
ZO1, 100

CONTENTS OF RECENT VOLUMES

Volume 33

Olfaction
S. G. Shirley

Neuropharmacologic and Behavioral Actions of Clonidine: Interactions with Central Neurotransmitters
Jerry J. Buccafusco

Development of the Leech Nervous System
Gunther S. Stent, William B. Kristan, Jr., Steven A. Torrence, Kathleen A. French, and David A. Weisblat

$GABA_A$ Receptors Control the Excitability of Neuronal Populations
Armin Stelzer

Cellular and Molecular Physiology of Alcohol Actions in the Nervous System
Forrest F. Weight

INDEX

Volume 34

Neurotransmitters as Neurotrophic Factors: A New Set of Functions
Joan P. Schwartz

Heterogeneity and Regulation of Nicotinic Acetylcholine Receptors
Ronald J. Lukas and Merouane Bencherif

Activity-Dependent Development of the Vertebrate Nervous System
R. Douglas Fields and Phillip G. Nelson

A Role for Glial Cells in Activity-Dependent Central Nervous Plasticity? Review and Hypothesis
Christian M. Müller

Acetylcholine at Motor Nerves: Storage, Release, and Presynaptic Modulation by Autoreceptors and Adrenoceptors
Ignaz Wessler

INDEX

Volume 35

Biochemical Correlates of Long-Term Potentiation in Hippocampal Synapses
Satoru Otani and Yehezkel Ben-Ari

Molecular Aspects of Photoreceptor Adaptation in Vertebrate Retina
Satoru Kawamura

The Neurobiology and Genetics of Infantile Autism
Linda J. Lotspeich and Roland D. Ciaranello

Humoral Regulation of Sleep
Levente Kapás, Ferenc Obál, Jr., and James M. Krueger

Striatal Dopamine in Reward and Attention: A System for Understanding the Symptomatology of Acute Schizophrenia and Mania
Robert Miller

Acetylcholine Transport, Storage, and Release
Stanley M. Parsons, Chris Prior, and Ian G. Marshall

Molecular Neurobiology of Dopaminergic Receptors
David R. Sibley, Frederick J. Monsma, Jr., and Yong Shen

INDEX

Volume 36

Ca^{2+}, *N*-Methyl-D-aspartate Receptors, and AIDS-Related Neuronal Injury
Stuart A. Lipton

Processing of Alzheimer Aβ-Amyloid Precursor Protein: Cell Biology, Regulation, and Role in Alzheimer Disease
Sam Gandy and Paul Greengard

Molecular Neurobiology of the $GABA_A$ Receptor
Susan M. J. Dunn, Alan N. Bateson, and Ian L. Martin

The Pharmacology and Function of Central $GABA_B$ Receptors
David D. Mott and Darrell V. Lewis

The Role of the Amygdala in Emotional Learning
Michael Davis

Excitotoxicity and Neurological Disorders: Involvement of Membrane Phospholipids
Akhlaq A. Farooqui and Lloyd A. Horrocks

Injury-Related Behavior and Neuronal Plasticity: An Evolutionary Perspective on Sensitization, Hyperalgesia, and Analgesia
Edgar T. Walters

INDEX

Volume 37

Section I: Selectionist Ideas and Neurobiology

Selectionist and Instructionist Ideas in Neuroscience
Olaf Sporns

Population Thinking and Neuronal Selection: Metaphors or Concepts?
Ernst Mayr

Selection and the Origin of Information
Manfred Eigen

Section II: Development and Neuronal Populations

Morphoregulatory Molecules and Selectional Dynamics during Development
Kathryn L. Crossin

Exploration and Selection in the Early Acquisition of Skill
Esther Thelen and Daniela Corbetta

Population Activity in the Control of Movement
Apostolos P. Georgopoulos

Section III: Functional Segregation and Integration in the Brain

Reentry and the Problem of Cortical Integration
Giulio Tononi

Coherence as an Organizing Principle of Cortical Functions
Wolf Singer

Temporal Mechanisms in Perception
Ernst Pöppel

Section IV: Memory and Models

Selection versus Instruction: Use of Computer Models to Compare Brain Theories
George N. Reeke, Jr.

Memory and Forgetting: Long-Term and Gradual Changes in Memory Storage
Larry R. Squire

Implicit Knowledge: New Perspectives on Unconscious Processes
Daniel L. Schacter

Section V: Psychophysics, Psychoanalysis, and Neuropsychology

Phantom Limbs, Neglect Syndromes, Repressed Memories, and Freudian Psychology
V. S. Ramachandran

Neural Darwinism and a Conceptual Crisis in Psychoanalysis
Arnold H. Modell

A New Vision of the Mind
Oliver Sacks

INDEX

Volume 38

Regulation of $GABA_A$ Receptor Function and Gene Expression in the Central Nervous System
A. Leslie Morrow

Genetics and the Organization of the Basal Ganglia
Robert Hitzemann, Yeang Olan, Stephen Kanes, Katherine Dains, and Barbara Hitzemann

Structure and Pharmacology of Vertebrate $GABA_A$ Receptor Subtypes
Paul J. Whiting, Ruth M. McKernan, and Keith A. Wafford

Neurotransmitter Transporters: Molecular Biology, Function, and Regulation
Beth Borowsky and Beth J. Hoffman

Presynaptic Excitability
Meyer B. Jackson

Monoamine Neurotransmitters in Invertebrates and Vertebrates: An Examination of the Diverse Enzymatic Pathways Utilized to Synthesize and Inactivate Biogenic Amines
B. D. Sloley and A. V. Juorio

Neurotransmitter Systems in Schizophrenia
Gavin P. Reynolds

Physiology of Bergmann Glial Cells
Thomas Müller and Helmut Kettenmann

INDEX

Volume 39

Modulation of Amino Acid-Gated Ion Channels by Protein Phosphorylation
Stephen J. Moss and Trevor G. Smart

Use-Dependent Regulation of $GABA_A$ Receptors
Eugene M. Barnes, Jr.

Synaptic Transmission and Modulation in the Neostriatum
David M. Lovinger and Elizabeth Tyler

The Cytoskeleton and Neurotransmitter Receptors
Valerie J. Whatley and R. Adron Harris

Endogenous Opioid Regulation of Hippocampal Function
Michele L. Simmons and Charles Chavkin

Molecular Neurobiology of the Cannabinoid Receptor
Mary E. Abood and Billy R. Martin

Genetic Models in the Study of Anesthetic Drug Action
Victoria J. Simpson and Thomas E. Johnson

Neurochemical Bases of Locomotion and Ethanol Stimulant Effects
Tamara J. Phillips and Elaine H. Shen

Effects of Ethanol on Ion Channels
Fulton T. Crews, A. Leslie Morrow, Hugh Criswell, and George Breese

INDEX

Volume 40

Mechanisms of Nerve Cell Death: Apoptosis or Necrosis after Cerebral Ischemia
R. M. E. Chalmers-Redman, A. D. Fraser, W. Y. H. Ju, J. Wadia, N. A. Tatton, and W. G. Tatton

Changes in Ionic Fluxes during Cerebral Ischemia
Tibor Kristian and Bo K. Siesjo

Techniques for Examining Neuroprotective Drugs *in Vivo*
A. Richard Green and Alan J. Cross

Techniques for Examining Neuroprotective Drugs *in Vitro*
Mark P. Goldberg, Uta Strasser, and Laura L. Dugan

Calcium Antagonists: Their Role in Neuroprotection
A. Jacqueline Hunter

Sodium and Potassium Channel Modulators: Their Role in Neuroprotection
Tihomir P. Obrenovich

NMDA Antagonists: Their Role in Neuroprotection
Danial L. Small

Development of the NMDA Ion-Channel Blocker, Aptiganel Hydrochloride, as a Neuroprotective Agent for Acute CNS Injury
Robert N. McBurney

The Pharmacology of AMPA Antagonists and Their Role in Neuroprotection
Rammy Gill and David Lodge

GABA and Neuroprotection
Patrick D. Lyden

Adenosine and Neuroprotection
Bertil B. Fredholm

Interleukins and Cerebral Ischemia
Nancy J. Rothwell, Sarah A. Loddick, and Paul Stroemer

Nitrone-Based Free Radical Traps as Neuroprotective Agents in Cerebral Ischemia and Other Pathologies
Kenneth Hensley, John M. Carney, Charles A. Stewart, Tahera Tabatabaie, Quentin Pye, and Robert A. Floyd

Neurotoxic and Neuroprotective Roles of Nitric Oxide in Cerebral Ischemia
Turgay Dalkara and Michael A. Moskowitz

A Review of Earlier Clinical Studies on Neuroprotective Agents and Current Approaches
Nils-Gunnar Wahlgren

INDEX

Volume 41

Section I: Historical Overview

Rediscovery of an Early Concept
Jeremy D. Schmahmann

Section II: Anatomic Substrates

The Cerebrocerebellar System
Jeremy D. Schmahmann and Deepak N. Pandya

Cerebellar Output Channels
Frank A. Middleton and Peter L. Strick

Cerebellar–Hypothalamic Axis: Basic Circuits and Clinical Observations
Duane E. Haines, Espen Dietrichs, Gregory A. Mihailoff, and E. Frank McDonald

Section III: Physiological Observations

Amelioration of Aggression: Response to Selective Cerebellar Lesions in the Rhesus Monkey
Aaron J. Berman

Autonomic and Vasomotor Regulation
Donald J. Reis and Eugene V. Golanov

Associative Learning
Richard F. Thompson, Shaowen Bao, Lu Chen, Benjamin D. Cipriano, Jeffrey S. Grethe, Jeansok J. Kim, Judith K. Thompson, Jo Anne Tracy, Martha S. Weninger, and David J. Krupa

Visuospatial Abilities
Robert Lalonde

Spatial Event Processing
Marco Molinari, Laura Petrosini, and Liliana G. Grammaldo

Section IV: Functional Neuroimaging Studies

Linguistic Processing
Julie A. Fiez and Marcus E. Raichle

Sensory and Cognitive Functions
Lawrence M. Parsons and Peter T. Fox

Skill Learning
Julien Doyon

Section V: Clinical and Neuropsychological Observations

Executive Function and Motor Skill Learning
Mark Hallett and Jordon Grafman

Verbal Fluency and Agrammatism
Marco Molinari, Maria G. Leggio, and Maria C. Silveri

Classical Conditioning
Diana S. Woodruff-Pak

Early Infantile Autism
Margaret L. Bauman, Pauline A. Filipek, and Thomas L. Kemper

Olivopontocerebellar Atrophy and Friedreich's Ataxia: Neuropsychological Consequences of Bilateral versus Unilateral Cerebellar Lesions
Thérèse Botez-Marquard and Mihai I. Botez

Posterior Fossa Syndrome
Ian F. Pollack

Cerebellar Cognitive Affective Syndrome
Jeremy D. Schmahmann and Janet C. Sherman

Inherited Cerebellar Diseases
Claus W. Wallesch and Claudius Bartels

Neuropsychological Abnormalities in Cerebellar Syndromes—Fact or Fiction?
Irene Daum and Hermann Ackermann

Section VI: Theoretical Considerations

Cerebellar Microcomplexes
Masao Ito

Control of Sensory Data Acquisition
James M. Bower

Neural Representations of Moving Systems
Michael Paulin

How Fibers Subserve Computing Capabilities: Similarities between Brains and Machines
Henrietta C. Leiner and Alan L. Leiner

Cerebellar Timing Systems
Richard Ivry

Attention Coordination and Anticipatory Control
Natacha A. Akshoomoff, Eric Courchesne, and Jeanne Townsend

Context–Response Linkage
W. Thomas Thach

Duality of Cerebellar Motor and Cognitive Functions
James R. Bloedel and Vlastislav Bracha

Section VII: Future Directions

Therapeutic and Research Implications
Jeremy D. Schmahmann

Volume 42

Alzheimer Disease
Mark A. Smith

Neurobiology of Stroke
W. Dalton Dietrich

Free Radicals, Calcium, and the Synaptic Plasticity–Cell Death Continuum: Emerging Roles of the Transcription Factor NFκB
Mark P. Mattson

AP-1 Transcription Factors: Short- and Long-Term Modulators of Gene Expression in the Brain
Keith Pennypacker

Ion Channels in Epilepsy
Istvan Mody

Posttranslational Regulation of Ionotropic Glutamate Receptors and Synaptic Plasticity
Xiaoning Bi, Steve Standley, and Michel Baudry

Heritable Mutations in the Glycine, GABA$_A$, and Nicotinic Acetylcholine Receptors Provide New Insights into the Ligand-Gated Ion Channel Receptor Superfamily
Behnaz Vafa and Peter R. Schofield

INDEX